# Signals and Communication Technology

This series is devoted to fundamentals and applications of modern methods of signal processing and cutting-edge communication technologies. The main topics are information and signal theory, acoustical signal processing, image processing and multimedia systems, mobile and wireless communications, and computer and communication networks. Volumes in the series address researchers in academia and industrial R&D departments. The series is application-oriented. The level of presentation of each individual volume, however, depends on the subject and can range from practical to scientific.

Indexing: All books in "Signals and Communication Technology" are indexed by Scopus and zbMATH

For general information about this book series, comments or suggestions, please contact Mary James at mary.james@springer.com or Ramesh Nath Premnath at ramesh.premnath@springer.com.

Shiban K. Koul · S. Swapna · G. S. Karthikeya

# Antenna Systems for Modern Wireless Devices

Springer

Shiban K. Koul
Centre for Applied Research in Electronics
Indian Institute of Technology Delhi
New Delhi, Delhi, India

S. Swapna
Centre for Applied Research in Electronics
Indian Institute of Technology Delhi
New Delhi, Delhi, India

G. S. Karthikeya
Department of Electronics
and Communication Engineering
B.M.S College of Engineering
Bengaluru, Karnataka, India

ISSN 1860-4862 ISSN 1860-4870 (electronic)
Signals and Communication Technology
ISBN 978-981-97-3371-2 ISBN 978-981-97-3369-9 (eBook)
https://doi.org/10.1007/978-981-97-3369-9



This Springer imprint is published by the registered company Springer Nature Singapore Pte Ltd.
The registered company address is: 152 Beach Road, #21-01/04 Gateway East, Singapore 189721, Singapore

*To baby Norah Koul*

*—Shiban K. Koul*

*To readers of the book*

*—S. Swapna*

*To the eternal friendship with Priyanka K C*

*—G. S. Karthikeya*

# Preface

Ever since the discovery of radio waves by Hertz in 1886, wireless communication technologies have continuously evolved. The expansion is so remarkable that today wireless technology touches every aspect of our lives. Wireless Local Area Network (WLAN) is a disruptive technology that is replacing most of the wired LANs and wired WANs. WLAN technology is based on the IEEE 802.11 standard which is continually expanding and evolving to support higher throughput capabilities, low-latency, enhanced positioning and location capabilities, and vehicle-to-vehicle capabilities. Personal workspaces are getting transformed to wireless through Wireless Personal Area Network (WPAN). WPAN enables adhoc connectivity among portable communication devices at a short range. A vital component in the wireless communication technology is the antenna integrated with the transceivers. Highly efficient antennas are essential to maintain and improve the signal quality. Consequently, antenna designs for WLAN are also going through lot of innovation with different functionalities incorporated into state-of-the-art antenna designs.

This book is aimed not only to describe up-to-date WLAN antenna designs, but also to provide enough information to anyone who wants to design a WLAN antenna or a radiating system to be deployed for practical coverage. The book primarily focuses on pattern diversity antennas. Pattern diversity antennas are very vital in wireless communication. High correlation between multiple signals can result in low data throughput which can be solved by using antennas with pattern diversity. Beam scanning antennas and their different types are also described in detail. Pattern diversity antenna systems with multiport feeds are also comprehensively discussed in this book. For a multiport system to maintain a reasonable link budget, equal antenna gains are preferred for the required antenna coverage. The book further describes latest techniques to enhance and equalize the antenna gain within a compact radiating system. With increasing demand for faster connectivity with minimum path loss, the demand for high-gain antennas is rapidly increasing. Detailed discussion on gain enhancement with latest high-gain antenna designs is required while describing WLAN antennas. Some antenna designs discussed in the book use additive manufacturing for their design and fabrication. Additive manufacturing is a much sought-after technology today that allows rapid development of antennas at an affordable cost.

Recently reported WLAN antennas make use of this technology to develop versatile antenna designs. Finally, the book includes a section on wideband antenna designs. Antenna designs that reduce the scanning loss are also illustrated towards the end of the book.

As mentioned, this book delves into the world of WLAN, exploring its working principle, its evolution, and its potential to shape the future. Chapter 1 lays the foundation by introducing the fundamental concepts of WLANs, laying the groundwork for deeper exploration. Chapter 2 discusses the fascinating realm of pattern diversity antennas, showcasing how they enhance signal quality and combat fading. Delving into the topic of beam steering, Chap. 3 unveils the capabilities of electronic beam scanning antennas, capable of directing signals with precision. Beyond electronic scanning, Chap. 4 explores diverse beam scanning methods, including mechanical approaches and using innovative methods such as using metasurfaces. Chapter 5 tackles the challenge of achieving wide-angle beam scanning, expanding coverage possibilities for WLANs. To maximize signal strength and reach, Chap. 6 delves into various antenna gain enhancement techniques, utilizing metamaterials and sophisticated cavity designs. Chapter 7 builds upon this foundation, for enhancing gain and equalizing signals in WLAN antennas, leading to more robust communication. Chapter 8 presents the exciting potential of multiport antennas, enabling the creation of multiple data streams and catering to multiple users simultaneously. Finally, Chap. 9 explains the transformative power of 3D printing, revealing its potential to revolutionize antenna design with customized and intricate solutions.

New Delhi, India Shiban K. Koul
New Delhi, India S. Swapna
Bengaluru, India G. S. Karthikeya

# Contents


**1 Introduction to Wireless Local Area Network** ..................... 1
1.1 Introduction ..................... 1
1.1.1 Radio Spectrum: A Key Resource of Wireless Networks ..................... 1
1.1.2 Types of Wireless Networks ..................... 3
1.1.3 Commercial Wireless Technologies ..................... 4
1.1.4 Benefits of Wireless Networks ..................... 6
1.1.5 Limitations of Wireless Networks ..................... 7
1.2 History of WLAN ..................... 7
1.3 WLAN Standards ..................... 9
1.3.1 IEEE 802 ..................... 9
1.3.2 IEEE 802.11 Architecture ..................... 12
1.3.3 IEEE 802.11 WLAN Standards ..................... 15
1.4 Cellular Generations ..................... 24
1.4.1 0G—Wireless Technology ..................... 25
1.4.2 1G—Analog Cellular Networks ..................... 25
1.4.3 2G—Digital Networks ..................... 25
1.4.4 3G—High-Speed IP Data Networks ..................... 27
1.4.5 4G—Growth of Mobile Broadband ..................... 28
1.4.6 5G Wireless Technology: The Next-Generation Network ..................... 29
1.4.7 6G Networks: Vision ..................... 29
1.5 WLAN Transceiver Architecture ..................... 30
1.6 Antennas for WLAN ..................... 31
1.6.1 Categorization of Wi-Fi Antennas ..................... 33
1.6.2 Typical Wi-Fi Antennas ..................... 34
1.6.3 Choosing the Right WLAN Antenna ..................... 35
1.6.4 General Requirements for Selecting a Wi-Fi Antenna ..................... 37
1.6.5 Antenna Diversity in Multipath Environments ..................... 37

1.7 Future of Wireless Communications ........ 38
1.7.1 6G—The Future Network Platform ........ 38
1.7.2 Wi-Fi 8 ........ 40
1.7.3 Wireless-Power Transfer ........ 41
1.7.4 Internet of Nanothings ........ 41
References ........ 42

**2 Pattern Diversity in WLAN Antennas** ........ 45
2.1 Introduction ........ 45
2.1.1 Radio Wave Propagation in Wireless Communication ........ 45
2.1.2 Fading in a Wireless Channel ........ 47
2.1.3 Multipath Fading ........ 47
2.2 Significance of Diversity in Wireless Communication ........ 48
2.2.1 Diversity Techniques ........ 48
2.2.2 Need for Pattern Diversity Antennas ........ 51
2.2.3 Significance of Pattern Diversity ........ 53
2.3 History of Pattern Diversity Antenna ........ 54
2.4 Various Pattern Diversity Techniques ........ 55
2.4.1 Integration of Multiple Radiators ........ 55
2.4.2 Different Phase Excitations ........ 56
2.4.3 Multiple Mode Excitation ........ 57
2.4.4 Even and Odd Mode Excitation ........ 58
2.4.5 CRLH-TL-Based Method ........ 59
2.4.6 Parasitic Antenna Arrays ........ 59
2.4.7 Other Pattern Diversity Methods ........ 60
2.5 Pattern Diversity Antennas with Polarization Diversity ........ 61
2.5.1 Polarization Diversity ........ 61
2.5.2 Polarization and Pattern Diversity Antennas ........ 62
2.6 Pattern Diversity Antennas with Reconfigurability ........ 65
2.6.1 Reconfigurable Antennas ........ 65
2.6.2 Reconfigurable Metamaterials ........ 66
2.7 Conclusion ........ 66
References ........ 67

**3 Electronic Beam-Scanning Antennas** ........ 71
3.1 Introduction ........ 71
3.1.1 Need for Beam-Scanning Antennas ........ 71
3.1.2 Applications of Beam-Scanning Antennas ........ 72
3.2 Different Techniques for Antenna Beam Scanning ........ 72
3.2.1 Electronic Scanning ........ 72
3.2.2 Mechanical Scanning ........ 74
3.2.3 Null Steering ........ 75
3.3 Frequency Beam Scanning ........ 75
3.3.1 Leaky-Wave Antenna ........ 75

3.4 Phased-Array Antenna ..... 77
3.4.1 Phase Shifters ..... 78
3.4.2 Phased-Array Scanning ..... 80
3.5 Electronic Scanning Using Discrete Devices ..... 81
3.5.1 Pin Diodes ..... 81
3.5.2 Varactor Diodes ..... 83
3.5.3 RF MEMS Switches ..... 85
3.6 Material Engineering for Beam Scanning ..... 86
3.6.1 Liquid Crystal ..... 87
3.6.2 Ferrite Material ..... 89
3.7 Conclusion ..... 90
References ..... 91

**4 Other Beam-Scanning Techniques** ..... 95
4.1 Introduction ..... 95
4.2 Mechanical Beam-Scanning Antennas ..... 95
4.2.1 Steering the Feed Antenna ..... 96
4.2.2 Steering the Focusing Aperture ..... 98
4.3 Mechanical Beam Scanning by Moving Antenna Parts ..... 98
4.3.1 Rotating Prisms ..... 98
4.3.2 Rotating Other Antenna Sub-components ..... 100
4.4 Beam Scanning with Lens ..... 103
4.4.1 Hemispherical Lens ..... 103
4.4.2 Dielectric Lens ..... 105
4.4.3 Luneburg Lens ..... 107
4.4.4 Inhomogeneous Flat Lens ..... 109
4.4.5 Integrated Lens Antenna ..... 110
4.5 Beam Scanning with Metasurface ..... 111
4.6 Beam Scanning with FSS ..... 114
4.7 Conclusion ..... 117
References ..... 118

**5 Wide-Angle Beam-Scanning Antennas** ..... 123
5.1 Introduction ..... 123
5.2 Challenges in Wide-Angle Beam Scanning ..... 125
5.3 Different Approaches for Wide-Angle Beam Scanning ..... 126
5.4 Wide-Angle Scanning Using Frequency Sweep ..... 129
5.5 Scan Loss ..... 131
5.5.1 Low Scan Loss Antennas ..... 132
5.6 Wideband Directional Cavity Antenna with Low Scan Loss for WLAN ..... 133
5.6.1 Narrow-Band Cavity Antenna ..... 134
5.6.2 Wideband Cavity Antenna ..... 136
5.6.3 Wideband Cavity Antenna with Slider ..... 138
5.6.4 Theoretical Analysis of the Slider ..... 140
5.6.5 Fabrication and Measurement ..... 141

5.6.6 Two-Element Cavity Antenna Array ... 143
5.6.7 Two-Element Cavity Antenna Array with Slider ... 144
5.7 Conclusion ... 148
References ... 148

**6 Antenna Gain Enhancement Techniques** ... 153
6.1 Introduction ... 153
6.1.1 Antenna Gain ... 153
6.1.2 Methods to Enhance Antenna Gain ... 154
6.2 Gain Enhancement Using Metamaterials ... 155
6.2.1 Metamaterial Classification ... 155
6.2.2 Metamaterial Integration for Gain Enhancement ... 157
6.3 Gain Enhancement Using Artificial Surfaces ... 165
6.3.1 Resonator Structures ... 165
6.3.2 Artificial Magnetic Conductors ... 167
6.3.3 Electromagnetic Bandgap Structures ... 168
6.3.4 Frequency-Selective Surface ... 169
6.4 Gain Enhancement Using Dielectrics ... 172
6.4.1 Dielectric Lens ... 172
6.4.2 Dielectric Superstrates ... 174
6.5 Gain Enhancement Using Fabry–Pérot Cavity ... 175
6.6 Other Gain Enhancement Techniques ... 180
6.6.1 Reflectors ... 180
6.6.2 Directors ... 181
6.6.3 Shorting Pins ... 182
6.6.4 Substrate Engineering ... 183
6.6.5 Orthogonal Radiating Choke ... 184
6.7 Conclusion ... 185
References ... 185

**7 Gain Enhancement and Equalization of WLAN Antennas** ... 189
7.1 Introduction ... 189
7.2 Gain Enhancement of 3D-Printed Wideband Cavity Antenna ... 190
7.2.1 Wideband Cavity Antenna Design ... 190
7.2.2 Metasurface Design ... 191
7.3 Common Aperture Two-Element Dipole Antenna Design ... 193
7.3.1 Antenna Design ... 194
7.3.2 Two-Port Antenna Designs ... 195
7.3.3 Metamaterial Design ... 197
7.3.4 Antenna Design with Metamaterials ... 198
7.3.5 Antenna Measurement ... 202
7.4 Common Aperture Three-Element Dipole Antenna Design ... 204
7.4.1 Antenna Design ... 204
7.4.2 Antenna Design Integrated with Metamaterials ... 205

7.5 Gain-Equalized Three-Port Antenna with Superstrate Loading .... 211
7.5.1 Motivation .... 211
7.5.2 Antenna Design .... 212
7.5.3 Antenna Design Theory .... 213
7.5.4 Metamaterial Design 1 .... 216
7.5.5 Metamaterial Design 2 .... 216
7.5.6 Antenna Design Using Metamaterials .... 218
7.6 Gain-Equalized Three-Port Antenna Using Fabry–Perot Cavity .... 227
7.6.1 Motivation .... 227
7.6.2 Antenna Design .... 228
7.6.3 Metamaterial Design .... 229
7.6.4 Antenna with Metamaterial Design .... 230
7.7 Conclusion .... 237
References .... 238

**8 Multiport Antennas for WLAN** .... 241
8.1 Introduction .... 241
8.2 Multiport Antenna Designs .... 242
8.3 Multibeam Antenna Designs .... 247
8.4 Multiport Antenna Design No. 1 with Gain Equalization .... 252
8.4.1 Micromachined Patch Antenna Design .... 253
8.4.2 Single Port Additively Manufactured Series-Fed Array .... 255
8.4.3 Two-Port Additively Manufactured Series-Fed Array .... 256
8.4.4 Three-Port Additively Manufactured Series-Fed Array .... 257
8.4.5 Gain Enhancement Using Parasitic Patch Superstrate .... 257
8.4.6 Three-Port Antenna with Gain Enhancement .... 259
8.5 Multiport Antenna Design No. II with Gain Equalization .... 262
8.5.1 Four-Element Antenna Design .... 265
8.5.2 Four-Element Antenna Design with 3D-Printed Dielectric Loading .... 268
8.6 Design Guidelines .... 276
8.7 Conclusion .... 278
References .... 279

**9 Additive Manufacturing in Antenna Development** .... 283
9.1 Introduction .... 283
9.1.1 3D Printing Methods .... 284
9.1.2 3D Printing Materials .... 286
9.1.3 3D Printing in RF Engineering .... 287
9.2 Classification of 3D-Printed Antennas .... 288
9.2.1 Intricately Shaped Antennas .... 289
9.2.2 Conformable Substrates .... 292
9.2.3 Flexible Substrates .... 293
9.2.4 Inhomogeneous Substrates .... 295

9.3 Metallization Techniques ........ 298
9.3.1 Conductive Spray Coating ........ 298
9.3.2 Electroless and Electroplating ........ 299
9.3.3 Liquid Metal ........ 301
9.3.4 Jet Metal Process ........ 302
9.3.5 Vacuum Filling ........ 302
9.3.6 Wire-Mesh Embedding ........ 303
9.4 3D Metal Printing ........ 304
9.4.1 Common Metal AM Processes ........ 304
9.4.2 Antenna Designs Using Metal AM Technology ........ 305
9.5 Latest Trends in 3D-Printed Antennas ........ 307
9.6 Enhancing Antenna Characteristics Through 3D Printing ........ 310
9.7 Conclusion ........ 315
9.8 Future Scope ........ 316
References ........ 316

**Index** ........ 321

# About the Authors

**Shiban K. Koul** received Bachelor of Technology degree in Electrical Engineering from the Regional Engineering College in 1977, Master of Technology degree in microwave engineering in 1979, and Ph.D. degree in the same field in 1983 from the Indian Institute of Technology Delhi, in India. He is currently an Honarary Professor at the Indian Institute of Technology, Delhi. His research interests include RF MEMS, high-frequency wireless communication, microwave engineering, microwave passive and active circuits, device modelling, millimetre and sub-millimetre wave IC design, body area networks, flexible and wearable electronics, medical applications of sub-terahertz waves, EMI/EMC suppression techniques, and reconfigurable microwave circuits including miniaturized antennas.

He has successfully completed 38 major sponsored projects, 52 consultancy projects, and 61 technology development projects. He has authored/co-authored 615 research papers, 23 state-of-the art books, four book chapters, and two e-books. He holds 26 patents, six copyrights, and one trademark. He has guided 30 Ph.D. theses and more than 120 master's theses.

He is a Life Fellow of IEEE, Fellow of INAE and IETE, and Honorary Fellow of BES. He is the Chief Editor of *IETE Journal of Research* and Associate Editor of the *International Journal of Microwave and Wireless Technologies*, Cambridge University Press. From 2012 to 2014, he served as Distinguished Microwave Lecturer (DML) for IEEE MTT-S. Previously, he served as Speaker Bureau Lecturer for IEEE MTT-S. In addition to serving as an AdCom member of the IEEE MTT Society from 2010 to 2018, he is currently a member of the society's Awards, Nominations and Appointments and Education Committees. Prior to this, he served for over a decade as regional coordinator for R-10. He has recently been selected as the Distinguished Lecturer (DL) of the IEEE EMC Society for 2 years (2024–2025).

He is recipient of numerous awards including Institution of Engineers Kolkata Gold Medal (1977), Indian National Science Academy (INSA) Young Scientist Award (1986); Top Invention Award (1991) of the National Research Development Council (NRDC) for his contributions to the indigenous development of ferrite phase shifter technology; VASVIK Award (1994) for the development of Ka-band components and phase shifters; Ram Lal Wadhwa Gold Medal (1995) from IETE; Academic Excellence Award (1998) from Indian Government for his pioneering contributions to phase control module development for Rajendra Radar; Shri Om Prakash Bhasin Award (2009) in the field of Electronics and Information Technology; VASVIK Award (2012) for the contributions made to the area of Information, communication Technology (ICT); Teaching Excellence Award (2012) from IIT Delhi; Award for contributions made to the growth of smart materials technology (2012) by Institute of Smart Materials and Systems(ISSS) Bangalore; MN Saha Memorial Award (2013) from IETE, IEEE MTT Society Distinguished Educator Award (2014), Lifetime Achievement Award (2023) from the IETE, India; and Lifetime Achievement Award (2024) from the Broadcast Engineering Society (BES), India . His name has recently been included in the "Year 2021" category of Scopus Elsevier, recognising him as one of the top 2% of scientists.

**S. Swapna** received Bachelor of Technology degree in Electronics and Communication Engineering from the University of Kerala in the year 2011 and Master of Technology degree in VLSI Design and Embedded Systems from the National Institute of Technology Rourkela, Odisha in 2013. She worked as an Assistant Lecturer in National Institute of Technology Karnataka, Surathkal from 2014 to 2015. She completed her Ph.D. from the Centre for Applied Research in Electronics, Indian Institute of Technology Delhi in 2022. She served as the Vice Chair of the IEEE MTT-S student branch chapter at IIT Delhi. She has authored/co-authored ten papers in peer-reviewed journals and international conference proceedings. Following her Ph.D., she worked as an IOE postdoctoral fellow at the Department of Electrical Communication Engineering at the Indian Institute of Science Bangalore, where she was part of the 5G Test Bed project. Her research interests include mm-wave antennas, pattern diversity antennas, and additive manufacturing.

**G. S. Karthikeya** received Bachelor of Technology degree in Electronics and Communication Engineering, with distinction, in the year 2010 from Visvesvaraya Technological University, Belgaum. He received Master of Technology degree, with distinction, in Microwave engineering from University of Kerala in 2012. He worked as an assistant professor in affiliated colleges of Visvesvaraya technological university from 2013 to 2016, where he established the Antenna Architects' lab. He joined the Centre for applied Research in Electronics (CARE), IIT Delhi, as a Ph.D. student in 2017 and defended his thesis in 2019. He worked as a Project Scientist at Foundation for Innovation and Technology Transfer (FITT) in association with Synergy Microwave Corporation, USA, from 2019 to 2021. From 2021 to 2022, he served as an Assistant Professor at the Centre for Antennas and Radio Frequency Systems (CARFS), Ramaiah Institute of Technology. Since 2023, he is working as an Assistant Professor at the Advanced RF and wireless communication lab (ARiWu), Department of Electronics and Communication, BMS College of Engineering, Bengaluru. He has authored or co-authored more than 85 articles in peer-reviewed journals and international conference proceedings. He holds a US patent and an Indian patent. He has also filed five Indian patents. He has co-authored three state of the art books and a book chapter. He also holds a trademark on 5G antenna module design. One of his articles has featured as the cover page in Microwave Journal, an industry leading magazine. His research interests include multi-octave antenna systems for portable devices, novel material design for antenna applications, mmWave antennas for mobile terminals and base stations, application of additive manufacturing in antenna designs. He serves as a research consultant to Liveon Technolabs, Alpha design Technologies, and Antsys Innovation Pvt. Ltd. He is a senior member of the IEEE-Antenna Propagation Society and a life member of Antenna Test and Measurement society. He serves as a reviewer of several IEEE and John Wiley journals. He is a recipient of Dr. N. Vasudevan Nair memorial award (2011) and Research Excellence Travel award (2019).

# Abbreviations

| | |
|---|---|
| 0G | Zeroth Generation |
| 1D | One-Dimensional |
| 1G | First Generation |
| 2D | Two-Dimensional |
| 2G | Second Generation |
| 2.5G | 2.5 Generation |
| 3D | Three-Dimensional |
| 3G | Third Generation |
| 3GPP | 3G technology under Generation Partnership Project |
| 4G | Fourth Generation |
| 5G | Fifth Generation |
| 6G | Sixth Generation |
| ABS | Acrylonitrile Butadiene Styrene |
| ADS | Antenna Decoupling Surface |
| AFR | Array Feeding Region |
| AFSS | Active Frequency-Selective Surfaces |
| AM | Additive Manufacturing |
| AMC | Artificial Magnetic Conductors |
| AMPS | Advanced Mobile Phone System |
| AMS | Anisotropic Metasurface |
| AMTS | Advanced Mobile Telephone System |
| AoA | Angle of Arrival |
| AP | Access Point |
| AR | Augmented Reality |
| AT&T | American Telephone and Telegraph |
| ATSA | Antipodal Tapered Slot Antenna |
| AZIM | Anisotropic Zero-Index Metamaterials |
| BCR | Beam Coverage Range |
| BJ | Binder Jetting |
| BOM | Bill of Materials |
| BSS | Basic Service Set |

| | |
|---|---|
| BST | Barium Strontium Titanate |
| CBSA | Cavity-Backed Slot Antenna |
| CCK | Complementary Code Keying |
| CCLR | Cross-Circular Loop Resonator |
| CDMA | Code Division Multiple Access |
| CEPT | European Conference of Postal and Telecommunications Administrations |
| CMS | Cylindrical MS |
| CNC | Computer Numerical Control |
| COTS | Commercial Off-The-Shelf |
| CP | Circular Polarization |
| CPFRA | Circularly Polarized Folded Reflectarray Antenna |
| CPW | Coplanar Waveguide |
| CPWG | Coplanar Waveguide Ground |
| CRLH TL | Composite Right/Left-Handed Transmission Lines |
| CSIRO | Commonwealth Scientific and Industrial Research Organization |
| CSIW | Corrugated Substrate-Integrated Waveguide |
| CSRR | Complementary Split-Ring Resonator |
| CTE | Coefficient of Thermal Expansion |
| DED | Directed Energy Deposition |
| DFS | Dynamic Frequency Selection |
| DL | Dielectric Lens |
| DMLS | Direct Metal Laser Sintering |
| DNG | Double-Negative Materials |
| DoS | Denial of Service |
| DP-CTS | Dual-Polarized Continuous Transverse Stub |
| DQDB | Distributed-Queue Dual-Bus |
| DS | Distribution System |
| DSS | Distribution System Service |
| DSSS | Direct Sequence Spread Spectrum |
| EBG | Electromagnetic Bandgap |
| EBM | Electron Beam Melting |
| ECC | Envelope Correlation Coefficient |
| EDGE | Enhanced Data Rates for GSM Evolution |
| EIRP | Effective Isotropic Radiated Power |
| EM | Electromagnetic |
| EMBB | Enhanced Mobile Broadband |
| ENG | Epsilon Negative |
| ENZ | Epsilon-Near-Zero |
| ERO | European Radiocommunications Office |
| ESS | Extended Service Set |
| ETSI | European Telecommunications Standards Institute |
| F/D | Focal-to-Diameter Ratio |
| FBR | Front-to-Back Ratio |
| FCC | Federal Communications Commission |

| | |
|---|---|
| FDM | Fused Deposition Modeling |
| FFF | Fused Filament Fabrication |
| FHSS | Frequency Hopping Spread Spectrum |
| FPC | Fabry–Perot Cavity |
| FPGA | Field Programmable Gate Array |
| FR2 | Frequency Range 2 |
| FRP | Flat Risley Prism |
| FSA | Frequency Scanning Antennas |
| FSS | Frequency-Selective Surface |
| GPR | Ground-Penetrating Radar |
| GPRS | General Packet Radio Service |
| GPS | Global Positioning System |
| GRIN | Gradient Index |
| GSM | Global System for Mobile Communications |
| HART | Highway Addressable Remote Transducer |
| HDICP-CVD | High-Density Inductively Coupled Plasma Chemical Vapor Deposition |
| HIPERLAN | High-Performance Radio LAN |
| HPBW | Half-Power Beamwidth |
| HPRS | Hybrid Partially Reflecting Surface |
| HRI | High Refractive Index |
| HS | Hybrid Surface |
| HSDPA | High-Speed Downlink Packet Access |
| HSR | H-Shaped Resonator |
| HSUPA | High-Speed Uplink Packet Access |
| HWMP | Hybrid Wireless Mesh Protocol |
| IBSS | Independent BSS |
| IEEE | Institute of Electrical and Electronics Engineers |
| IF | Intermediate Frequency |
| IFA | Inverted-F Antenna |
| ILA | Integrated Lens Antennas |
| IMS | Impedance Modulation Surface |
| IMT 2000 | International Mobile Telephone 2000 |
| IMTS | Improved Mobile Telephone Service |
| IoNT | Internet of Nano Things |
| IP | Internet Protocol |
| IRS | Intelligent Reflecting Surface |
| IS-95 | Interim Standard 95 |
| ISLAN | Integrated Services LAN |
| ISM | Industrial, Scientific, and Medical |
| ITS | Intelligent Transportation Systems |
| ITU | International Telecommunication Union |
| LAN | Local Area Network |
| LC | Liquid Crystals |
| LCD | Liquid Crystal Display |

| | |
|---|---|
| LCRA | Liquid-Crystal Reflect Array |
| LEO | Low Earth Orbit |
| LHCP | Left-Hand Circular Polarization |
| LHMs | Left-Handed Materials |
| LIMs | Low-Index Materials |
| LLC | Logical Link Control |
| LMSC | LAN/MAN Standards Committee |
| LNA | Low-Noise Amplifier |
| LO | Local Oscillator |
| LOS | Line-of-Sight |
| LPWAN | Low-Power Wide Area Network |
| LSAP | Link Service Access Point |
| LTCC | Low-Temperature Co-fired Ceramic |
| LTE | Long-Term Evolution |
| LWA | Leaky-Wave Antenna |
| MAC | Medium Access Control |
| MBSA | Mechanical Beam Steering Array |
| MDCM | Metal-Dielectric Composite Metasurfaces |
| MEMS | Micro-Electro-Mechanical Systems |
| MEO | Medium Earth Orbit |
| MIMO | Multiple Input Multiple Output |
| MJ | Material Jetting |
| ML | Machine Learning |
| MMIC | Monolithic Microwave Integrated Circuit |
| MMS | Multimedia Messaging Service |
| MMTC | Massive Machine Type Communication |
| MNG | Mu-Negative |
| MNZ | Mu-Near-Zero |
| MPA | Microstrip Patch Antenna |
| MRC | Maximum Ratio Combining |
| MSAP | MAC Service Access point |
| MTM | Metamaterial |
| MTS | Mobile Telephone System |
| MUSA | Multiple-Unit Steerable Antenna |
| NFC | Near-Field Communication |
| NFMS | Near-Field Metasteering |
| NMT | Nordic Mobile Telephone |
| NPJ | Nanoparticle Jetting |
| NRI | Negative Refractive Index |
| NRI-TL | Refractive Index Transmission Line |
| NTT | Nippon Telephone and Telegraph |
| NZIM | Near Zero-Index Materials |
| OFDM | Orthogonal Frequency Division Multiplexing |
| OMT | Orthogonal-Mode Transducer |
| ORC | Orthogonal Radiating Choke |

| | |
|---|---|
| OSB | Open-stop Band |
| OSI | Open Systems Interconnection |
| P2MP | Point-to-Multipoint |
| P2P | Point-to-Point |
| PA | Power Amplifier |
| PAA | Phased-Array Antennas |
| PC | Polycarbonate |
| PDC | Personal Digital Cellular Telecommunication System |
| PEC | Perfect Electric Conductor |
| PECVD | Plasma Enhanced Chemical Vapor Deposition |
| PhSAP | Physical Service Access Point |
| PHY | Physical Specification |
| PIFA | Planar Inverted-F Antenna |
| PLA | Polylactic Acid |
| PLPDA | Printed Log-Periodic Dipole Array Antenna |
| PLYDA | Printed Log-Yagi Dipole Array Antenna |
| PMC | Perfect Magnetic Conductor |
| PMM | Phase Matching Method |
| PPW | Parallel-Plate Waveguide |
| PR | Polarization Rotator |
| PRS | Partially Reflecting Surface |
| PS | Portable Stations |
| PSS | Phase-Shifting Surface |
| PTT | Push to Talk |
| QoS | Quality of Service |
| RA | Reflect Array |
| RCA | Resonant Cavity Antenna |
| RF | Radio Frequency |
| RFI | Radio Frequency Interference |
| RFID | Radio Frequency Identification |
| RHCP | Right-Hand Circular Polarization |
| RP | Rapid Prototyping |
| RPA | Risley-Prism Antenna |
| RS | Reflective Surface |
| RSN | Robust Security Network |
| SATCOM | Satellite Communication |
| SFF | Solid-Freeform |
| SIW | Substrate-Integrated Waveguide |
| SIW-SA | Substrate-Integrated Waveguide Fed-Slot Antenna |
| SLA | Stereolithography |
| SLL | Sidelobe Level |
| SLM | Selective Laser Melting |
| SLS | Selective Laser Sintering |
| SMS | Short Message Service |
| SNIR | Signal-to-Noise Plus Interference Ratio |

| | |
|---|---|
| SNR | Signal-to-Noise Ratio |
| SPDT | Single-Pole Double-Throw |
| SRR | Split-Ring Resonator |
| SS | Station Service |
| SSPP | Spoof Surface Plasmon Polariton |
| STA | Stations |
| SW | Surface Waves |
| SWL | Surface-Wave Launchers |
| TA | Transmit Array |
| TACS | Total Access Communications Systems |
| TDMA | Time Division Multiple Access |
| TE | Transverse Electric |
| TM | Transverse Magnetic |
| TPC | Transmit Power Control |
| TRAI | Telecom Regulatory Authority of India |
| UAV | Unmanned Aerial Vehicle |
| UC | Unit Cell |
| UHF | Ultra-High Frequency |
| UMTS | Universal Mobile Telecommunications System |
| URLLC | Ultra Reliable Low Latency Communications |
| UV | Ultraviolet |
| UWB | Ultra-Wideband |
| VGA | Variable Gain Amplifier |
| VICTS | Variable Inclination Continuous Transverse Stub |
| VLE | Very Large Epsilon |
| VLF | Very Low Frequency |
| VoIP | Voice Over Internet Protocol |
| VP | Vertically Polarized |
| VR | Virtual Reality |
| VSRR | Vertical Split-Ring Resonator |
| WABS | Wide-Angle Beam Scanning |
| WAIM | Wide-Angle Impedance Matching |
| WAP | Wireless Application Protocol |
| WCDMA | Wideband Code Division Multiple Access |
| WEP | Wired Equivalent Privacy |
| Wi-Fi | Wireless Fidelity |
| WiMAX | Worldwide Interoperability for Microwave Access |
| WLAN | Wireless Local Area Network |
| WMAN | Wireless Metropolitan-Area Network |
| WPA | Wi-Fi Protected Access |
| WPAN | Wireless Personal-Area Network |
| WPT | Wireless Power Transfer |
| WWAN | Wireless Wide-Area Network |
| XR | Extended Reality |
| ZIM | Zero-Index Material |

# Chapter 1
# Introduction to Wireless Local Area Network

## 1.1 Introduction

We are in an era where everyone and everything is interconnected. Along with people using wireless networks to talk to one another, all possible electronic devices also talk to one another. The traditional wired networks are now becoming more and more obsolete with the onset of wireless networks. If users must be connected to a network by physical cables, their movement would be dramatically limited, which is inconceivable in today's fast-paced life. Wireless connectivity, however, poses no such restriction and allows a great deal of free movement on the part of the user. There are many benefits to wireless communication like lower maintenance costs, more flexibility, and better adaptability for challenging remote environments like isolated deserts and steep valleys where wired networks cannot be installed [1]. As a result, wireless technologies are becoming ubiquitous, replacing wired networks. The use of Wireless Local Area Network (WLAN) systems is rapidly growing in the communications industry. A WLAN enables devices to connect and communicate based on radio transmission rather than wired connections. WLANs transmit data using high-frequency radio waves and often include an access point to the Internet.

### *1.1.1 Radio Spectrum: A Key Resource of Wireless Networks*

Audio frequencies extend from 3 to 20 kHz in the very low-frequency (VLF) band, whereas radio frequency (RF) occupies a very wide range of spectrum (20 kHz–300 GHz). Depending on the nature and requirement of the application, suitable RF frequency can be selected [2].

The spectrum is divided into different frequency bands as shown in Fig. 1.1 and each band has a specific application. This spectrum would be typically allocated to companies or organizations depending on the state policies. For most countries,

 
S. K. Koul et al., *Antenna Systems for Modern Wireless Devices*, Signals and Communication Technology, https://doi.org/10.1007/978-981-97-3369-9_1

the RF (Radio Frequency) spectrum is a national resource, like water, land, gas, and minerals [3]. The use of radio spectrum is rigorously controlled by regulatory authorities through licensing processes. In the USA, regulation is done by the Federal Communications Commission (FCC). European allocation is performed by the European Radiocommunications Office (ERO) of the European Conference of Postal and Telecommunications Administrations (CEPT). In India, the spectrum allocation is performed by the Telecom Regulatory Authority of India (TRAI). Different national bodies are coordinated by an international body, the International Telecommunication Union (ITU) for the purpose of standardization of global telecommunications. Table 1.1 lists some common frequency bands used.

In this book, the focus is more on the Industrial Scientific and Medical (ISM) bands as these are popular for numerous commercial applications. ISM bands are

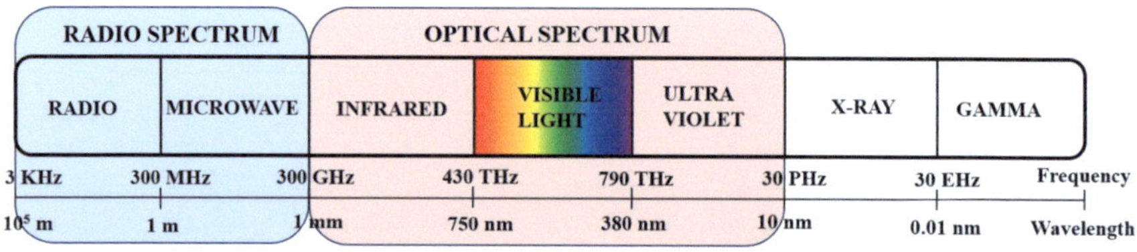


**Fig. 1.1** Electromagnetic spectrum

**Table 1.1** Common frequency bands [4]

| Band | Frequency range | Typical applications |
|---|---|---|
| HF | 3–30 MHz | Short wave broadcasting |
| VHF | 30–300 MHz | Mobile radio communication |
| UHF | 300–1000 MHz | TV broadcasting |
| UHF ISM | 902–928 MHz | GSM |
| L | 1–2 GHz | GPS |
| S | 2–4 GHz | 4G LTE |
| S Band ISM | 2.4–2.5 GHz | Wi-Fi/Bluetooth |
| C | 4–8 GHz | Satellite communication |
| C Band ISM | 5.725–5.875 GHz | Wi-Fi |
| X | 8–12 GHz | Defense |
| Ku | 12–18 GHz | Satellite communication |
| K | 18–26.5 GHz | Satellite communication |
| Ka | 26.5–40 GHz | 5G (FR2) |
| Q | 32–50 GHz | Radio astronomy |
| U | 40–60 GHz | Band currently unused |
| V | 50–75 GHz | Millimeter wave radar |
| W | 75–100 GHz | Military radar |

set aside for equipment related to industrial, scientific, or medical areas and can be operated without requiring any licenses.

## *1.1.2 Types of Wireless Networks*

Wireless networks are categorized into four types according to the geographic area where a signal and its services are available [5]. A complete wireless illustration is shown in Fig. 1.2.

### 1.1.2.1 Wireless Personal-Area Network (WPAN)

WPAN uses low-power transmitters to create a short-range network, typically 20–30 feet (7–10 m). It is used to communicate among a group of private devices such as wireless mouse, wearable devices, USB flash drives, digital cameras, thermostats, lighting controls, and motion sensors. It has no connection to the outside world and is only among the selected devices. This includes technologies like Bluetooth®, Zigbee®. The unlicensed 2.4 GHz ISM band frequencies between 2.402 GHz and 2.480 GHz are used for the protocols.

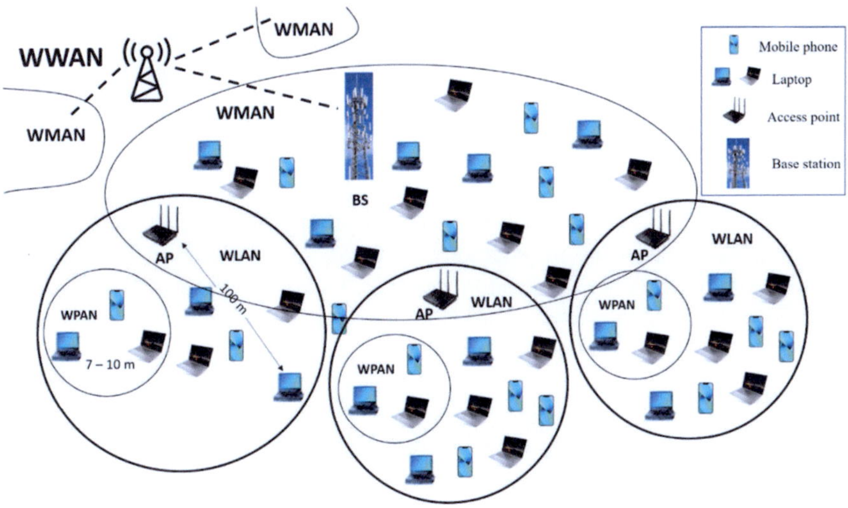


**Fig. 1.2** Wireless network illustration

#### 1.1.2.2 Wireless Local-Area Network (WLAN)

WLAN is a wireless service that connects multiple devices over a medium-sized range up to 100 m. It eliminates the need for an access point by enabling users in a small area to create a temporary network. However, through the usage of an access point, it can be used to gain access to the Internet. The unlicensed 2.4 GHz and 5 GHz ISM band frequencies with frequency bandwidth of 5.15–5.85 GHz are used.

#### 1.1.2.3 Wireless Metropolitan-Area Network (WMAN)

WMAN is a wireless service over a large geographic area of more than 100 m but smaller than a Wireless Wide-Area Network (WWAN). Licensed frequencies are commonly used, and these spectra would be allocated by the state.

#### 1.1.2.4 Wireless Wide-Area Network (WWAN)

WWAN is a wireless data service for mobile phones that is offered over a very large geographic area (regional, national, and even global) by telecommunications carriers. Only licensed frequencies are used here.

### *1.1.3 Commercial Wireless Technologies*

Wireless connectivity technology like Bluetooth®, Wi-Fi, ZigBee® [6], RFID, NFC, and Z-wave® [7] are used in short-range communication, while Cellular communication (2G and beyond) [8, 9], and LPWAN [10] are used for large-range communication. Bluetooth® uses radio signals to connect nearby devices like wireless speakers and smartphones and share information between them. As a result, Bluetooth® can be used in both Wireless LANs and Wireless PANs. ZigBee® and Z-Wave® specifications are also used to create wireless PANs primarily for smart home devices like programmable light bulbs and security sensors.

***Bluetooth®***

The Bluetooth technology operates in the ISM band at 2.4 GHz. The Bluetooth devices will connect and communicate through short-range networks known as piconets. A piconet consists of a minimum of two and a maximum of eight peer devices. The piconets are dynamically established as and when a Bluetooth device enters and leaves a network. Bluetooth range is around 5–10 m and it consumes very low power.

***Wi-Fi***

Wi-Fi is a wireless technology that allows computers, mobile phones, and other devices to interface with the Internet. The wireless access point allows the wireless device to connect to Wi-Fi and it has a range of 20–100 m which could be extended with repeaters. The ISM spectrum of 2.4 GHz and 5 GHz is used in the Wi-Fi. The protocols that enable Wi-Fi communication are defined in the IEEE 802.11 standard, which will be discussed in a later section.

***ZigBee®***

ZigBee is a wireless technology for low-power and low-data rate applications because its focus is on wireless monitoring and control. The frequency range supported in ZigBee is 2.4 GHz and it has a range of 10–30 m. When a device using ZigBee finds another active device in its vicinity, it initiates communication without any interruptions. ZigBee is used in smart home applications, healthcare devices, and sensors.

***Radio Frequency Identification (RFID)***

RFID is a wireless technology that can identify, locate, track, and monitor people and objects without a clear line of sight between the tag and the reader. The main frequencies used by passive RFID tags are 125 kHz, 134 kHz, 13.56 MHz, and 860–956 MHz and active tags use 433 MHz and 2.45 GHz. The read range of an RFID tag varies from 1 to 10 m based on the frequency and tag type used. RFID technology has the potential to replace bar codes in the future.

***Near-Field Communication (NFC)***

NFC is a short-range, high-frequency, low-bandwidth wireless technology between two NFC-enabled devices. It integrates RFID technology and a contactless smartcard interface with mobile phones. The communication between NFC devices occurs at 13.56 MHz as in RFID but at a very close range of 4 cm or less. In the case of contactless smart cards, the communication is performed when the devices are in close proximity, and for contact smart cards, the communication is performed when the devices are in contact with each other.

***Z-wave***

Z-wave is a wireless communication technology for home automation. It interconnects lighting, heating, security, and safety devices with devices from different vendors able to work together. Z-wave uses the sub-1 GHz frequency with a range of 30–40 m. A Z-Wave mesh network can have interconnections of up to 232 devices. The devices can be controlled via a Z-Wave remote control, a cell phone, or a PC. Z-Wave devices can also be programmed to turn up the heater or turn off all lights with one button.

***Cellular Communication***

Cellular communication is a wireless technology that enables users to use voice and data communication through mobile phones. Mobile communication technology has continuously evolved in the past four decades, from the second generation to the fifth generation. The 2G is designed to operate at the 900/1800 MHz frequency band, 3G at 2100 MHz and 1920–2170 MHz, 4G at 700–800 MHz and 1710–2700 MHz, and 5G at 26/39 GHz band. Cellular communication is facilitated through various base stations installed at fixed locations at various geographic locations by various service providers.

***Low-Power Wide-Area Network (LPWAN)***

LPWAN is a wireless communication technology designed to allow long-range communications with low-power consumption, low-cost interface, low transmission data rates due to small packet sizes, and a long battery life. LPWAN-based applications are expected to be one-third of all IoT applications. LPWAN can provide long-range communication up to 10–40 km in rural/desert zones and 1–5 km in urban zones. LPWANs use sub-GHz band to achieve robust and reliable communication as the sub-GHz frequencies have better propagation characteristics.

### *1.1.4 Benefits of Wireless Networks*

- ***Mobility***: The most obvious advantage of wireless networking is mobility. WLAN allows users to move around the coverage area, often a home or small office while maintaining a network connection. A mobile telephone user can drive miles during a single conversation because the phone connects the user through cell towers. The access to the network is not bounded by the length of the cables.
- ***Device adaptability***: Wireless networks typically have a great deal of adaptability, which can translate into rapid deployment. It supports the use of a wide range of devices, such as computers, phones, tablets, gaming systems, and IoT devices.
- ***Easier installation***: The back-end infrastructure side of a wireless network is qualitatively the same whether you are connecting one user or a million users. A WLAN requires less physical equipment than a wired network, which saves money, reduces installation time, and takes up a lesser footprint in typical office settings.
- ***Economical***: The equipment and setup costs are reduced. To offer service in an area, you need base stations and mobile terminals in place. Adding a device to the network is a matter of configuring the infrastructure, and does not involve running cables, punching down terminals, and patching in a new jack.
- ***Scalability***: Wireless networks are scalable in nature. Once the initial infrastructure is built, adding a user to a wireless network is mostly a matter of authorization, which does not require additional infrastructure.

***Flexibility***: Wireless networks provide clutter-free homes, offices, and other networked places. Flexibility may be particularly important in older historical buildings because it reduces the need for further construction. Wireless networks can be deployed extremely rapidly in such environments [11].

### *1.1.5 Limitations of Wireless Networks*

- In wireless networks, radio waves are used for communication. This makes signals noisier due to more interference from nearby devices. External factors like dust, fog, and rain can also interfere with the wireless communication.
- Wireless LANs are slower than wired LANs. The speed of wireless networks is constrained by the available bandwidth. Unless the regulatory authorities can make the spectrum bands wider, there is an upper limit on the speed of wireless networks.
- Security on any wireless network is of prime concern. The information must be encrypted for data security. This is critical as the network transmissions are available to anyone within range of the transmitter with the appropriate receiver operating on the same protocol. On a wired network, the signals stay in the wires and can be protected by strong physical access control. The data is also prone to errors. Hence, the bandwidth requirement for wireless LANs is more than for wired LANs.
- Like all networks, wireless networks transmit data over a network medium. Here medium is in the form of electromagnetic radiation. To be well-suited for use on mobile networks, the medium must be able to cover a wide area so clients can move around in certain area.

## 1.2 History of WLAN

Wireless communication has a fascinating and lengthy history. The humble beginnings of wireless services take us back to the nineteenth century at a time when Guglielmo Marconi, "the father of radio" made his mark on the world of wireless technology [12]. In 1894, Marconi started experimenting with radio waves (Hertzian Waves). His goal was to produce and detect radio waves over long distances. After his successful patent application in 1896, Marconi founded the Wireless Telegraph and Signal Company Limited, the world's first radio factory. In 1901, radio signals were received across the Atlantic, and in 1905, the first wireless distress signal was sent using Morse Code.

The US Military eventually developed wireless technology into a useful instrument. They set up wireless signals to transfer information across a medium with sophisticated encryption, which virtually eliminated unauthorized access to network traffic. This type of technology was first introduced during World War II when the

Army began sending battle plans over enemy lines and when Navy ships instructed their fleets from shore to shore.

For over a century, wireless services remained synonymous with "radio" Eventually, wireless proved so valuable as a secure communication medium that many businesses and schools thought they could expand their wired local area networks (LAN) using wireless LANs. The first wireless LAN came in existence in 1971 when Norman Abramson, a Professor at the University of Hawaii, developed the world's first wireless computer communication network, ALOHAnet [13]. Seven computers spread over four islands made up the system's bi-directional star topology, allowing them to connect with the Oahu island central computer without the need for phone lines. Here, a UHF wireless packet network called ALOHAnet linked the Hawaiian Islands. ALOHAnet and the ALOHA protocol were early alternatives to Ethernet, and later the IEEE 802.11 protocols, respectively.

Vic Hayes is frequently referred to as the "father of Wi-Fi" [14]. He began working on various packet-level protocols when he joined National Cash Registers Corporation in 1974, which is now a division of Agere® Systems, a manufacturer of semiconductor components. One of the scientists that helped form the IEEE 802.11 Working Group on Wireless LANs in 1990 was Hayes.

In 1985, the US Federal Communications Commission decided to release the ISM band for unlicensed use. These are centered around 2.4 GHz band frequencies. However, these frequency ranges are prone to interference because they are also used by devices such as microwave ovens.

In 1991, NCR Corporation with AT&T Corporation invented the precursor to 802.11, created for the use in checkout systems. The WaveLAN brand was used for the initial wireless products. They are the ones who are given credit for creating Wi-Fi.

As a byproduct of a Commonwealth Scientific and Industrial Research Organization (CSIRO), Australia, research project, "a failed experiment to detect exploding mini black holes the size of an atomic particle," the Australian radio astronomer John O'sullivan and his colleagues Terence Percival, Graham Daniels, Diet Ostry, and John Deane developed a key patent used in Wi-Fi [15]. In 1992 and 1996, CSIRO obtained patents for a technique to "unsmear" the signal, which was later used in Wi-Fi.

Initially, wireless LAN hardware was only utilized as a backup to cabled LAN in locations where cabling was impractical or impossible due to its high cost. Early development included industry-specific solutions and proprietary protocols, but by the end of the 1990s, these were replaced by IEEE 802.11 technical standards. The 802.11 standard first came out in 1997 and offered connectivity rates of up to 2 Mbit/s. This was modified in 1999 with 802.11b, to allow for 11 Mbit/s link speeds, and this proved to be popular.

The European Telecommunications Standards Institute (ETSI) pursued High-Performance Radio LAN (HiperLAN) in Europe around the same time in 1991, and the first version was authorized in 1996. Then, in February 2000, a HiperLAN/2 functional specification including ATM effects was completed. Neither European standard achieved the commercial success of 802.11, although much of the work on HiperLAN/2 has survived in the physical specification (PHY) for IEEE 802.11a, which is nearly identical to the PHY of HiperLAN/2.

The Wi-Fi Alliance was established in 1999 as a trade organization to manage the Wi-Fi trademark, which is used to sell the majority of products [16]. The name, Wi-Fi, was coined by the brand-consulting firm Interbrand® which was hired to create a name that was more appealing than IEEE 802.11b Direct Sequence. The initial Wi-Fi standards of 802.11a and 802.11b had very low speeds (up to 54 Mbps & 11 Mbps, respectively). It was acceptable back then because there weren't many laptops or Wi-Fi-enabled handheld mobile phones.

By 2003, Wi-Fi-capable mobile devices started to appear, and portable laptops started to be used more often for both professional and personal purposes. At that time, 802.11g was approved for delivering up to 54 Mbps in the 2.4 GHz space. In 2007, the emergence of the smartphone really came about, and along with it, the 802.11n standard was adopted. The "n" standard supported both 2.4 GHz and 5 GHz bands at a maximum data transfer rate of 600 Mbps. Most newer routers are dual-band and able to utilize both wireless bands. This allows data communication to avoid the crowded 2.4 GHz band, which is also shared with Bluetooth devices and microwave ovens. The 5 GHz band also has more channels than the 2.4 GHz band, permitting a greater number of devices to share the space.

## 1.3 WLAN Standards

Wireless standards are a set of conventions and protocols that determine how a wireless network behaves. This includes the protocols to be followed to set up a wireless network, the establishment of a connection between the access point and a client, and so on. The most common set of wireless standards is the IEEE 802.11 WLAN standard.

### *1.3.1 IEEE 802*

The first meeting of the IEEE Computer Society's "Local Network Standards Committee" Project 802 was held in February 1980. The number 802 does not have any significance. It was simply the next number in the sequence being issued by the IEEE for standards projects. IEEE 802 is a family of Institute of Electrical and Electronics Engineers (IEEE) standards for developing, maintaining, and recommending the networking standards for local, personal, and metropolitan and other area networks. The IEEE 802 LAN/MAN Standards Committee (LMSC) maintains these standards using an open and accredited process and advocates them on a global basis. LMSC is organized into several working groups and the focus area for each group as shown in Table 1.2.

**Table 1.2** IEEE 802 working groups [17]

| Name | Description | Status |
|---|---|---|
| IEEE 802.1 | Higher layer LAN protocols working group | Active |
| IEEE 802.2 | LLC | Disbanded |
| IEEE 802.3 | Ethernet | Active |
| IEEE 802.4 | Token bus | Disbanded |
| IEEE 802.5 | Token ring MAC layer | Disbanded |
| IEEE 802.6 | MANs (DQDB) | Disbanded |
| IEEE 802.7 | Broadband LAN using coaxial cable | Disbanded |
| IEEE 802.8 | Fiber optic TAG | Disbanded |
| IEEE 802.9 | Integrated services LAN (ISLAN or iso-Ethernet) | Disbanded |
| IEEE 802.10 | Interoperable LAN security | Disbanded |
| IEEE 802.11 | Wireless LAN (WLAN) and Mesh (Wi-Fi certification) | Active |
| IEEE 802.12 | 100BaseVG | Disbanded |
| IEEE 802.13 | Unused | Reserved for fast Ethernet development |
| IEEE 802.14 | Cable modems | Disbanded |
| IEEE 802.15 | Wireless PAN | Active |
| IEEE 802.15.1 | Bluetooth certification | Disbanded |
| IEEE 802.15.2 | IEEE 802.15and IEEE 802.11 coexistence | Hibernating |
| IEEE 802.15.3 | High data rate wireless PAN (e.g., UWB, etc.) | – |
| IEEE 802.15.4 | Low data rate wireless PAN (e.g., ZigBee, Wireless HART, Mi Wi, etc.) | Active |
| IEEE 802.15.5 | Mesh networking for WPAN | – |
| IEEE 802.15.6 | Body area network | Active |
| IEEE 802.15.7 | Visible light communications | – |
| IEEE 802.16 | Broadband wireless access (WiMAX certification) | Hibernating |
| IEEE 802.16.1 | Local multipoint distribution service | Hibernating |
| IEEE 802.16.2 | Coexistence wireless access | Hibernating |

(continued)

**Table 1.2** (continued)

| Name | Description | Status |
|---|---|---|
| IEEE 802.17 | Resilient packet ring | Disbanded |
| IEEE 802.18 | Radio regulatory TAG | – |
| IEEE 802.19 | Wireless coexistence working group | – |
| IEEE 802.20 | Mobile broadband wireless Access | Disbanded |
| IEEE 802.21 | Media independent handoff | Hibernating |
| IEEE 802.22 | Wireless regional area network | Hibernating |
| IEEE 802.23 | Emergency services working group | Disbanded |
| IEEE 802.24 | Vertical applications TAG | – |

#### 1.3.1.1 IEEE 802 LAN/MAN Reference Model

The IEEE 802 LAN/MAN reference model is emulated after the Open Systems Interconnection (OSI) Basic Reference Model [18]. The detailed explanation of each layer of the OSI reference model is beyond the scope of this book. Here, we will only focus on the datalink layer as that falls within the scope of IEEE 802 standards. Figure 1.3 shows the architectural view of LAN/MAN reference model and its relation to the OSI reference model.

The Logical Link Control (LLC) sublayer controls the synchronization, flow control, and error-checking functions of the data link layer. The Medium Access Control (MAC) sublayer provides an addressing mechanism and channel access so that each node available on a network (source) can communicate with other nodes (destination) available on the same or other networks. Multiple-link service access points (LSAPs) provide interface ports to support multiple higher layer users above the LLC sub-layer. The MAC sub-layer provides a single MAC service access point (MSAP) as an interface port to the LLC sub-layer in an end station. The physical layer provides an interface port to a single MAC station and may optionally offer isochronous bearer services at multiple physical service access points (PhSAPs).

#### 1.3.1.2 General Requirements for an 802 LAN/MAN

1. An IEEE 802 LAN is required, at minimum, to support information exchange between LLC entities and the construction of MAC addresses through MAC Internal Sublayer Service.

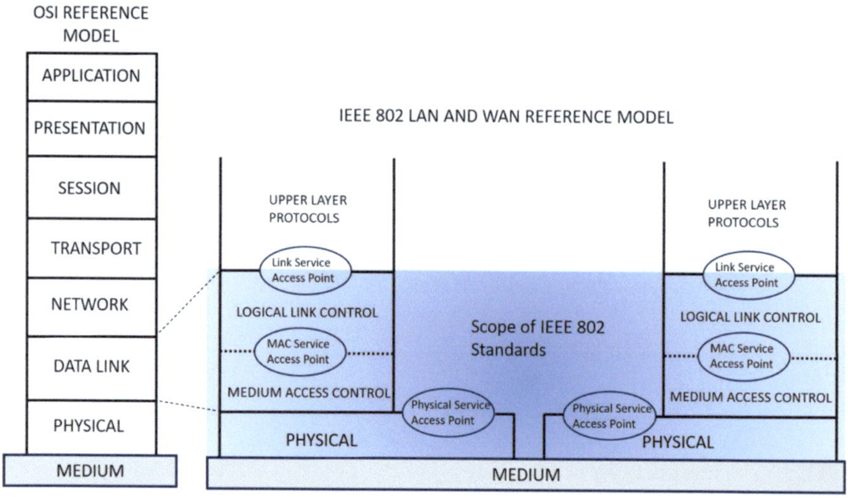


**Fig. 1.3** IEEE 802 reference model for end stations (LAN and MAN/RM)

2. The original IEEE 802 LAN/MAN is required to support at least 200 end stations within a geographical extent of at least 2 km for LANs and 50 km for MANs. Subsequent developments in LAN technology have eased the necessary geographical extent of LAN as the stations can be interconnected using bridged LANs.
3. For wireless physical media, the probability that a data unit is not received correctly shall be less than $8 \times 10^{-8}$ per octet of MAC Service Data Unit length.
4. Insertion or removal of a device from LAN/MAN shall not cause a transient loss of availability of the access domain for not more than 1 s.

***IEEE 802.11 Family***

In 1985, the Federal Communications Commission (FCC) of the USA authorized the industrial, scientific, and medical (ISM) frequency bands [19]. The ISM band refers to that part of the radio spectrum that is internationally reserved for the industrial, scientific, and medical requirements rather than for communications. The ISM bands expedited the WLAN development as vendors no longer needed to apply for licenses to operate their products.

## *1.3.2 IEEE 802.11 Architecture*

The components of WLAN architecture as laid down in IEEE 802.11 are—

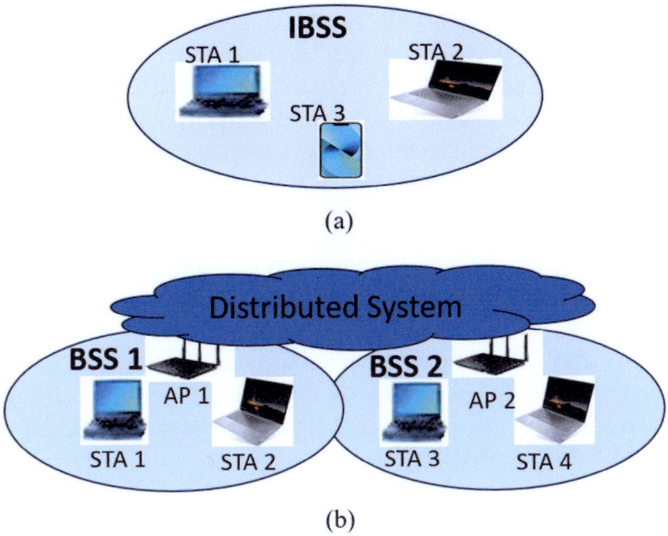


**Fig. 1.4** **a** Ad-hoc network architecture, **b** infrastructure network architecture

1. Stations (STA)—Stations comprise of all devices and equipment that are connected to the wireless LAN. Each station has a wireless network interface controller. A station can be of two types:
  - Wireless Access Point (WAP or AP)
  - Client
2. Basic Service Set (BSS)—The Basic Service Set (BSS) is the basic building block of an IEEE 802.11 LAN. It is a collection of stations that communicate together within an 802.11 network. Figure 1.4 [20] shows different components of 802.11 LAN architecture. The BSS has two or more fixed, portable, and/or moving nodes or stations (STA) that can communicate with each other over the air in a geographically limited area.

- Independent BSS (IBSS)—The most basic type of IEEE 802.11 LAN is the IBSS. In this mode of operation, the stations communicate directly without the use of an access point as shown in Fig. 1.4a. This type of 802.11 LAN is formed only as long as the LAN is needed. Hence, it is also referred to as the ad hoc network. If a station moves out of its BSS, it can no longer directly communicate with other members of the BSS.
- Infrastructure BSS—In the infrastructure mode, there are access points (AP) within a BSS as shown in Fig. 1.4b. An access point acts like a station along with providing access to the Distribution System (DS) which normally is a LAN. Handoff will happen when a station moves from the coverage area of one access point to another access point. The DS enables mobile device support by providing the necessary services to handle address to destination mapping and seamless integration of multiple BSSs.

- Extended Service Set (ESS)—The combination of DS and BSSs allows IEEE 802.11 to create a wireless network of arbitrary size and complexity. IEEE 802.11 refers to this type of network as the extended service set (ESS) network. The ESS network appears like a IBSS network to an LLC layer. Stations within an ESS may communicate and mobile stations may move from one BSS to another (within the same ESS) transparently to LLC.

The IEEE 802.11 architecture is integrated with a traditional wired network through an architectural component called portal. A portal is the logical point at which data units from an integrated non-IEEE 802.11 LAN enter the IEEE 802.11 DS.

IEEE 802.11 also specifies the services associated with different components of the architecture. There are two categories of IEEE 802.11 service—the station service (SS) and the distribution system service (DSS). Both these categories of service are used by the IEEE 802.11 MAC sublayer.

The service provided by stations is known as the station service. The SS is present in every IEEE 802.11 station (including APs, as APs include station functionality). The SS is specified for use by MAC sublayer entities. The service provided by the DS is known as the distribution system service. The DSSs are provided by the DS. They are accessed via a STA that also provides DSSs. A STA that provides access to DSS is an AP. Figure 1.5 combines the different LAN components and both types of services to show the complete IEEE 802.11 architecture [21].

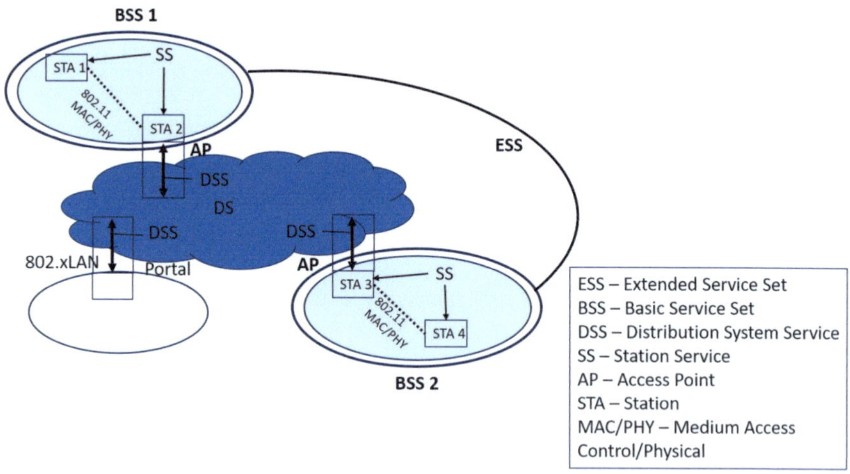


**Fig. 1.5** Complete IEEE 802.11 architecture

### 1.3.3 IEEE 802.11 WLAN Standards

IEEE 802.11-based WLANs [22–24] are becoming popular in various environments primarily due to their lower cost of implementation and higher data rates. Also, the availability of a number of variants of this standard can help users to select a technology suitable for their needs. Thus, it provides a flexible way of achieving data communication among various devices.

#### 1.3.3.1 IEEE 802.11 Technical Specifications

***802.11***

In 1989, the IEEE 802.11 Working Group began elaborating on the Wireless LAN Medium Access Control and physical layer specifications. The final draft was ratified on June 26, 1997. The devices that are compliant with the IEEE 802.11 standards operate at the 2.4 GHz ISM band with 20 MHz channel bandwidth. The Direct Sequence Spread Spectrum (DSSS) physical layer provides 2 Mbps of peak rate and optional 1 Mbps in extremely noisy environments. The Frequency Hopping Spread Spectrum (FHSS) physical layer operates at 1 Mbps with optional 2 Mbps in very clean environments. The LAN can operate within a range of 20–100 m, and it has applications in RF heating, microwave oven, and medical diathermy machines.

***802.11a***

The 2.4 GHz frequency band is used by many applications. Therefore, the IEEE 802.11a standard was approved in September 1999 that uses the 5 GHz frequency band. Consequently, to reduce the signal interference, the 802.11a standard moved to the 5 GHz band with 20 MHz channel bandwidth. The 802.11a uses Orthogonal Frequency Division Multiplexing (OFDM), which provides higher data rates up to 54 Mbps. The LAN can operate within a range of 30–120 m and supports video streaming and Voice Over Internet Protocol (VoIP). To reduce the signal penetration through various obstacles, the 5 GHz carrier frequency operates in almost line of sight. This mode of operation reduces the signal range, increasing the number of access points required. However, this helps in more reliable communication. The higher number of access points increases the overall system cost. Another disadvantage of 802.11a is that it is only available in half the bandwidth in Japan, and not available for use in Europe. The 5 GHz spectrum is not license-free in every country. 802.11a standard is not compatible with 802.11b as they operate in different frequency bands.

***802.11b***

Two years after the legacy 802.11 standard was proposed, it was officially revised on 16 September 1999. The new standard, named 802.11b or 802.11 high data rate, still operated in the 2.4 GHz frequency band and 20 MHz channel bandwidth with DSSS modulation scheme. The 802.11b standard provides a data rate up to 11 Mbps, which is comparable to a fixed Ethernet. The new standard is more robust but still accommodates the same 802.11 protocols, ensuring backward compatibility. This promise encouraged the manufacturing of 802.11b products replacing the older standard. The IEEE 802.11b physical layer adopts Complementary Code Keying (CCK) technology with data rate of 11 Mbps. The LAN can operate within a range of 38–140 m and supports Internet access and the DSL or cable modem services. Although 802.11b products have conquered the WLAN market, signal interference at 2.4 GHz is a major issue. Microwave ovens, cordless telephones, Bluetooth devices, many medical equipment, and household appliances use 2.4 GHz frequency band causing severe interferences.

***802.11c***

The IEEE 802.11c standard deals with bridge operation procedures. A bridge is a device that links to LANs with a similar or identical MAC protocol. It performs functions that are similar to those of an Internet Protocol (IP)-level router, but at the MAC layer. A bridge is simpler and more efficient than an IP router. In 2003, the 802.11c task group completed its work on this standard. Some protocols associated with bridging 802.11 wireless client devices were added. Now, the IEEE 802.11c standard is part of IEEE 802.1d.

***802.11d***

IEEE 802.11d is an amendment to the IEEE 802.11b standard with further support for additional regulatory domains. This standard resolves the regulatory differences of various countries by adding a country information element to the probe requests and responses. In the absence of IEEE 802.11d, equipment manufacturers have to produce country-specific products and users that travel to different countries have to carry country-specific network cards. This standard supports international roaming and will make devices automatically adjust based on geographic requirements. This country information element greatly simplified the creation of 802.11 wireless APs and client devices that meet the different regulations enforced in various parts of the world. In 2015, the FCC ruled that new devices are not allowed to rely solely on 802.11d for setting country-specific radio parameters.

***802.11e***

802.11e is an enhancement to 802.11b specifications revised in November 2005. It operates on 2.4 GHz and 5.8 GHz frequency bands and has a bandwidth of up to 100 MHz, and this high-frequency range accounts for more channels, fast data transfer speeds, and less chances of interference. 802.11e standard defines a set of quality of service (QoS) enhancements for wireless LAN applications through modifications to the media access control layer. IEEE 802.11e has the ability to accommodate time-scheduled and polled communication during null periods when no other data is moving through the system. This improves the polling efficiency and channel robustness. These enhancements should provide the quality which is necessary for services such as VoIP and video streaming. 802.11e offers subscribers high-speed Internet access with VoIP, full motion video, and high-quality audio.

### *802.11f*

The IEEE 802.11f or Inter-Access Point Protocol is an extension to IEEE 802.11b. It provides wireless access point communication among multivendor systems. An access point also functions as a bridge that connects two different 802.11 LANs across another type of a network, such as Ethernet, LAN, or a WMAN. IEEE 802.11f handles the registration of APs within a network and the exchange of information when a user is roaming among coverage areas supported by APs from different vendors. This standard, introduced in 2003, was never implemented by any of the vendors, and it was withdrawn by IEEE in Feb 2006.

### *802.11g*

IEEE proposed the 802.11g standard in June 2003 to enhance the 2.4 GHz 802.11b technology. It provides all the functionalities of 802.11b along with higher data performance of 54 Mbps compared to 11 Mbps in 802.11b. It also has the advantage of backward compatibility with 802.11b. It uses DSSS and OFDM modulation schemes and operates for a range of 38–140 m. All devices that were previously functioning under 802.11b can be easily migrated to the new standard resulting in significant cost reduction engineering. Customers get increased performance, more robust security, and better range for exactly the same price. Despite all the advantages, 802.11g suffers from the same interference issues that affected the 802.11b in the already crowded 2.4 GHz range. The success of this standard resulted in high density of usage in the urban areas leading to further interferences.

### *802.11h*

The 802.11a uses the 5 GHz frequency band, which is also used for satellite and radar communication. This resulted in interference with other wireless networks. Therefore, IEEE 802.11a was amended to a new standard called IEEE 802.11h, to address the interference problems with satellite and radar. To handle such situations, 802.11h introduced two mechanisms. First, the Dynamic Frequency Selection (DFS) scheme, where the AP detects other networks operating in the same frequency band and changes the operating frequency to prevent interference. Second, the Transmit Power Control (TPC) scheme, that keeps the signal level below a certain preset level if there is a satellite signal in the nearby channels. It was originally designed to address the European regulation but it is now applicable to other countries as well.

### *802.11i*

802.11i is a standard that provides enhanced encryption for networks that use the 802.11a, 802.11b, and 802.11g standards. The aim of the IEEE 802.11i standard is to address security issues. This standard defined the security and authentication mechanisms at the MAC layer.

Wired Equivalent Privacy (WEP) was a security algorithm introduced as part of 802.11 when it was initially launched in 1997. The intention was to provide data security that was comparable to a wired network, but this authentication mechanism was not very strong and it was possible to intercept the data. Wi-Fi Protected

Access (WPA) was initially introduced as an intermediate solution for the security vulnerabilities faced by WEP. It was further improved to WPA2, also called Robust Security Network (RSN), and it is referred as 802.11i standard. The standard consists of two new protocols called the four-way handshake and the group key handshake to establish more effective data authentication services. IEEE 802.11i also incorporates Temporal Key Integrity Protocol and Counter Mode/CBC-MAC Protocol for data transmission confidentiality, protection, packet authentication, and encryption.

***802.11j***

802.11j standard was introduced in 2004, designed especially for the Japanese market. The standard works in the 4.9–5 GHz band to conform to the Japanese rules for radio operation for indoor, outdoor, and mobile applications.

***802.11k***

IEEE 802.11k standard defines enhancements for radio resource management. It provides information to discover the best available access point for mobile devices and improves the way traffic is distributed within a network. Every device connects to the AP that provides the strongest signal. In situations where a higher number of users are accessing a particular AP, this can cause significant overcrowding resulting in performance degradation. In a network conforming to 802.11k, if the AP having the strongest signal is loaded to its full capacity, the device will be connected to any of the nearby underutilized APs. Even though the signal may be weaker, the overall throughput is greater because more efficient use is made of the network resources.

***802.11l***

This standard is reserved and will not be used. Moreover, no task group was named IEEE 802.11l to avoid confusion with another well-known standard IEEE 802.11i.

***802.11m***

IEEE 802.11m is an initiative to provide continuous monitoring, editorial maintenance, management, improvements, interpretations, clarifications, and corrections for the documentations relevant to 802.11 family. IEEE 802.11m is based on two documented initiatives—IEEE 802.11ma and IEEE 802.11mb. The IEEE 802.11ma is comprised of the eight previous standards (a, b, d, e, g, h, i, j) which were editorially applied through 2003. The IEEE 802.11mb is the current maintenance draft version.

***802.11n***

The purpose of 802.11n standard is to improve network throughput and data rates over the 802.11a and 802.11g standards. To increase the MAC layer efficiency and throughput, two new mechanisms are introduced called frame aggression and block acknowledgement. The frame aggregation technique enabled STAs to aggregate small frames into larger ones. In the previous standards, the STA had to wait for some time after sending a MAC frame which resulted in under-utilization if the MAC frame was small. In block acknowledgement, instead of transmitting individual acknowledgement for every frame, multiple frames can acknowledge together in a single BA frame. This resulted in a significant increase in the data rates from 54 to 72 Mbps at a channel width of 20 MHz in the 2.4 GHz frequency band and to 600 Mbps at a channel width of 40 MHz in the 5 GHz frequency band. It has backward compatibility as well. 802.11n introduced various enhancements to the MAC layer. These enhancements include support for multiple-input multiple-output antennas, multiple antennas, smart antennas, security improvements, and changes to signal encoding schemes. It uses multiple antennas to increase the data rates. These features can be used in the 2.4 or 5 GHz frequency bands.

***802.11o***

IEEE 802.11o is reserved and it will not be used to avoid confusion.

***802.11p***

IEEE 802.11p is an amendment to the IEEE 802.11a standard to add wireless access in vehicular environments. It defines enhancements to 802.11a required to support Intelligent Transportation Systems (ITS) applications. It includes data exchange between vehicles and the roadside infrastructure in the licensed ITS band of 5.9 GHz (5.85–5.925 GHz). The vehicles send the information like speed, distance to nearby vehicles so that all the vehicles in the neighborhood are aware about the traffic status.

***802.11q***

802.11q is reserved and it will not be used. It can be confused with 802.1q.

***802.11r***

IEEE 802.11r is an amendment to realize continuous connectivity for mobile devices through fast and secure handoffs between STAs. This standard enables fast roaming and reduced the roaming delay between two basic service sets (BSSs) to less than 50 ms. It reduced the delay by redefining the security key negotiation protocol by allowing requests and negotiations for wireless resources to occur in parallel. 802.11r facilitates the deployment of portable phones with VoIP via mobile phones designed for wireless Internet networks instead of or in addition to standard cellular networks.

***802.11s***

IEEE 802.11s is an amendment to define how wireless devices interconnect to create a WLAN mesh network. Mesh STAs use mesh services to communicate with other devices in the network. 802.11s provides a proxy mechanism to provide addressing support for non-mesh 802 devices, enabling the end-points to be aware of the external addresses. 802.11s standard defines a routing protocol for the network devices connected to the mesh network, called Hybrid Wireless Mesh Protocol (HWMP). This is a mesh routing protocol that combines the flexibility of on-demand routing with proactive topology tree extensions. The combination of reactive and proactive elements of HWMP enables optimal and efficient path selection in a wide variety of mesh networks. 802.11s also provides a framework for congestion control and power saving. Peer-to-peer security protocols that allow for simultaneous initiation from both sides or from one side to the other are used in mesh networks. This establishment protocol called "Simultaneous Authentication of Equals" is a secure password-based authentication technique. After a SAE exchange, the two peers establish a cryptographically strong key, which is then used with the "Authenticated Mesh Peering Exchange" to protect mesh traffic.

***802.11t***

The IEEE 802.11t, also known as Wireless Performance Prediction, recommends testing and measurement practices to be followed in wireless networks. It provides a set of measurement methodology and performance metrics recommendations that enable the service providers, testing labs, manufacturers, and end users to measure the performance of IEEE 802.11 standard equipment and networks. However, this is only a set of recommendations and not a standard.

***802.11u***

IEEE 802.11u is an amendment to add features for improved Internet working with external non-802.11 networks. The standard enables sending and requesting the information relating to a wireless network. This is achieved through the network discovery and selection where the AP displays pre-association information to the clients. The AP describes the type of network offered through a pre-defined list, venue information, roaming information, and other optional information. Another feature in 802.11u is the Access Query Network Protocol, which describes the services offered by an AP. The clients can send a query to the AP for this information. 802.11u also enables the clients to contact emergency services. This standard can also be used to create a network dedicated to emergency services. It can provide location information of the client to assist the emergency services. There is also provision to send alert messages to the clients.

***802.11v***

IEEE 802.11v is an amendment that allows configuration of clients while they are connected to the wireless network. It allows client devices to exchange information about the network topology and the RF environment. This information exchange makes the clients more aware of their surrounding and enables an overall improvement of the network. 802.11v standard also describes enhancements to wireless network management, like the network-assisted power savings that help the clients to improve battery life by enabling them to sleep longer and network-assisted roaming so that better APs are able to associate with various clients.

***802.11w***

IEEE 802.11w is an amendment to increase the management frame security. It introduces protected management frames that enable data origin authenticity, data integrity, and replay protection. This standard aims to mitigate certain types of WLAN Denial of Service (DoS) attacks like de-authentication attack and dis-association attack. DoS attacks based on management frames are not protected by 802.11w. One issue with this standard is that 802.11w requires a software update in both the client and the AP.

### *802.11x*

802.11x is reserved, and it will not be used to avoid confusion with 802.1x.

### *802.11y*

The frequency channels at the 2.4 GHz and 5 GHz frequency bands are already occupied with WLANs. The 802.11y is a standard developed for the USA, and it operates in the 3650–3700 MHz band. The standard enables the operation of high-powered wireless equipment, and thus, it provides services to more users and serves longer ranges.

The standard introduces three new concepts—Contention-Based Protocol that allows many users to use the same channel, Extended Channel Switch Announcement for the AP to notify the connected stations about its intentions to change a channel and Dependent Station Enablement by which an enabling station grants permission and dictates operational procedures to dependent stations.

### *802.11z*

In a wireless network, the data transfer from one client to another occurs through the AP. The 802.11z standard makes it possible to transfer data between two clients that are part of the same network. This mechanism is called Tunneled Direct Link Setup and sets up a direct link between the clients while they remain connected to the AP. This helps in reducing the network traffic and prevents congestion at the AP, in addition to adding power save capabilities.

### *802.11aa*

The 802.11aa is a standard for improving the delivery of audio–video streams of WLAN without any interference with other types of traffic. This allows efficient transmission of various multimedia streams through the network with a stronger QoS provisioning mechanism. For example, a video conference can be served with higher quality than a lower priority video streaming.

### *802.11ab*

802.11ab is reserved, and it will not be used to avoid confusion with devices using 802.11a and 802.11b standards, which are often abbreviated as 802.11ab.

### *802.11ac*

The 802.11ac was introduced to improve the speed of the network. It was an improvement over the 802.11n standard. 802.11ac was able to achieve a very high throughput data speeds of 1 Gbps. It has a wider channel bandwidth of 80 MHz. The routers of 802.11ac adopted multiple-user MIMO which the routers can use to send information to multiple devices at the same time. 802.11ac standard also consists of standardized beamforming. This also helps in increased throughput as the radio signals are directed at specific devices during transmission.

***802.11ad***

The 802.11ad is a wireless standard that uses the 60 GHz spectrum. As it uses the V-band of mmW frequency, the communication range is limited as it is difficult to pass through obstacles. As the frequency is higher the frequency channels are wider with a bandwidth of 1760 MHz. This is helpful for users in dense networks as they can maintain high-speed performance without having to share bandwidth with the legacy bands. As it is of 60 GHz frequency some modifications in the physical layer need to be considered for optimal operations.

***802.11ae***

The IEEE 802.11ae describes the Quality of Service for Management frames. The management frames are used to manage the BSS. The standard introduces a mechanism to determine the prioritization of the Management frames. In the absence of such prioritizations, irrelevant or latency-insensitive management traffic receives priority over data traffic and VoIP traffic is also severely affected.

***802.11af***

The IEEE 802.11af standard is also called Super Wi-Fi or White-Fi as it utilizes the unused radio spectrum or white spaces in the television bands. It operates in the frequency range of 54–790 MHz. Cognitive radio technology is used to identify the white spaces it can use without any interference to the television users. It is based on a geolocation database, which provides information on frequencies, as it can vary in different locations. The frequency channels are 6–8 MHz wide. The physical layer is based on orthogonal frequency division multiplexing (OFDM).

***802.11ag***

802.11ag is reserved and it will not be used to avoid confusion with devices using 802.11a and 802.11g standards, which are often abbreviated as 802.11ag.

***802.11ah***

The 802.11ah standard, also called Wi-Fi HaLow, provides WLAN operations in the unlicensed industrial, scientific, and medical (ISM) bands that are available at 900 MHz. This allows devices to gain access to short-term transmissions in these frequency bands that are less congested. The utilization of lower frequencies, compared to the conventional frequencies of 2.4 or 5 GHz, provides an improved coverage due to better propagation ranges. It consumes low power and provides coverage for applications like wide area-based sensor networks, sensor backhaul systems, and potential Wi-Fi off-loading functions.

***802.11ai***

The IEEE 802.11ai standard provides fast initial link setup methods that enable a wireless LAN client to achieve a secure link setup within 100 ms. It improves user experience in dense environments, as well as allowing a more secure and reliable connection. It enables more efficient spectrum utilization by optimizing the protocol overhead in an area where many mobile users are continually entering and leaving an existing extended service set (ESS) coverage area.

***802.11aj***

The IEEE 802.11aj standard enables operations in the 45 GHz unlicensed spectrum available in some regions of the world, especially China. It is also known as China Millimeter Wave. The amendment also maintains backward compatibility with 802.11ad when it operates in the 59–64 GHz frequency band.

***802.11ak***

The IEEE 802.11ak-2018 is a standard to simplify the expansion and functionality of mixed IEEE 802.11 wireless networks for different network applications. IEEE 802.11ak provides protocols and procedures that enhance the ability of IEEE 802.11 networks to provide internal connections as transit links within IEEE 802.1Q bridged networks, and wireless connectivity for fixed, portable, and moving stations within a local area.

***802.11aq***

The IEEE 802.11aq-2018 is an amendment to IEEE 802.11 that addresses the discovery of available services in WLANs, benefitting network operators and end users alike. The IEEE 802.11aq specifies parameters for pre-association queries between wireless networks and devices. By facilitating such an exchange of information between the wireless access point and the user's device, users can discover which types of services are supported before making the decision to connect to a WLAN. It delivers a critical competitive advantage through service differentiation in crowded market environments.

***802.11ax***

The IEEE 802.11ax, marketed by the Wi-Fi alliance as Wi-Fi 6, is an improvement over the IEEE 802.11ac standard. The IEEE 802.11ax, also known as High-Efficiency Wi-Fi, adds to the overall efficiency, flexibility, and scalability with increased speeds for the next-generation applications. It also resolves the mutual interference between the access points in densely covered areas that are used to slow down the wireless networks. The 802.11ax uses OFDMA technology which increases the capacity of the system.

***802.11ax (extended)***

Wi-Fi 6E takes the efficiency features from Wi-Fi 6 like OFDMA, and WPA3 and improves upon the current Wi-Fi 6 (802.11ax) standard. The 1200 MHz of spectrum between 5.925 and 7.125 GHz, which is known collectively as the 6 GHz band is allocated for Wi-Fi 6E providing a more contiguous spectrum and less interference. Wi-Fi 6E eliminates multiprotocol complexity by restricting access to the 6 GHz band to Wi-Fi 6 and later protocols only [25]. Firmware that requires multigigabit rates, such as high-definition video, can be supported by organizations using Wi-Fi 6E.

### *802.11ay*

The IEEE 802.11ay is an enhancement of 802.11ad standard in the unlicensed 60 GHz millimeter-wave band of spectrum. It has a transmission rate of 20–40 Gbit/s and an extended transmission distance of 300–500 m. The 802.11ay standard has mechanisms for channel bonding with four channels of 802.11ad bonded together for a bandwidth of 8.64 GHz. The applications of the standard include replacement for Ethernet and other cables, and providing backhaul connectivity outside for service providers.

## 1.4 Cellular Generations

Cellular mobile communication started in the late 1970s and has been around for 40 years now. Wi-Fi has been around for 20 years since the IEEE passed the 802.11 standard. The wireless communication has evolved rapidly over the years to become the ultra-fast, highly convenient communication standard that we know today.

The evolution of wireless communication is broadly categorized into 0G, 1G, 2G, 3G, 4G, 5G, and 6G. The G stands for Generation and the number is just a representation of the evolution of the technology. These representations began after the development of 2G technology and 0G and 1G were named to represent the pre-2G era. Figure 1.6 gives a broad picture of the different stages of the cellular evolution.

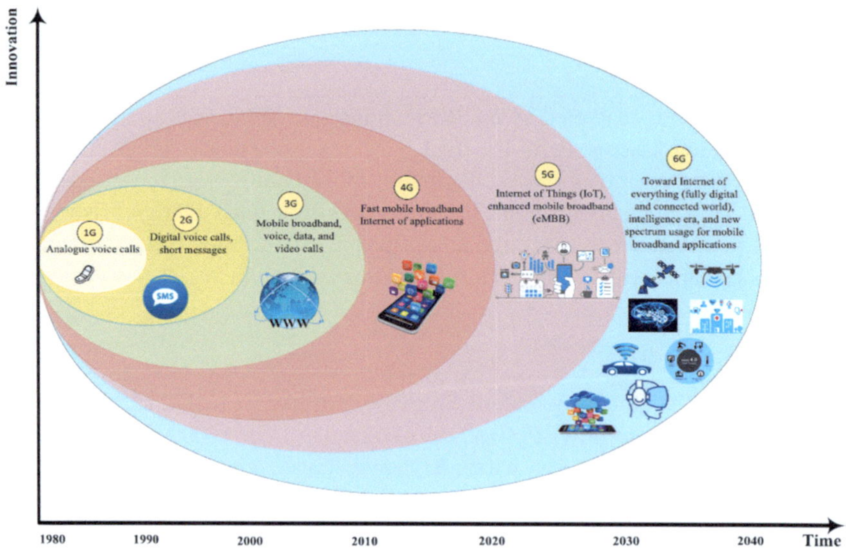


**Fig. 1.6** Evolution of cellular mobile network [26]

### 1.4.1 0G—Wireless Technology

Zeroth-generation mobile technology (0G) refers to pre-cellular phone technology, such as radio-telephone, which some had in cars before the arrival of cell phones. Technologies used in 0G systems included PTT (Push to Talk), MTS (Mobile Telephone System), IMTS (Improved Mobile Telephone Service), and AMTS (Advanced Mobile Telephone System).

### 1.4.2 1G—Analog Cellular Networks

1G technology was the first generation of mobile network technology. It supported only voice call capabilities. The first cellular system in the World was launched in Japan by Nippon Telephone and Telegraph (NTT) DoCoMo firm in 1979. Two years after 1G was launched, in 1981 Nordic countries Norway, Sweden, Denmark, and Finland, along with Switzerland, Netherlands, Eastern Europe, and Russia got a new standard known as NMT (Nordic Mobile Telephone) which was based on 1G technology. Similarly, AMPS (Advanced Mobile Phone System) started to be used in North America and Australia, TACS (Total Access Communications Systems) in the UK, and many others [27]. The main technological development in this generation that distinguished the first-generation mobile phones from the previous generation was the use of multiple cell sites, and the ability to transfer calls from one site to the next site as the user traveled between cells during a conversation. All the systems offered handover and roaming capability but the cellular networks were unable to interoperate between countries. 1G is an analog technology and uses analog signals at 150 MHz frequency. The maximum speed of 1G was 2.4 kbps. 1G's inadequate voice link quality, low capacity, and erratic handoff made calls less reliable. The low-frequency bandwidth for 1G, and a low data transmission speed of 3–4 kbps made 1G technology feasible only for phone calls as they take low data to carry the signals. In addition, it uses lot of energy to transmit data to far-off distances and signals would get disturbed while reaching the recipient. Most users often experience dropped calls. The 1G phones were big and thick in size due to bigger antennas that were required for the chosen operating frequency. Also, mobile phones suffer from lower battery life because data transmission consumes lot of energy, thanks to the powerful wireless transceivers.

### 1.4.3 2G—Digital Networks

2G or second generation of wireless technology was first launched commercially in Finland in 1991. It utilized digital signals for voice transmission with bandwidth of 30–200 kHz. 2G provided services such as text message (SMS), picture messages, and Multimedia Messaging Service (MMS), but not videos. 2G offered greater security for both sender and receiver. Data transmission for phone conversation got fully digitized and encrypted, which ensured that only the intended receiver can receive and read it. Data transmission speed was also increased to 14.4 kbps. In 2G technology, the voice quality improved significantly. Since analog components were no longer needed, the mobile phone size was also reduced. Battery life improved due to digitization of data transmission which usually took less energy to transmit. This happened due to two different multiplexing standards on which 2G technology relied on such as TDMA (Time Division Multiplexing Access) and CDMA (Code Division Multiplexing Access). Group Special Mobile (GSM), PDC, and IS-136 are the different 2G mobile telecommunications standards based on TDMA and Interim Standard 95 (IS-95) is based on CDMA technology. GSM, which has its origins in Europe, is the most admired standard of all the mobile technologies, and was used in more than 212 countries in the world. GSM standard makes international roaming very common between mobile phone operators, enabling subscribers to use their phones in many parts of the world. GSM operates in the 900 and 1800 MHz bands. IS-136 is used particularly in the USA and Canada. New technologies have been developed based on the original GSM system, leading to some advanced system, known as 2.5 generation (2.5 G) Systems.

#### 1.4.3.1 2.5G—GPRS (General Packet Radio Service)

2.5G is used to describe 2G-Systems that have implemented a packet switched domain in addition to circuit switched domain [28]. This enabled the sending of any request to network bases in the form of small packets rather than continuous signals in real-time connection. It helped to save battery even more as the transmissions were in small packet form. If network went off, and came back on, the request would continue from where it stopped when the network went off and there would not be any need to re-initiate everything. GPRS provided data rates of 50–60 kbps. It provides services such as Wireless Application Protocol (WAP) access, Multimedia Messaging Service (MMS) and Internet communication services. GPRS data transfer is typically charged per megabyte of traffic transferred, while data communication via traditional circuit switching is billed per minute of connection time, independent of whether the user is utilizing the capacity or is in an idle state. The G symbol we see on our mobile phones while using Internet is a notation of GPRS.

#### 1.4.3.2 2.75G—EDGE (Enhanced Data Rates for GSM Evolution)

GPRS networks evolved to EDGE networks with the introduction of 8PSK encoding. This further improved voice conversation quality. EDGE was deployed on GSM networks beginning in 2003 initially by Cingular (now AT & T) in the United States. EDGE is an upgrade that provides a potential three-fold increase in the capacity of GSM/GPRS networks. The specification achieves data rates up to 236.8 kbps. The biggest advantage of using EDGE technology is that one does not need to install any additional hardware and software to make use of it. EDGE is a part of third-generation technology. It is denoted by the E symbol near the mobile network.

### *1.4.4 3G—High-Speed IP Data Networks*

In 1998, 3G was pre-commercially launched in Japan by NTT DoCoMo for testing purposes and branded as FOMA. Until 2001, 3G technology was not released for commercial use due to unreliability. 3G is based on the International Telecommunication Union (ITU) plan called International Mobile Telephone 2000 (IMT-2000), Standard. This is a wireless communication standard for all countries throughout the world in the 2000 MHz range. In October 2001, it was widely launched commercially on W-CDMA standard which is based on GSM. 3G evolution for CDMA systems lead to CDMA 2000. In European countries, the UMTS (Universal Mobile Telecommunications System) network spread 3G technology under Generation Partnership Project (3GPP) standards. W-CDMA standard is widely used in 3G technology as it provides better data transmission speed than UMTS. 3G technologies achieve greater network capacity through improved spectral efficiency. Also, better encrypting/encoding technique used to transmit data offered better data security. 3G devices became smaller in size and also battery life improved with better frequency and encoding technique. The maximum speed of 3G was estimated to be around 2 Mbps for non-moving devices and 384 kbps in moving vehicles [29]. 3G technology offers users more advanced services like wide-area wireless voice telephony, video calls, enhanced audio and video streaming, higher data speed, IPTV (TV through Internet support), high-speed web and WAP browsing, mobile television, and GPS (global positioning system).

#### 1.4.4.1 3.5G—HSDPA (High-Speed Downlink Packet Access)

High-speed packet access or HSPA is a 3.5G technology which increases the data rate by adding transport channels and improves latency and increases throughput. It is an enhanced version of WCDMA technology based on GSM standards. The major advancements of HSPA are the HSDPA in the downlink and HSUPA in the uplink direction. HSDPA is a packet-based data service with data transmission up to 8–10 Mbps over a 5 MHz bandwidth in WCDMA downlink. Video streaming improved in 3.5G and the network latency reduced to a greater extent. The "H" symbol on the mobile network represents HSPA or 3.5G network.

#### 1.4.4.2 3.75G—HSUPA (High-Speed Uplink Packet Access)

An improved version of HSDPA was launched as HSUPA+ or Evolved High-Speed Packet Access, also termed as 3.75G. HSUPA is a UMTS/WCDMA uplink evolution technology with advanced encryption/encoding techniques and using multiple antennas to transmit and receive data requests. Data transmission rates increased up to a peak speed of 42 Mbps uplink and 22 Mbps downlink theoretically. The high data rates introduced stable mobile broadband services. It also enhanced various business and consumer applications and real-time person-to-person gaming. Any mobile device could be used as a Wi-Fi hotspot and via a dongle Internet could be accessed on the computer.

### *1.4.5 4G—Growth of Mobile Broadband*

The speed of the 3G network quickly became insufficient as technology and smartphones evolved. 4G came around 2010 and drastically improved data transfer speeds. It supports mobile web access like 3G but also gaming services, HD mobile TV, Video Conferencing, 3D TV, and other services that demand higher speeds. The maximum speed of a 4G network when the device is moving fast is 100 Mbps or 1 Gbps for low mobility communication when stationary or walking. A new technology named OFDMA is introduced in 4G. In OFDMA, the data is transmitted in pure packet form, bringing better quality for voice calls. The channel is divided into narrow bands through which data packets are sent and the network latency is reduced. Two different standards were proposed under 4G—LTE and WiMAX. WiMAX was first released commercially in June 2006 in South Korea with peak data speed for download at 128 Mbps and peak data speed for upload at 56 Mbps. The WiMAX network does not support legacy systems like 2G and 3G and it was not adopted by enough carriers to make it commercially successful. In Europe and North America, most carriers dropped WiMAX by the end of 2017.

#### 1.4.5.1 4G Long-Term Evolution (LTE)

The 4G LTE standard was first launched in Stockholm and Oslo in 2009, and other countries slowly joined them in the following years. It was a complete redesign and simplification of 3G HSPA network architecture, resulting in a significant reduction in transfer latency and thus, increasing efficiency and speeds on the network. 4G VoLTE stands for Voice Over Long-Term Evolution is an extension in LTE network to fully digitize voice telephony in packet form from circuit switched technology. In this protocol, voice quality is enhanced much more along with better data security and at higher data speed.

### 1.4.6 5G Wireless Technology: The Next-Generation Network

The 5G network was first launched in 2019 in some parts of the United States. There are three categories of 5G—enhanced Mobile Broadband (eMBB), Ultra Reliable Low Latency Communications (URLLC), and massive machine type communication (mMTC) [30]. The biggest advantage of 5G over 4G is the increased bandwidth. It also provides significantly faster data rates with maximum speed of 10 Gbps, higher connection density, and much lower latency. 5G network can connect to thousands of devices at high speeds that can move computing and processing power away from devices and into the network. The shift from a hardware to a software-based network environment will bring about lower overheads for the mobile operators. This can make IoT devices much smaller. 5G will also become the underlying wireless infrastructure to support autonomous vehicles, virtual reality (VR)/augmented reality (AR) headsets, and smart cities. 5G networks with reliable and fast network ensure stable connectivity during transport.

### 1.4.7 6G Networks: Vision

The goal of 6G networks is to handle the massive demands for data traffic while significantly improving the quality of services in conventional wireless communications systems. The goals of 6G networks are to increase data rates, decrease energy consumption, improve broadband connectivity and coverage, boost link dependability, lower latency, and accomplish intelligent communication. 6G may be able to support a very high data rate of above 100 Gbps with an end-to-end delay of less than 1 ms. 6G is also anticipated to provide extremely high levels of communication reliability. Additionally, we anticipate that future 6G networks will usher in an ultra-era. Ultra-low latency and ultra-high reliability wireless communications are predicted to be supported by 6G networks. Future 6G networks want to allow extremely quick mobility as well. Ultra large-scale Multiple Input Multiple Output (MIMO) systems and ultra-high frequencies are anticipated to be employed by 6G networks to support ultra-high-speed wireless data transmission. Additionally, 6G networks hope to handle streaming of ultra-high-definition video and offer ultra-high broadband access.

A brief history of the development of mobile communications networks from the first generation (1G) to the fifth generation (5G) is given in Table 1.3. There have been five distinct generations of mobile communications systems developed to date, each with its own set of standards, methods, and capabilities. It is noteworthy that a new generation of mobile communications networks has generally been created every decade.

**Table 1.3** Key evolution indication of 1G to 5G mobile wireless communications [26]

| Key indication | 1G | 2G | 3G | 4G | 5G |
|---|---|---|---|---|---|
| Period | 1980–1990 | 1990–2000 | 2000–2010 | 2010–2020 | 2020–2030 |
| Technology | Analog voice | GSM | CDMA-2000 | Wi-Fi, WiMax, LTE | 5G NR, IPv6, LAN, WAN, PAN |
| Multiplexing | FDMA | TDMA, CDMA | CDMA | CDMA, OFDM | OFDM, BDMA |
| Data rate | 2.4–14.4 kb/s | 14.4–64 kb/s | 3.1–14.7 Mb/s | 100 Mb/s–1 Gb/s | 1 Gb/s and above |
| Bandwidth | 150 kHz | 5–20 MHz | 25 MHz | 100 MHz | 1–2 GHz |
| Architecture | SISO | SISO | SISO | MIMO | Massive-MIMO |
| Main network | PSTN | PSTN | Packet | Internet | Internet |
| Features | Voice | Voice, SMS | Voice, data | Video | VoIP, ultra HD |
| Band-type | Narrow | Narrow | Broad | Ultra-broad | Ultra-wide |
| Highlight | Mobility | Digitization | Internet | Real-time streaming | Extra-high-rate |
| Switching | Circuit | Circuit, packet | Packet | All packet | All packet |
| Handoff | Horizontal | Horizontal | Horizontal | Horizontal, vertical | Horizontal, vertical |

## 1.5 WLAN Transceiver Architecture

A wireless transceiver is a central component of any wireless communication system. A wireless transceiver consists of two functional layers: a physical (PHY) layer and a media access control (MAC) layer. The PHY layer consists of an RF front end and a baseband processor. The baseband processor modulates a bitstream to a constellation symbol stream for transmission. The receiver demodulates a symbol stream to recover the transmitted signal. The wireless transceiver RF front end adds an RF carrier to the baseband symbol stream for transmission and down-converts the received RF signal to baseband. The MAC layer provides link traffic control for the wireless transmitter to access the wireless links, avoid collisions, and optimize data throughput. There are many factors involved in the design of WLAN Transceivers, [31] which include:

- Integration level.
- 802.11 support.
- Process/Design Characteristics.
- Die size.
- Radio architecture.

These factors come into play when dealing with WLAN transceiver design. The most challenging designs in a wireless transceiver integrated circuit are the low-noise amplifier (LNA) at the receiver side and the power amplifier (PA) at the transmitter side. The use of a Superheterodyne radio architecture versus a direct conversion architecture also becomes paramount when trading off design specifications with a lower bill of materials (BOM). As integration levels increase, the designers are trying to build all radio components into a single chip. Although it is sometimes more beneficial to design separate components of a radio to maximize the performance of the overall system, many companies are increasingly pushing to integrate. When formerly external components are integrated into a WLAN transceiver, additional design or process-related issues could surface, necessitating some innovative design approaches.

***Transceiver Architecture***

Since the WLAN market is very cost-sensitive, for a competitive radio, it must be low cost and low power for the consumer. Choosing the transceiver architecture wisely is thus critical. Superheterodyne radios require two synthesizers, but have a high level of transceiver performance. Their inherent frequency diversity reduces DC offset and VCO pulling effects as compared to direct conversion receivers. Although by comparison direct conversion receivers require only one local oscillator, oftentimes mixing schemes are used to achieve similar frequency diversity, resulting in a more complicated transceiver. A compromise between the two approaches is to use a walking intermediate frequency (IF) architecture where the RF local oscillator (LO) is four times the IF LO [32]. Thus, both LOs can be derived from a single synthesizer obtaining the benefit of a direct down conversion radio, but retaining the advantages of the superheterodyne radio.

Shi et al. [33] implemented the transceiver RFIC illustrated in Fig. 1.7. The walking IF architecture is chosen because, to supply both RF and IF LO frequencies, a single Integer-N frequency synthesizer with a 16 MHz reference would be sufficient. In addition, only one VCO with a wide tuning range is needed to cover the 802.11a low and middle bands from 5.18 to 5.32 GHz. The receiver includes an LNA followed by an image filter to reject the RF image tone at about 3.2 GHz. The RF mixer down-converts the received signal to IF frequency at about 1.05 GHz. The IF mixers further demodulate the IF signal to baseband quadrature signals. The baseband block includes the variable gain amplifiers (VGAs) for gain tuning and the baseband filter is off-chip. The transmitter starts with the baseband VGA blocks, followed by IF up-converters. The IF signal is further up-converted to a 5.2 GHz signal by an RF mixer followed by an image filter. A pre-driver amplifies the 5.2 GHz signal with −5 dBm output power for driving the off-chip power amplifier.

## 1.6 Antennas for WLAN

Wireless Local Area Networks (WLANs) provide wireless network communication between computers and other portable devices with fixed access points over a short distance. WLANs include Wireless Fidelity (Wi-Fi) and High-Performance Radio LAN (HIPERLAN). The WLANs operate in the unlicensed Industrial, Scientific,

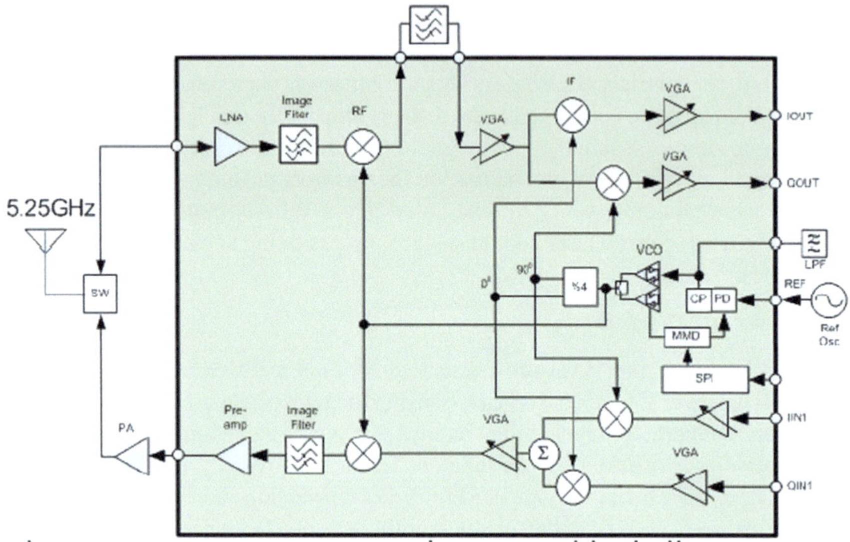


**Fig. 1.7** WLAN 802.11a transceiver block diagram [33]. Reprinted with permission from IEEE

and Medical (ISM) bands, namely the 2.4-GHz band with a frequency range of 2.4–2.485 GHz and the 5-GHz bands with frequency ranges of 5.150–5.350 GHz, 5.470–5.725 GHz, and 5.725–5.850 GHz with channel bandwidth ranging from 5 to 20 MHz. The data rate of a wireless communication link increases with a larger bandwidth, but decreases with mobility. Antennas play an important role in WLAN systems. The antennas for the client devices have critical constraints in terms of cost and size, which severely limits the antenna performance. The antennas for base stations in point-to-point (P2P) and/or point-to-multipoint (P2MP) systems face different challenges such as the performance, cost, size, and integration of multiple functions into one antenna design, as well as the integration of antennas into radios. Wi-Fi connectivity is maintained by the radio-frequency energy that is transmitted and received over antennas. The rapid growth of laptops and smartphones paved the way for a swift growth of WLAN standards. However, the antenna designs did not keep pace with the WLAN communication system development. The antenna operates as a transition between the transceiver and its surrounding environment. To attract a wide market, designed antenna must be of low profile and low cost. Wi-Fi is widely regarded as a wireless communication technology that can be used to create local networks across short distances. But with the right antenna, Wi-Fi signals can reach a higher range. Standard wireless Access Points (APs) can create WLANs that traverse distances of 100–300 m with a clear line of sight. These ranges can be greatly increased by replacing the typical omnidirectional antennas with directional antennas. It is simple to establish a network "bridge" that spans 2–4 km or more by using a matched pair of APs with directed antennas [34]. Wireless links employing 2.4 GHz equipment and dish antennas have achieved distances up to 50 km. Wi-Fi signals can pass through isolated patches of vegetation or thin walls, but not through large towers or dense trees. Such obstructions can cause signal loss and the transmission can suffer from losses of data packets, which would be higher in the 5 GHz band.

There are three general requirements for transmission of Wi-Fi signals over long distances:

- A direct line of sight connection between the transmitting and receiving antennas.
- The transmitting and receiving antennas should be preferably on the same horizontal plane.
- Directional antennas need to be used on the sending and receiving sides. Antennas with a smaller aperture angle need more accurate orientation but can transmit over longer distances.

The long-distance Wi-Fi antennas come with trade-offs between aperture angles and transmission distances. Omnidirectional antennas radiate power equally in all directions in a uniform plane. The omnidirectional antenna has a doughnut-shaped radiation pattern. Omni-directional antennas are commonly used in Wi-Fi Routers and Wi-Fi chipsets of mobile phones, laptops, television, etc. Semi-directional antennas transmit radio waves in particular patterns depending up on the design. Directional antennas radiate power only in one direction. It is most commonly used in point-to-point communication. The focused beaming allows increased performance by reducing interference due to unwanted noises, coming from various angles.

### *1.6.1 Categorization of Wi-Fi Antennas*

Wi-Fi antennas can be categorized into three groups—designs for outdoor, indoor, and mobile applications [35].

***Outdoor Antennas***

Outdoor Antennas include backhaul and fixed base-station antennas. They provide directional point-to-point link with gain up to 29 dBi. For point-to-multipoint coverage, gain from 3 to 16 dBi for omnidirectional coverage, and 15–20 dBi for sectoral coverage, with typical azimuth-plane half-power beamwidths of 60, 90, or 120 is desired. They are usually mounted on towers or buildings for line-of-sight (LOS) communication with reduced multipath interference. The antennas can be directional/sectorized patch antenna arrays, omnidirectional dipole-type arrays, highly directional Yagi antennas, dish/grid reflector antennas, or circularly polarized endfire helix antennas. Antennas with polarization diversity require high isolation, typically around 30–40 dB between the orthogonal polarizations with a compact form factor.

***Indoor Antennas***

Indoor Antennas or access point antennas provide omnidirectional coverage with gain upto 6 dBi. These patch or monopole antennas are usually mounted on walls,

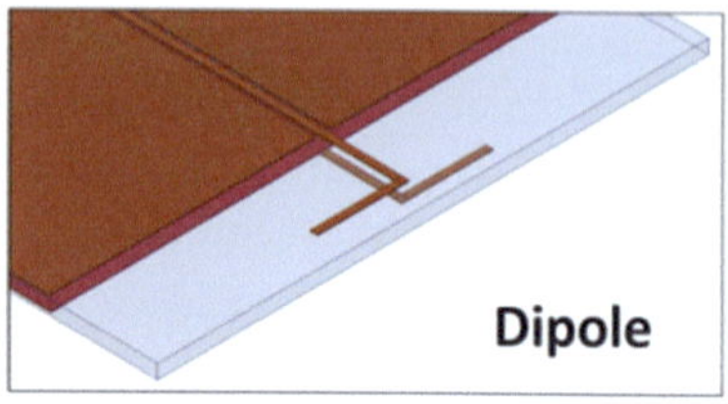


**Fig. 1.8** Schematic of a dipole antenna [38]

ceilings, or computers [36]. A small-form factor or low-profile design is important because of the limited space and aesthetical concerns.

***Antennas for Mobile Devices***

Antennas for mobile devices provide omnidirectional coverage with an average gain of 5 dBi. These Wi-Fi antennas have been extensively embedded into the mobile devices such as laptops, personal computers, video game consoles, mobile/smart phones, and digital audio players. As the multiband and diversity antennas operate within a very limited space in the devices, physically and electrically small antennas are required.

## *1.6.2 Typical Wi-Fi Antennas*

### 1.6.2.1 Dipole Antenna

Dipole Wi-Fi Antennas consist of two identical metal rods that are separately fed using two different wires or feed lines. The dipole radiation pattern is 360° in the horizontal plane and 75° in the vertical plane (assuming the dipole antenna is standing vertically) and resembles a doughnut in shape [37]. Dipole antennas are said to have a gain of 2.14 dBi in comparison with an isotropic antenna. These antennas are commonly used for indoor purposes such as Wi-Fi Routers, USB Wi-Fi adaptors, and television. A dipole antenna illustration is shown in Fig. 1.8.

### 1.6.2.2 Panel Antennas

A panel antenna is like a sector antenna, which is a directional antenna used in telecommunications for cell phone connectivity and wireless Internet. A panel antenna is shaped like a panel and is usually attached to a mounting bracket, which may then be affixed to the side of a building or inside a building [39]. Additionally, it may be mounted on a swivel mount, which allows the user to position the antenna for optimal concentration of signal strength. Panel antennas are used to increase the range and capability of wireless Internet networks. Panel antennas can extend wireless connectivity up to 15 km. The antenna can be connected to a Wi-Fi router or can be fixed on a pole on the roof of the house, or to a building where the user wants to receive the signal at a distance very far away from the source.

#### 1.6.2.3 Parabolic Grid

Parabolic refers to the shape of the reflector on the device. These include a small antenna that is backed with a parabolic reflector that will direct Wi-Fi waves to the antenna. This is a long-range antenna and can provide point-to-point link between the source and the receiver. These antennas have high gain but a very narrow beamwidth which makes it highly directional with a radiating angle between 3° and 12°. These antennas are designed to withstand extreme weather conditions. It can be used in Wi-Fi or other cellular devices like routers, modems, and signal boosters.

#### 1.6.2.4 Patch Antenna

Patch antennas are very popular in wireless communications because they are easy to design and easy to integrate with other microwave-integrated circuits. A single patch antenna provides a maximum gain of 6–9 dBi. An inherent advantage of the patch antennas is its ability to have diversity of polarization. They can be easily designed to have vertical, horizontal, right-hand circular (RHCP) or left-hand circular (LHCP) polarizations, by means of use of multiple feed points or one. This property allows patch antennas to be used in several communication areas, such as in the design of personal communication equipment. A patch antenna illustration is shown in Fig. 1.9.

#### 1.6.2.5 Omnidirectional Antenna

Omnidirectional antennas provide a 360° donut-shaped radiation pattern to provide the widest possible signal coverage in indoor and outdoor wireless applications. Types of omnidirectional antennas include "rubber duck" antennas often found on access points and routers. Omnidirectional antennas are also used in base stations and mobile devices that use radio.

### *1.6.3 Choosing the Right WLAN Antenna*

Usually, two kinds of antennas are considered for use in WLAN. One is associated with a mobile user (or a peripheral device), while the other one concerns the network side (Access Point—AP). The first type of antenna needs to be small, to minimize the overall size of the mobile or portable unit [41]. A number of antennas capable of fulfilling such requirement include printed planar monopoles, such as the Planar

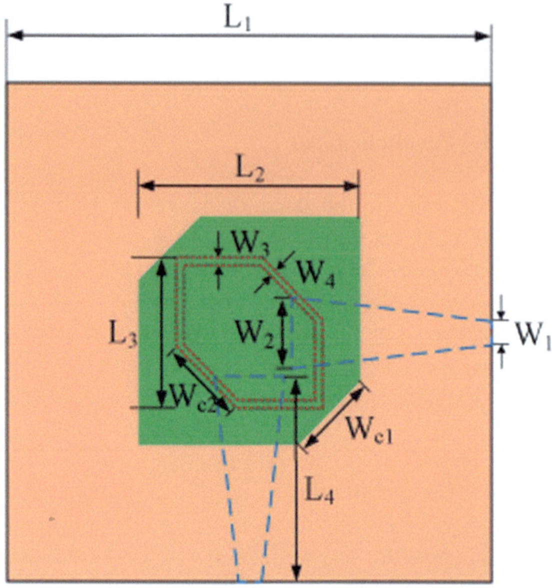


**Fig. 1.9** Schematic of a patch antenna [40]

Inverted F antenna, and reduced-size microstrip patches usually involving a shorting pin as the size reduction mechanism.

The access point antennas do not introduce stringent limitations on the antenna size and its appearance. The primary performance objective for these antennas is to spread the signal to mobile or stationary devices within an enclosed space like a room. Perhaps due to these non-specific objectives, the area of designing antennas for WLAN AP has been largely overlooked. This contrasts with the mobile units, for which antennas have been rigorously researched in recent years. The simplest and most frequently utilized antenna for WLAN AP is a quarter-wavelength wire monopole positioned on a finite ground plane. This antenna, usually located at the ceiling of a room, exhibits an omnidirectional radiation pattern. Although very simple in construction, which is advantageous from the development point of view, it suffers from a considerable exposure to an accidental damage. This possible environmental hazard leads to necessity of employing a radome, which significantly increases profile of the antenna, and may be unacceptable in some applications. Another problem, which creates concerns among network users, is because of a possible intrusion into the network, as this antenna features a widely spread radiation pattern. This is unwelcome in applications requiring high level of security.

### 1.6.4 General Requirements for Selecting a Wi-Fi Antenna

Major considerations for WLAN antenna design include electrical properties such as frequency range/bandwidth in terms of gain, impedance matching, and polarization, as well as other factors such as size, cost-effectiveness, and mechanical robustness. Fulfilling the electrical properties of the antenna is a priority, which stems from system requirements. In addition to an antenna's electrical performance, other factors are also vital for a successful design. For example, a low-cost antenna product is always preferable for the commercial market. The relevant cost factors include the material used, manufacturing, mechanical tolerance, installation, switching from one standard to another, and integration with the radio. In addition, the form factor of the antenna is also an important consideration. Indoor antennas are generally designed to be compact and less protrusive.

### 1.6.5 Antenna Diversity in Multipath Environments

Antenna diversity generally refers to using two antennas for each radio, increasing the probability of one of the two receiving a better signal. The antennas must be separate but similar at each location. Diversity addresses challenges experienced by wireless networks in multipath environments. The antennas are physically separated from each other to ensure that each one encounters different multipath propagation effects. Signal null points are located throughout multipath environments. In some instances, it's necessary to move antennas around to avoid those null points and enable adequate reception. Diversity is intended to maximize throughput by minimizing the number of dropped packets.

Delay spread is a measure of multipath performance and is defined as the time between when the main signal arrives and when the last reflected signal arrives. The amount of delay is related to the environment and size of the network. For residences, it's typically under 50 ns; in offices, it can be 100 ns, and in manufacturing and warehouse environments, it can be 300 ns or more.

A multipath signal can have good RF signal strength yet have poor signal quality. In a typical antenna diversity application, each access point samples the signals from the two transmitting antennas and uses the best combination of signal strength and quality. Antenna diversity is not meant to extend a Wi-Fi router's coverage range or bandwidth but to enhance the coverage by eliminating multipath distortion and signal nulls.

Spacing is an important consideration in antenna diversity designs. If the antennas are separated by more or less than one wavelength distance, the coverage area of each antenna differs. If the coverage differs too much, the benefit of antenna diversity can be lost. While antenna diversity relies on the use of two antennas, many Wi-Fi routers have three or more antennas. Antenna diversification is meant to improve coverage by eliminating multipath distortion and signal nulls; the use of three or more antennas can improve the speed and reliability of transmissions.

## 1.7 Future of Wireless Communications

Over the past several years, the devices for wireless networks have made tremendous progress in terms of increased data speed per user, lower cost, reduced latency, and increased efficiency. The most prominent wireless networks: cellular communication and Wi-Fi have already undergone several generational advancements over the years and the transition to the next generation of wireless technology is already underway. The deployment of 5G has introduced an ultra-broadband and low-latency network. Wi-Fi has evolved from a way to connect to the Internet, to a serious competitor to other forms of wireless networks. Wi-Fi and cellular data services are competing technologies today, particularly given the massive rise in data usage per user brought about by social media, streaming services, online gaming, video calls, and frequent application updates. Due to this increase in traffic, cellular and Wi-Fi providers are competing more and more to capture their market share.

### *1.7.1 6G—The Future Network Platform*

For the 6G networks of 2030, a broad range of promising technologies should be considered. The study of these potential elements of 6G, shown in Fig. 1.10, will be a key topic of research in the coming years [42]. The 6G era is expected to be a hyper-collaboration society with various services and research findings converging to open a new era in communication. In the deployment of the 6G network, it is expected that software-based network architecture using software-defined radio/infrastructure will be widely employed.

The four main challenges for the future 6G era are the trustworthiness of the systems, sustainability through the efficiency of mobile technology, accelerated automation, and limitless connectivity meeting the demands for communication anywhere, anytime, and for anything. Triggered by these challenges, new application areas will appear in the coming days, calling for new capabilities in future networks.

***Research Directions in 6G Communications*** [43]

1. THz Channel Modeling: The ultra-high-frequency band to be used in 6G THz communication, 0.1–10 THz, is sensitive to atmospheric attenuation, weather effects, such as rain or fog, and high molecular absorption loss. Therefore, it is important to model the THz communication channel considering the effects of reflection, diffraction, and scattering in the atmosphere.

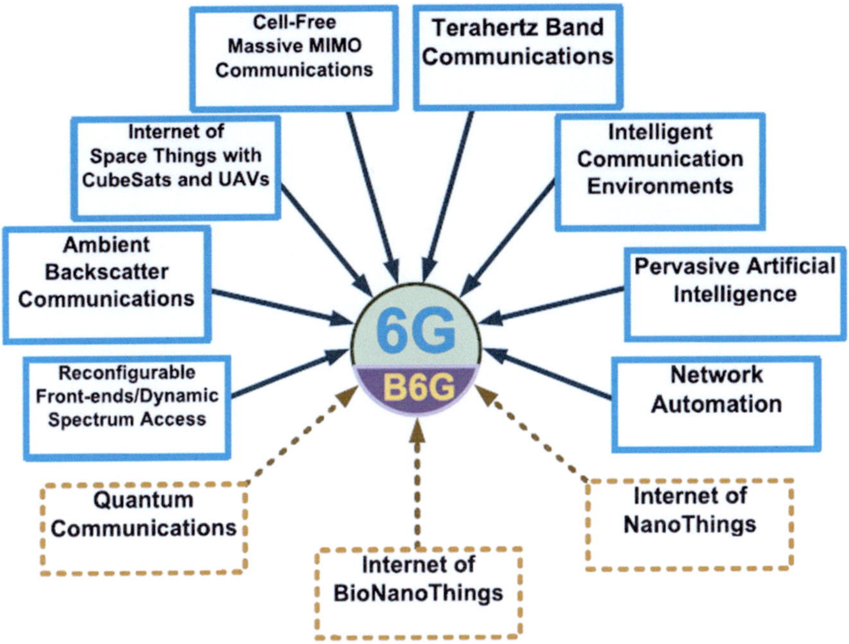


**Fig. 1.10** Key candidate technologies of 6G [42]

2. Ultra-massive MIMO and Intelligent Surface: An ultra-massive, multiple-input, multiple-output (MIMO) technique is being investigated to combat the THz channel's significant free-space path loss. Due to the THz band's short wavelength, a very large antenna array can be created by integrating more than 10,000 extremely small antennas, which allows for the formation of an ultra-narrow beam to reduce path loss. The use of intelligent reflecting surface (IRS) technology is being investigated as a potential solution to the THz channel's significant route loss. By overcoming path loss and securing line-of-sight (LoS), IRS technology can enhance communication propagation distance. This can be achieved by regulating the wireless channel environment with an additional intelligent antenna between base stations [44].
3. Signal Processing for Ultra-Broadband Communication: High-speed transmission in 6G THz communication can be accomplished using a broad bandwidth of 10 GHz or more, but doing so will greatly raise the noise floor and provide technical difficulties for the implementation of radio frequency (RF) equipment. Hence, a technique for configuring ultra-wideband in THz communication is taken into consideration. Hardware calibration of phase noise and the carrier frequency of numerous carriers is required to do this.

The performance requirements of 6G communications [45] are as follows:

1. High-Performance Networking—6G is a complex networking system with many heterogeneous space-air-ground-underwater communication networks. With the help of massive-scale heterogeneous networks, 6G communications can achieve up to 1 Tbps data rate per user [46] with ultra-low end-to-end delay.
2. Higher Energy Efficiency—Long battery life and low energy consumption are two popular research topics for 6G communications. Various studies have proposed energy harvesting technology, and wireless power transfer technology to improve energy efficiency and extend the working time of wireless devices [47].
3. High Security and Privacy—Federated Learning, an emerging decentralized machine learning (ML) solution can be used to achieve privacy-enhanced deep learning in 6G networks.
4. High Intelligence—For 6G networks, intelligent applications are one baseline of application requirements. The high intelligence of 6G will be beneficial to provide users with high-quality, personalized, and intelligent services like intelligent voice assistants to complete daily tasks.
5. Increased Device Density—6G networks are made of smaller but large number of radio cells that quickly transmit data. These cells consist of high-performance transmission links with high transmission rates, shorter delay, greater device density, and the integration of artificial intelligence.
6. Green Communication—Green communication techniques optimize energy usage and energy scheduling for wireless devices. Advanced machine learning techniques like deep reinforcement learning optimizes the computation tasks of a wireless device lowering the energy consumption and enhancing the energy efficiency.

### *1.7.2 Wi-Fi 8*

As the product development of Wi-Fi 7 is underway, the industry has already begun the discussions for Wi-Fi 8 specifications. Starting with Wi-Fi 8 [48], ways to improve the spectral efficiency and reduce latency will become a major focus area for development. The focus will also be on delivering increased network capacity, better quality of service and higher reliability. It will be achieved through a collaborative multidevice operation mode, new transmission schemes, and increased access point managed network operation. Wi-Fi 8, while coexisting with legacy devices, will be optimized for Extended Reality (XR) usages, Industrial IoT, metaverse, improvements in P2P operations, and in overall networking performance. The discussion to define the Wi-Fi 8 in the 60 GHz band or to align with sub-8 GHz band is still underway. The 802.11bf task group is assigned to review the Wi-Fi 8 specifications and see how to best enable mmWave Wi-Fi sensing as part of Wi-Fi 8.

### 1.7.3 Wireless-Power Transfer

Today, wired Internet connections are being replaced by wireless LANs. The wireless connectivity has exploded with the advent of new generations of mobile and wireless devices. However, one aspect of our technology usage that is still tied to cables is in the need for power, even while using battery-operated devices. For charging the batteries we still must connect it to the socket through a wire. The wireless power transfer technology has the potential to enable charging without the use of wires. WPT technology allows mobile devices to be charged at any location and in any direction, even if the devices are away from the power source. With this technology, so long as mobile users stay in a designated area where the charging is available, the device will pick up power automatically, as needed. The WPT system is capable of charging multiple mobile devices concurrently even while holding the devices in midair or 0.5 m away from the power source, which is a transmitter [49]. This technology is not new and it was initially proposed by Nikola Tesla decades ago. Currently, there has been a resurgence of interest on wireless power transfer. The resurgence is due to the prospect that if massively deployed, they could open the door to transferring enough electrical energy to efficiently power mobile electronic devices without having to locate where the electrical power outlet is in relation to the mobile device [50]. Unlike the power wireless communication uses, the transmitted electromagnetic power required for commercial implementation of wireless power transfer could be substantial. More research is also needed to understand the biological effects and human safety of such implementation.

### 1.7.4 Internet of Nanothings

As the concept of wireless is becoming ubiquitous, there are certain situations where electromagnetic waves do not yield acceptable performance. In applications with THz band communications, the operating frequency falls to the nanometer range, triggering the need for nanonetwork communications. The devices and transceivers used in the Internet of Nanothings (IoNT) [42] are in the scale of nanometers and behave differently from the classical wireless communication systems that operate at lower frequencies in the microwave range. Given the smaller size, each nanothing consumes much less energy and is envisioned to be self-powered via vibrational energy harvesting using piezoelectric nanogenerators. Communications in the paradigm of nanonetworks are implemented through plasmonic radiation on metamaterial-based antennas including graphene and carbon nanotubes operating in the THz band. In the realm of IoNT, applications can be primarily found in body area networks, short-distance local environments, nanocameras, on-chip networks, and nanorobots for IoNT.

## References


1. Hellstrom, H., Luvisotto, M., Jansson, R., Pang, Z.: Software-defined wireless communication for industrial control: a realistic approach. IEEE Ind. Electron. Mag. **13**(4), 31–37 (2019)
2. Hamza, A.S., Deogun, J.S., Alexander, D.R.: Wireless communication in data centers: a survey. IEEE Commun. Surv. Tutorials **18**(3), 1572–1595 (2016)
3. Madjar, H.M.: Radio Spectrum Management: Policies, Regulations and Techniques. Wiley, United Kingdom (2016)
4. Whitaker, J.C.: The Resource Handbook of Electronics. CRC Press, United States (2018)
5. Hucaby, D.: CCNA Wireless 640–722 Official Cert Guide. Pearson Education (2014)
6. Labiod, H., et al.: Wi-Fi™, Bluetooth™, Zigbee™ and WiMax™. Springer (2007)
7. Mishra, A.R.: Fundamentals of Network Planning and Optimisation 2G/3G/4G: Evolution to 5G. John Wiley & Sons Ltd. (2018)
8. Zhang, Z.: Antenna Design for Mobile Devices. Wiley, Singapore (2017)
9. Hong, W., Sim, C.Y.D.: Microwave and Millimeter-Wave Antenna Design for 5G Smartphone Applications. Wiley-IEEE Press (2023)
10. LPWAN Technologies for IoT and M2M Applications. Elsevier Science, United Kingdom (2020)
11. Gast, M.: 802.11 Wireless Networks_The Definitive Guide-O'Reilly Media (2002)
12. Salazar-Palma, M., Sengupta, D.L., Sarkar, T.K., Oliner, A.A., Mailloux, R.: History of Wireless. Wiley, Germany (2006)
13. Meinel, C., Sack, H.: Internetworking: Technological Foundations and Applications. Springer, Berlin Heidelberg, Germany (2013)
14. Youssef, A., et al.: Wi-Fi Enabled Healthcare. CRC Press, United States (2014)
15. O'sullivan, J.D., Daniels, G.R., Percival, T.M.P., Ostry, D.I., Deane, J.F.: Wireless LAN, US5487069A, 23 Jan 1996
16. Sherlock, I.: Wi-Fi alliance: connecting everyone and everything, everywhere. IEEE Commun. Stand. Mag. **1**(3), 6–6 (2017)
17. https://www.ieee802.org/
18. IEEE standard for telecommunications and information exchange between systems—LAN/ MAN specific requirements—Part 11: Wireless medium access control (MAC) and physical layer (PHY) specifications: high speed physical layer in the 5 GHz band. In: IEEE Std 802.11a-1999, pp. 1–102, 30 Dec 1999
19. Tuch, B.: Development of WaveLAN, an ISM band wireless LAN. AT&T Tech. J. **72**(4), 27–37 (1993)
20. Yeh, J.-H., Chen, J.-C., Lee, C.-C.: WLAN standards. IEEE Potentials **22**(4), 16–22 (2003)
21. IEEE standard for wireless LAN medium access control (MAC) and physical layer (PHY) specifications. In: IEEE Std 802.11-1997, pp. 1–445, 18 Nov 1997
22. Labiod, H., et al.: Wi-FiTM, BluetoothTM, ZigbeeTM and WiMaxTM. Springer Netherlands, Germany (2007)
23. Reynolds, J.: Going Wi-Fi: Networks Untethered with 802.11 Wireless Technology. CRC Press, United States (2003)
24. IEEE Standard for Information technology—Local and metropolitan area networks—Specific requirements—Part 11: Wireless LAN Medium Access Control (MAC) and Physical Layer (PHY) specifications amendment 1: Radio resource measurement of wireless LANs. In: IEEE Std 802.11k-2008 (Amendment to IEEE Std 802.11-2007), pp. 1–244, 12 June 2008
25. Wi-Fi 6E: The Next Great Chapter in Wi-Fi, Cisco, White Paper (2021)
26. Alsabah, M., et al.: 6G wireless communications networks: a comprehensive survey. IEEE Access **9**, 148191–148243 (2021)
27. Wang, S., Cai, Y., Xu, Y., Cai, Y.: Wireless Communication Network Technology and Evolution. World Scientific Publishing Company (2022)
28. Karmakar, G., Dooley, L.S.: Mobile Multimedia Communications: Concepts, Applications, and Challenges. Information Science Reference, Ukraine (2008)

29. Mishra, A.R.: Fundamentals of Network Planning and Optimisation 2G/3G/4G: Evolution to 5G. Wiley, Germany (2018)
30. White Paper 5G Evolution and 6G NTT DOCOMO, INC. (January 2020)
31. WLAN Transceiver Whitepaper A Technology Review and Comparison of Five Manufacturers, Semiconductor insights (2003)
32. Rogers, J.W.M., Dai, F.F., Cavin, M.S., Rahn, D.G.: A multi-band AT fractional-N frequency synthesizer for a MIMO WLAN transceiver RFIC. J. Solid-State Circ. **40**(3), 678–689 (2005)
33. Shi, Y., et al.: A 5.2 GHz low power transceiver RFIC for WLAN applications. In: 2007 IEEE Bipolar/BiCMOS Circuits and Technology Meeting (2007)
34. Wireless Networking over Extended Range, US Robotics. White Paper (2005)
35. Chen, Z.N., Qing, X., See, T.S.P., Toh, W.K.: Antennas for WiFi connectivity. Proc. IEEE **100**(7), 2322–2329 (2012)
36. Zhou, L., Jiao, Y., Qi, Y., Weng, Z., Lu, L.: Wideband ceiling-mount omnidirectional antenna for indoor distributed antenna systems. IEEE Antennas Wirel. Propag. Lett. **13**, 836–839 (2014)
37. Cisco Aironet and Catalyst Antennas and Accessories (2019)
38. Jang, T.H., et al.: Wideband wide beam-width modified angled dipole antenna for 5G millimeter-wave IoT applications. IEEE Access **11**, 63324–63332 (2023)
39. Choosing the Right WiFi Antenna for your Application, L-com, Infinite Electronics, White Paper (2019)
40. Zhang, E., Michel, A., Pino, M.R., Nepa, P., Qiu, J.: A dual circularly polarized patch antenna with high isolation for MIMO WLAN applications. IEEE Access **8**, 117833–117840 (2020)
41. Bartz, R.J.: Mobile Computing Deployment and Management: Real World Skills for CompTIA Mobility Certification and Beyond. Wiley (2015)
42. Akyildiz, I.F., Kak, A., Nie, S.: 6G and beyond: the future of wireless communications systems. IEEE Access **8**, 133995–134030 (2020)
43. Hong, E.-K., et al.: 6G R&D vision: requirements and candidate technologies. J. Commun. Netw. **24**(2), 232–245 (2022)
44. Tataria, H., Shafi, M., Molisch, A.F., Dohler, M., Sjoland, H., Tufvesson, F.: 6G wireless systems: vision, requirements, challenges, insights, and opportunities. Proc. IEEE **109**(7), 1166–1199 (2021)
45. Liu, Y., Yuan, X., Xiong, Z., Kang, J., Wang, X., Niyato, D.: Federated learning for 6G communications: challenges, methods, and future directions. China Commun. **17**(9), 105–118 (2020)
46. Xiao, Y., Shi, G., Krunz, M.: Towards ubiquitous AI in 6G with federated learning. arXiv preprint arXiv:2004.13563 (2020)
47. Dang, S., Amin, O., et al.: What should 6G be? Nat. Electron. **3**(1), 20–29 (2020)
48. Reshef, E., Cordeiro, C.: Future directions for Wi-Fi 8 and beyond. IEEE Commun. Mag. **60**(10), 50–55 (2022)
49. Gozalvez, J.: Advances in wireless power transfer [Mobile radio]. IEEE Veh. Technol. Mag. **10**(4), 14–32 (2015)
50. Lin, J.C.: Wireless power transfer for mobile applications, and health effects [Telecommunications health and safety]. IEEE Antennas Propag. Mag. **55**(2), 250–253 (2013)

# Chapter 2
# Pattern Diversity in WLAN Antennas

## 2.1 Introduction

This chapter introduces the concept of pattern diversity and different techniques to achieve the same. The discussion on pattern diversity, from evolution to the latest trends is covered over the course of this chapter. To understand the concept of pattern diversity, one needs to understand how the waves travel from the transmitter antenna to the receiver antenna. So, a discussion on radio wave propagation is also included in this chapter. Diversity in terms of antenna design is comprised of antennas with different radiation patterns, polarizations, frequencies, and so on. The concept of diversity is very advantageous in antenna design as the electromagnetic (EM) waves after traveling long distances get affected by signal fading. As more antennas are placed at the receiver, it can collect more accurate information coming from different paths from the transmitter antenna. As we go further into the chapter, pattern diversity antennas with polarization diversity and reconfigurability are also discussed as higher diversity delivers a higher throughput of the system.

### *2.1.1 Radio Wave Propagation in Wireless Communication*

One of the prerequisites for understanding diversity is the knowledge of the propagation of EM waves, which are used for the transmission of information. When a signal propagates between a base station and a mobile station, it comes across various obstacles in its path. Outdoors, the obstacles consist of buildings, and trees, and indoors the obstacles consist of office machines, furniture, walls, etc. These obstacles can have a considerable influence on the propagation of radio waves. The mobile radio environment has five unique factors [1]. These factors are significant because the mobile antenna is very close to the ground.



S. K. Koul et al., *Antenna Systems for Modern Wireless Devices*, Signals and Communication Technology, https://doi.org/10.1007/978-981-97-3369-9_2

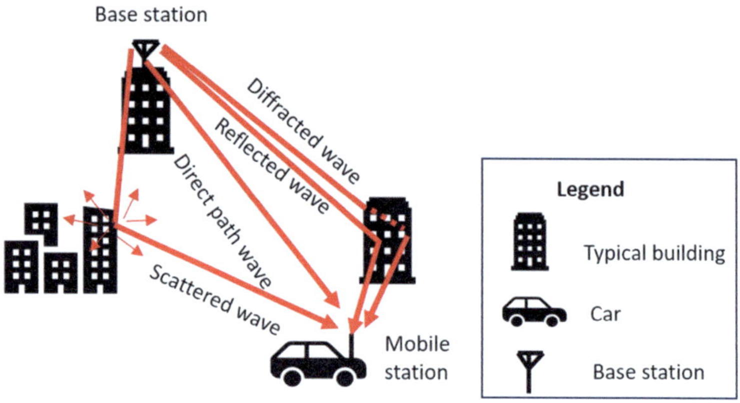


**Fig. 2.1** Radio wave propagation in mobile communication

- Natural terrain configurations such as flat ground, hills, water, mountain, valley, and desert.
- Manmade structures such as open areas, suburban and urban areas, and metropolitan areas.
- Manmade noise such as automotive ignition noise and machine noise.
- Moving medium brought about by the mobility of mobile and portable units.
- Dispersive medium causing frequency-selective fading and time-delay spread

Figure 2.1 illustrates typical conditions for radio wave propagation in mobile systems [2]. In wireless communications, the radio waves transmitted from an outdoor base station are radiated in all directions and do not necessarily reach the receiver directly. When radio waves propagate in space it undergoes several different physical mechanisms: such as free-space propagation or line-of-sight propagation, reflection, transmission, diffraction, scattering, and waveguiding.

In free-space propagation, a wave propagates without encountering any obstacle. The attenuation in free space results from the scattering of energy which occurs as the wave propagates away from the transmitter. Free-space attenuation is a function of the distance and the frequency. The phenomenon of reflection occurs when a propagating wave impinges upon a surface with large dimensions compared to the wavelength like walls of buildings or other ground obstacles. The phenomenon of transmission is the process whereby waves propagate through a medium, like vacuum, air, or an obstacle, without a change of frequency, but with a drop in magnitude. The phenomenon of diffraction is the bending of the radio waves when the obstacles have sharp corners and edges. It occurs when waves impinge upon an obstacle or an aperture with large dimensions compared to the wavelength. Scattering is the phenomenon where the energy of an electromagnetic wave is distributed in a propagation medium along several directions when the wave hits a rough, non-uniform surface or heterogeneities with small dimensions compared to the wavelength. Certain environments, for instance, streets flanked by tall skyscrapers, corridors, or tunnels behave like

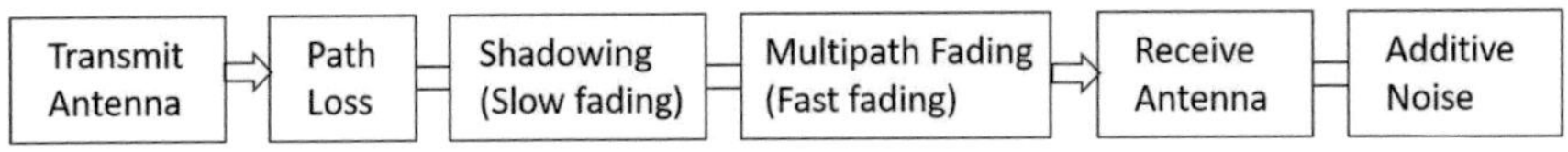


**Fig. 2.2** Contributions to noise in the wireless channel

waveguides with respect to the propagation of radio waves. The wave propagates by successive reflections off the surfaces, similar to the propagation inside a waveguide.

During its propagation, a wave follows different paths between the transmitter and the receiver. Depending on the nature of the obstacles, the wave encounters during its propagation, it is subjected to different types of phenomena. This results in a multitude of elementary paths and the signals reach the mobile station at different times carrying the same signal information. Each such path is characterized at the receiver level by an attenuation, a delay, and a specific phase difference. This mode of propagation is referred to as multipath propagation [3]. Different waves propagated along multiple paths interfere with each other at the receiver.

### *2.1.2 Fading in a Wireless Channel*

Figure 2.2 shows various types of fading an EM wave experiences during its propagation from a transmitter antenna to a receiver antenna. The entire fading process can be subdivided into path loss, shadowing, and fast fading. Path loss is the reduction in the power of an electromagnetic wave as it travels through space. Path loss leads to an overall decrease in signal strength as the distance between the transmitter and the receiver increases. Shadowing is caused by obstacles between the transmitter and receiver that absorb power [4]. When the obstacle absorbs all the power, the signal is blocked. Fast fading or multipath fading results from the constructive and destructive interference between multiple waves reaching the mobile from the base station [5].

### *2.1.3 Multipath Fading*

When signals reach the receiver through multiple paths, the signals arriving through these paths add in phase. As the signal travels through the intervening medium the signal power is lost due to attenuation and absorption by the medium resulting in a reduction in the received power at the receiver. This random fluctuation of power is identified as “fading” in wireless systems [6]. As the signal reaches through multiple paths, this phenomenon is referred to as multipath fading. This multipath fading results in errors in the signal reception.

Multipath fading is also called Rayleigh fading in the mobile radio environment. When these multipath waves bounce back and forth due to the buildings and houses,

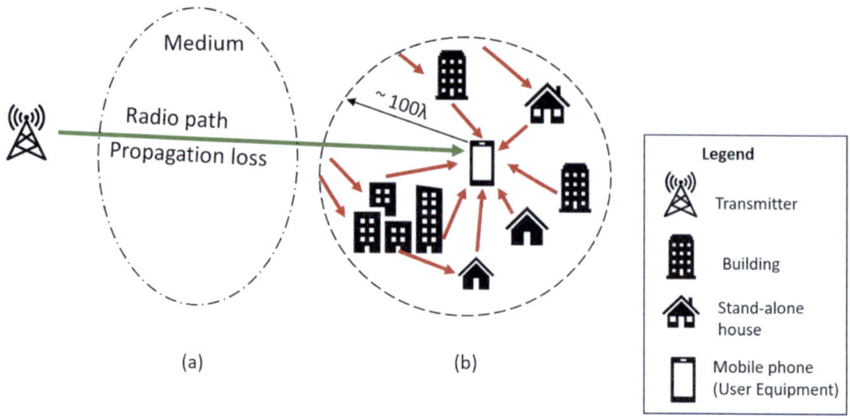


**Fig. 2.3** A mobile radio environment **a** propagation loss **b** multipath fading

they form many standing-wave pairs in space, as shown in Fig. 2.3. Those standing-wave pairs are summed together and become an irregular wave-fading structure [7]. When a mobile unit is standing still, its receiver only receives a signal strength at that spot, so a constant signal is observed. When the mobile unit is moving, the fading structure of the wave in the space is received. The recorded fading increases with the vehicle's speed.

## 2.2 Significance of Diversity in Wireless Communication

### *2.2.1 Diversity Techniques*

Diversity reception is used for mobile terminals and base stations to help compensate for very weak signal reception caused by multipath fading. The base station transmits tens of watts, while the mobile terminal radiates power can be less than 1 W in current cellular systems. The main purpose of using diversity reception is to compensate for the reduction in signal level due to fading. There are several antenna diversity techniques to help reduce fading. Based on the propagation mechanism and physical arrangement, the diversity schemes may be categorized into six major types: space, polarization, pattern, frequency, and time, as shown in Fig. 2.4.

The basic principle of antenna diversity is that multiple antenna outputs experience different signals due to different channel conditions, and these signals are only partially correlated. During transmission in a multipath propagation condition, the signals encounter blocked or shadowed direct line-of-sight (LOS) path, and each receiver antenna experiences a different fading environment. Thus, it is highly likely that if one antenna is in a deep fade, then the other one is not, and provides sufficient signal [8].

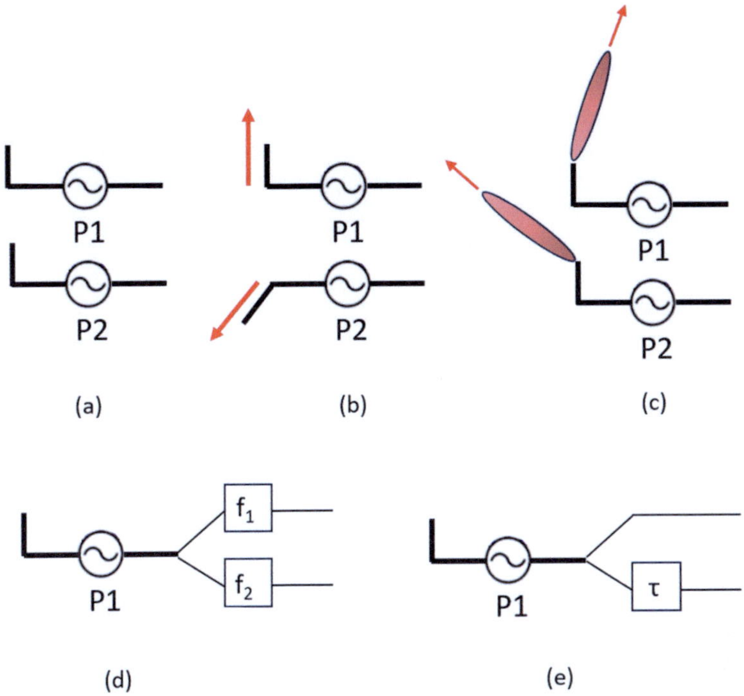


**Fig. 2.4** Various diversity reception schemes **a** space diversity, **b** polarization diversity, **c** antenna pattern diversity, **d** frequency diversity, and **e** time diversity

#### 2.2.1.1 Space Diversity

Antenna diversity or Space diversity or Spatial diversity is a wireless diversity scheme that uses two or more antennas separated in space for reception or transmission to improve the system reliability. It uses several independent signals from multiple antenna elements placed at separate locations on the receiver. The space diversity concept relies on the fact that if the antennas are sufficiently spaced, the paths coming to various receivers are independent of each other. If the spacing between the receivers is reduced, the signals received by the receivers will become correlated. The independence of these signals is crucial for an acceptable diversity performance. A spacing of $\lambda/2$ ($\lambda$ is the mean wavelength of the transmitted signal) between antennas is sufficient to produce uncorrelated signals from each of them [9]. However, space diversity systems are widely used at the cellular base station antennas and the antenna gain of the base station is 10 dBi or more. This requires antennas to be spaced by more than 10–20 wavelengths apart to decrease the mutual interference between the antennas [10].

#### 2.2.1.2 Polarization Diversity

In polarization diversity, multiple antennas provide two or more different polarizations. It assumes that the signals transmitted or received in two orthogonal polarizations exhibit uncorrelated behavior and hence are independent of each other [11]. Polarization diversity also takes advantage of the normal and cross-polarization components of the received signal. When the transmitter operates in a single (vertical or horizontal) polarization, the scattering takes place in the propagating channel which can couple some energy from one polarization to the other. Thus, if we have two antennas at the receiver, one for each polarization, polarization diversity of order two can be achieved. As the two polarizations are orthogonal, positioning and locating the antennas is easier with polarization diversity. The polarization diversity concept can be extended further by using two transmitting antennas and two receiving antennas providing pairs of orthogonal polarization at both transmitter and receiver. The polarization diversity is a form of space diversity since multiple antennas are used.

#### 2.2.1.3 Antenna Pattern Diversity

Pattern diversity, also known as angle diversity, consists of two or more antennas that are physically separated by some short distance with different radiation patterns. It utilizes the signals from multiple antennas with non-overlapping radiation patterns [12]. Pattern diversity occurs in many instances at the mobile terminals because the antennas will pick up signals coming from different angles. In addition to the pattern diversity antennas, a type of directional antenna called sector antenna with a sector-shaped radiation pattern is widely used for base stations. This antenna divides the coverage area into various sectors. It helps to eliminate the signal delay due to multiple reflections and can be used for applications like wireless LANs or mobile communication systems.

#### 2.2.1.4 Frequency Diversity

Frequency diversity uses two or more widely spaced frequency channels to send the same message, usually at the same time. The separation between the carrier signals must be larger than the coherence bandwidth of the channel so that the signals can be treated as uncorrelated. Even though a higher separation between the carrier frequencies assures uncorrelated signals, increasing separation will be at the cost of reduced bandwidth available for data transmission. Frequency diversity can be implemented using one antenna at the transmitter and receiver. But because some of the transmit power is used to send data in bands that are in deep fades, the power efficiency is less compared to space diversity, where the transmitter sends all its power in one channel. For simultaneous transmission and reception, multiple transmitters, antennas, and receivers are required, all of which can be costly in size, cost, power, and management.

#### 2.2.1.5 Time Diversity

Time diversity is based on the premise that if the information is transmitted sequentially through a random fading channel, the multiple replicas of the signal will be uncorrelated if the time separation among the samples is sufficiently large. The coherence time is dependent on the speed of the transmitter/receiver. If there is no motion, time diversity ideally does not produce uncorrelated replicas. In practice and with time, even in the absence of any motion, one can treat the channel as dynamic from motions of the scatterers from wind, change in temperature, and so on. Time diversity also requires that information be stored at the transmitter and receiver which makes it rather inconvenient to use time diversity. Additionally, repeated transmission of the same information and subsequent processing produces time delay (latency effects). Because of this, time diversity is often used sparingly.

### 2.2.2 *Need for Pattern Diversity Antennas*

In a pattern diversity communication system, all the antennas are placed in the same place pointing to different directions in space or having a phase shift between the co-located antennas, so that their radiation patterns are distinctly directed in space. An illustration of the pattern diversity system is given in Fig. 2.5 which shows three antennas placed distinctly at the same location, which are radiating in three different directions. Since their radiation patterns are directional, there is no coupling between the antennas, which further improves the envelope correlation coefficient (ECC). The envelope correlation between the received signals is different if the directions of the patterns are different because the relative amplitudes of the received signals are different for each antenna even though these antennas are co-located [13].

The radiation pattern can be distorted because of the mutual coupling between the antennas. The effectiveness of the pattern diversity in handsets should be high because, in a mobile system, signals can come from any direction over a narrow-angle spread, and this could improve the performance of the mobile handsets. This method reduces the antenna spacing, still maintaining a high diversity gain.

An attractive feature in many communication systems is the capability to support multidirectional far-field radiation patterns. These systems face the challenge of maintaining identical characteristics in different radiation patterns. Pattern diversity antennas can provide the required coverage in such cases.

1. Reduce multipath fading—Pattern diversity has multiple antennas with different antenna patterns receiving different sets of multipath waves. As the antennas receive waves from different directions it reduces the fading effect.

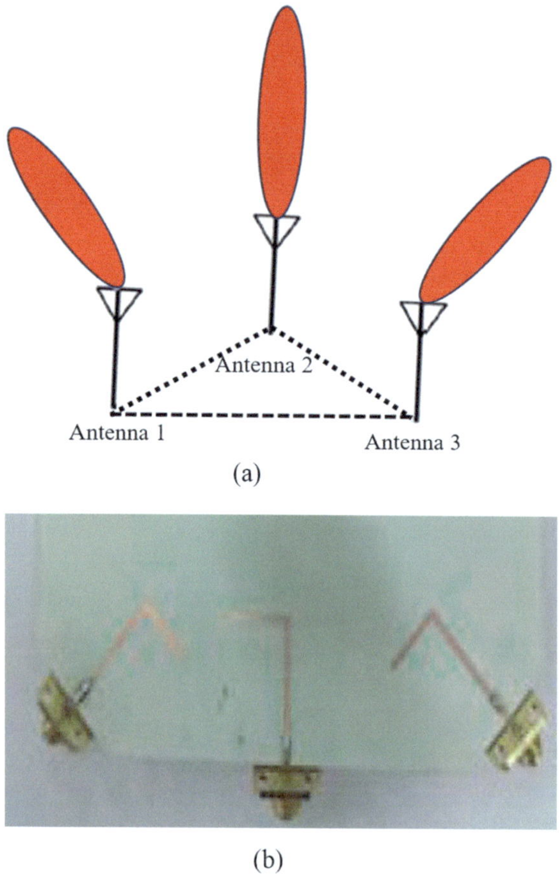


**Fig. 2.5** **a** Antenna configuration showing pattern diversity **b** example of a pattern diversity system

2. Improved reliability in multipath channels—Pattern diversity helps to reduce disturbances from adjacent channels onto the required network. The interferences from reflected signals in multipath signals causes fading of the received signal. The fade level experienced on an average for a given outage probability (percent down time) is decreased through diversity. Systems that use diversity combining can provide 10 dB or more diversity gain for the worst 1% of cases [14].
3. Increase in overall average received power—Systems that use polarization or angle diversity automatically adjust the antenna characteristics to produce a near-maximum received signal, increasing the efficiency of the radio link. For example, a radio without polarization diversity can easily experience 10—20 dB decrease in mean received signal power due to polarization mismatch, but with a simple polarization diversity system the worst-case mismatch is 3 dB. Handheld radios that use diversity antennas have system-level benefits. Diversity systems permit the use of lower transmit power for a given level of reliability. This decreases

interference, increases battery life of handheld radios, and reduces the probability that a hostile party will intercept the signals.

4. Wider antenna beam coverage—Today, as many state-of-the-art communication technologies are being used, the coverage provided by a single antenna becomes insufficient. Therefore, we need antennas that can provide wider beam coverages. But as the beamwidth of the antenna increases, its gain reduces. Pattern diversity antennas provide efficient and sustained communication without suffering any signal drops.
5. Higher antenna diversity gains—WLAN antennas are often small and require an omnidirectional radiation pattern due to area limitations in wireless devices [15]. In handheld devices, the antenna spacing is relatively small which results in a high correlation between the received antenna signals, resulting in a lower diversity gain. This problem can be solved by using antenna-pattern diversity [16].
6. Pattern diversity antenna configuration is also an effective solution for RF harvesting because the real environment is full of multipath signals [17].

### 2.2.3 Significance of Pattern Diversity

For the next-generation wireless systems to deliver high data speeds and expanded channel capacity, diversity antenna structures are necessary. The pattern diversity antennas that are described in the subsequent chapters of the book consist of multiple radiators and each radiator is designed to focus on different directions to ensure adequate beam coverage. In the case of a phased-array antenna each radiator has to be fed with different phases for the beam to be steered electronically and to ensure the adequate beam coverage for the beams coming from different directions. The addition of the phase shifters makes the antenna bulky and costly compared to a pattern diversity antenna. A reconfigurable pattern diversity antenna would require active devices for the antenna to steer the beam in the required directions. The bias lines needed for the active devices could complicate the design and affect the overall antenna performance. The pattern diversity antennas that will be described make use of individual antennas placed close together in multiple ports and designed to contribute to the different antenna beams. The number of ports can be increased or decreased based on the gain and beam coverage requirement of a particular design. The designs consist of only passive structures which eliminate the use of bias lines and phase shifter networks.

## 2.3 History of Pattern Diversity Antenna

One of the fundamental motives for employing pattern diversity antennas is to overcome the multipath fading effects. The multipath fading and the flat fading experienced by the two received signals can be decorrelated if the antennas are sufficiently separated. One of the earliest experiments to understand the relationship between the received signal fluctuation and the direction of the main lobe of the receiver antenna pattern was conducted by Bruce [18]. Bruce and Beck [19] found a remarkable reduction in fading by using a horizontal rhombic antenna with an extremely sharp directional pattern. The reduction was based on the existence of stable angular separation among signal components following different paths, thus enabling a steerable antenna with sufficiently sharp directivity to accept only one of them, maybe the best after some calibration, at any time.

The qualitative relation between angles of arrival and propagation delays of the received signal components was demonstrated by Friis [20]. He concluded that greater the delay in the arrival of a signal, the greater the angle above the horizontal (or azimuth) plane. Friis and Feldman [21] exploited this phenomenon with the aid of a special arrangement whereby individual wave groups arriving at different vertical angles were received separately and, after delay equalization, combined coherently to offer directional diversity or directivity diversity. They accomplished that with the multiple-unit steerable antenna (MUSA) system. According to Gregory and Newsome, the MUSA array was the last major technological development in the short-wave communication era, representing the ultimate receiving system [22]. It was probably the most complex radio receiver ever built and gave valuable service between 1940 and 1960s. Later, Vogelman [23] extended the concept of a single beam to a multibeam steerable antenna, where signal components associated with individual beams were particularly combined to offer angle diversity or angle-of-arrival diversity. However, the effectiveness of this system depends upon the antenna characteristics, which have to be properly chosen so that the angle between adjacent beams is always smaller than the angular spread of the channel and that low correlation among different beams is ensured.

Pattern diversity can be provided via several antennas with diverse patterns spaced apart from each other. When antennas with similar patterns are placed close to each other, the mutual coupling effects alter their antenna pattern, thus yielding an effective pattern diversity scheme. The concept of the energy density antenna to mitigate signal fading in mobile radio channels was first proposed by Gilbert [24]. It utilizes sophisticated energy density antenna arrangements to receive uncorrelated electric and magnetic field components of the transmitted signal so as to reduce fading in mobile radio.

One of the earliest methods to achieve pattern diversity uses two horizontally separated antennas which either have different patterns or are purposely misaligned in the elevation plane relative to boresight and to each other. Another method is to provide one antenna with two separate main beams which can be generated, for example, by using two misaligned feeds. This technique would result in an antenna with deteriorated sidelobe performance and higher cross-polarization. In [25] the use of mode separation technique in a horn reflector antenna to achieve pattern diversity is discussed. Significant improvements are shown in system performance in both Rayleigh and Rician fading environments when antennas with different radiation patterns are employed [26]. To introduce significant subchannel decorrelation to the MIMO system that has insufficient space available, the MIMO designer can make use of pattern diversity as proposed by Vaughan [27]. The proposed design has switched parasitic elements that can either be terminated to the ground or left open, to realize antenna-pattern diversity. In [28] pattern diversity as the natural generalization of conventional orthogonal polarization diversity is defined since the orthogonality between patterns decorrelates the signals in highly scattering environments. Pattern diversity having better performance than spatial diversity when used in handsets is identified in [29]. Diversity gain is explained by the fact that different antenna patterns receive different sets of multipath waves. This gain will be largest if the two sets are completely independent that can, in theory, be realized with two antenna patterns that have zero overlaps.

## 2.4 Various Pattern Diversity Techniques

Pattern diversity is one of the techniques to achieve antenna diversity. When excited by two or more co-located antennas, different radiation patterns can be combined. Different methods to achieve pattern diversity are discussed below.

### *2.4.1 Integration of Multiple Radiators*

A traditional method to realize pattern diversity is to integrate two or more separate radiators with different radiation patterns as shown in Fig. 2.6. Though this approach is easy for understanding and implementation, it results in large size of the antenna, which adds to the structural complexity of the system. In [30], two-port pattern diversity antenna is presented by combining a low-profile mono-cone antenna and a broadband microstrip patch antenna. The cup-shaped microstrip patch is loaded on top of the mono-cone antenna and shares part of its radiation structure. The lower antenna has a monopole-type pattern, and the upper antenna has a microstrip-antenna-type pattern. It can be seen that the radiation pattern of the upper antenna compensates for the null of the radiation pattern of the lower antenna. The radiation patterns of the two antennas are complementary and pattern diversity is achieved in 3G and 4G bands. The planar pattern diversity antenna in [31] is composed of two parts, an annular ring, and a circular patch. The outer annular ring results in broadside radiation patterns, while the inner center-fed circular patch provides omnidirectional patterns in the horizontal plane, due to its physical circular symmetry. Both the

Fig. 2.6 Multiple radiators with different radiation patterns

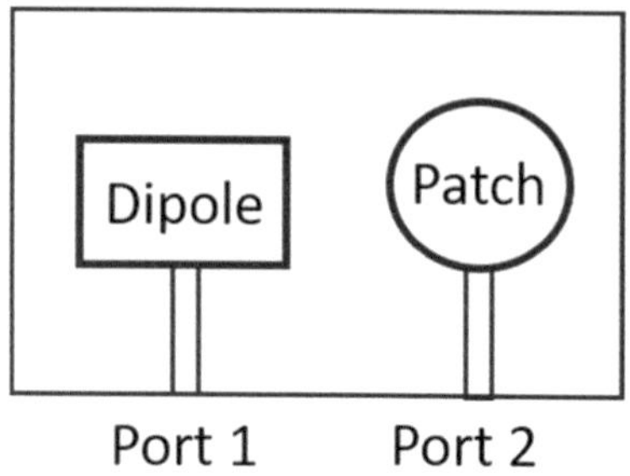


coupling and the correlation coefficient between the two parts are quite low, which increases the diversity performance in a MIMO system.

### 2.4.2 *Different Phase Excitations*

Another method to realize pattern diversity is to excite two or more separate radiators with different phase excitations as shown in Fig. 2.7. Pattern diversity is achieved in [32] by controlling the phase relationship between the two monopole antenna elements through in-phase and out-of-phase signals. The design consists of two closely spaced elements that are excited simultaneously through a compact two-port feed network. The mechanism of providing in-phase and out-of-phase excitations is through varying the lengths of the current paths to the elements. When the current paths to the two elements are equal, it provides in-phase excitation and when the current paths to the two elements have a length difference of $\lambda/2$, where $\lambda$ is the wavelength in the substrate at 3.5 GHz, it provides an out-of-phase excitation. Since the antennas are close to the ground, the direction of the mirrored currents on the ground is opposite to that of the horizontal currents on the monopoles. The radiation ability of the horizontal segments is poor. Thus, the radiation pattern is mainly contributed by the vertical segments.

Fig. 2.7 Multiple radiators with different radiation phase feed

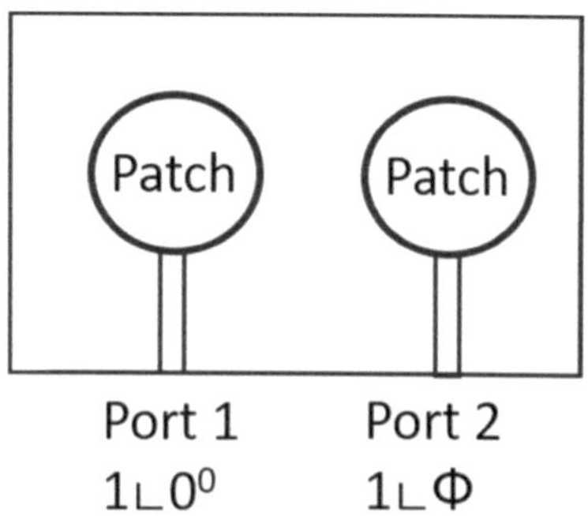

In the in-phase excitation, the currents on the two monopoles are the same. The vertical segments can be viewed as a two-element array with in-phase excitation. The array has maximal gain in the zenith direction. On the other hand, when the two elements are fed with out-of-phase signals, the currents on the two monopoles have opposite direction. Here the two elements can be viewed as an array with 180° out-of-phase excitation and the main beam is spitted into two parts. Thus, the two radiation patterns excited by in-phase and out-of-phase signals can achieve pattern diversity.

### *2.4.3 Multiple Mode Excitation*

To simplify the radiated structures, another approach is to excite multiple modes of a single radiator, corresponding to different radiation patterns as shown in Fig. 2.8. Since different modes of a single radiator are orthogonal, the antenna shows different radiation patterns with different modes.

A wideband circular patch antenna with the loop-fed technique for pattern diversity is proposed in [33]. The $TM_{11}$ mode of the patch is excited by two open feeding loops with in-phase current, which radiates a broadside beam. The $TM_{01}$ mode is excited by another set of feeding loops with an anti-phase current, which radiates conical beam. To excite the $TM_{11}$ mode of the patch, the two loops should provide two equivalent magnetic currents in phase. To excite the $TM_{01}$ mode of the circular patch, the two loops should provide two equivalent magnetic currents out of phase. Two different T-junction power dividers are used for exciting the two current loops. The difficulty in achieving pattern diversity is that the two beams share a common aperture for impedance matching at the same frequency. Two symmetrical printed shorted posts are used for impedance matching in the design.

Circular patch antennas excited in $TM_{11}$ mode lead to broadside radiation, and $TM_{01}$ mode leads to conical radiation. But the pattern diversity element suffers from narrow bandwidth. L-shaped probe feeding technique is used for enhancing the bandwidth of such antennas. By supporting a L-probe coupled circular patch antenna with four metallic posts at appropriate locations, it is demonstrated in [34] that the conical

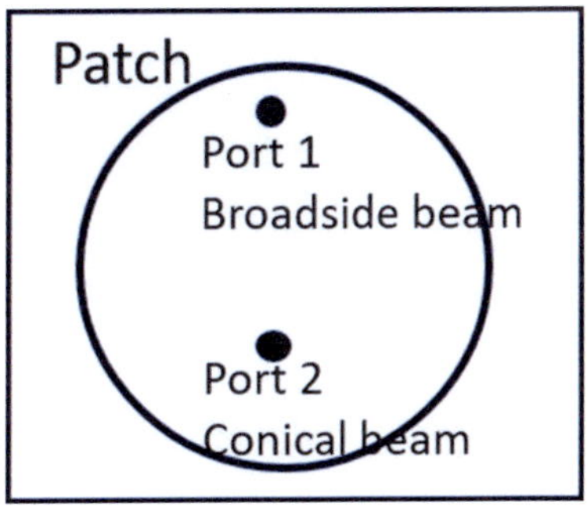


**Fig. 2.8** Exciting multiple modes of a patch

and the broadside modes which are excited separately by two different sets of L-shaped probes, can be operated at the same frequency range, thus improving the overall antenna bandwidth. With suitable switching network, the antenna can be used as a pattern diversity element.

A conformal dual-port diversity patch antenna, operating with a broadside and/or conical radiation pattern that is excited separately by a hybrid feed network is demonstrated in [35]. In conventional circular patches, the $TM_{01}$ mode is a higher-order mode compared to $TM_{11}$ mode. So the two modes are operated at different frequencies. In the proposed design, the $TM_{11}$ mode with broadside radiation is proximity-coupled fed by an H-shaped slot, while a driven rod and the capacitive feed technique are applied to shift down the resonant frequency of $TM_{01}$ mode. By making use of a hybrid feed network, the conical $TM_{01}$ mode and the broadside $TM_{11}$ mode of the antenna can be excited at an overlapped frequency band, making the resonant modes of the antenna to operate at the same frequency.

### 2.4.4 Even and Odd Mode Excitation

Another method for pattern diversity utilizes the even and odd modes of a single radiator as shown in Fig. 2.9. The two modes are excited by 180° out-of-phase and in-phase excitations, which can be delivered by a single feeding network.

A compact dual-mode metamaterial loop antenna for pattern diversity is proposed in [36]. By making use of a hybrid feed network, the even and odd modes of the proposed antenna are exploited for pattern diversity. For odd mode radiation, magnetic dipole-like horizontally polarized omnidirectional pattern in the *xy*-plane is obtained and for even-mode radiation, an omnidirectional pattern in the *yz*-plane is obtained.

A pattern diversity antenna, capable of radiating broadside as well as conical radiation patterns using quarter mode substrate-integrated waveguide subarray is proposed in [37]. The pattern diversity is achieved by in-phase and out-of-phase excitation of the subarray using a rat race coupler feed network. The currents in

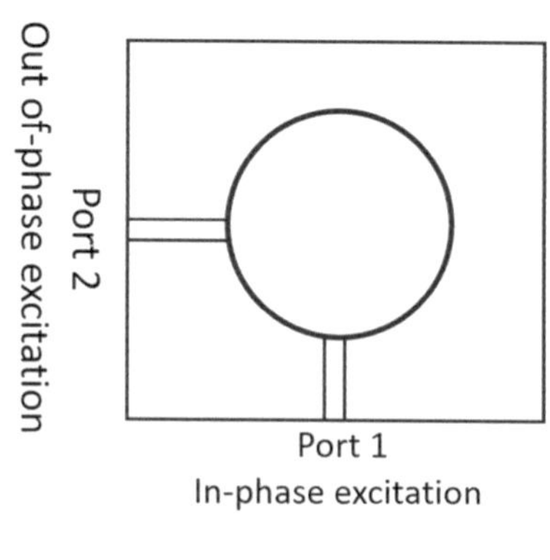


**Fig. 2.9** Exciting even and odd modes of a radiator

the case of in-phase excitation have the opposite direction in the subarray patches, resulting in a null in the broadside, while they are in the same direction in the patches for out-of-phase excitation giving maximum radiation at the broadside.

### 2.4.5 CRLH-TL-Based Method

Composite right/left-handed transmission lines (CRLH-TL) have been widely used in the design of microstrip antennas, due to their ease of fabrication and good performance [38]. CRLH-TL is a subtype of planar metamaterial that can be systematically designed to have both positive and negative phase velocities, which allows antennas to work not only in the positive mode but also in the zero modes and the negative mode. The use of CRLH-TL can thus be quite beneficial to achieve pattern diversity since it has the ability to achieve both broadside and omnidirectional patterns by working with different modes. By reconfiguring the dispersion properties, it is possible to choose the resonant mode of the CRLH-TL-based radiator in certain frequency bands. Thus, by using the CRLH-TL-based method, more freedom and flexibility in the design can be obtained compared to other approaches.

The reconfigurable dual-band antenna proposed in [39] can realize a broadside pattern and an omnidirectional pattern in both the ISM band and the WiMAX band. The antenna switches its radiation pattern from an omnidirectional pattern in the lower band and a broadside pattern in the higher band to a broadside pattern in the lower band and an omnidirectional pattern in the higher band. The antenna is able to switch its radiation pattern by reconfiguring the resonant modes through switches in the reconfigurable metamaterial structure. These structures contain a combination of a CRLH-TL part and a RH-TL part. By changing the states of the switches, the structure is tuned to change the dispersion properties of the CRLH and RH combined TL and the radiation performance in the two bands can be reconfigured. The use of the RF switches along with the CRLH-TL and RH-TL combined structure makes it possible to accurately control the radiation pattern of the antenna simultaneously in two bands.

### 2.4.6 Parasitic Antenna Arrays

Parasitic antenna arrays have been considered to generate various radiation patterns with less complexity than regular phased arrays as shown in Fig. 2.10. With the advent of numerous wireless low-cost systems, several teams started to investigate the reconfiguration of parasitic antenna arrays for pattern diversity [40]. Although parasitic antenna arrays can produce a better pattern diversity characteristic, they have some drawbacks including comparatively complex structures and large sizes.

In [41], a micro-electro-mechanical system (MEMS) switched parasitic antenna array with radiation pattern diversity is presented. Using stand-alone packaged series

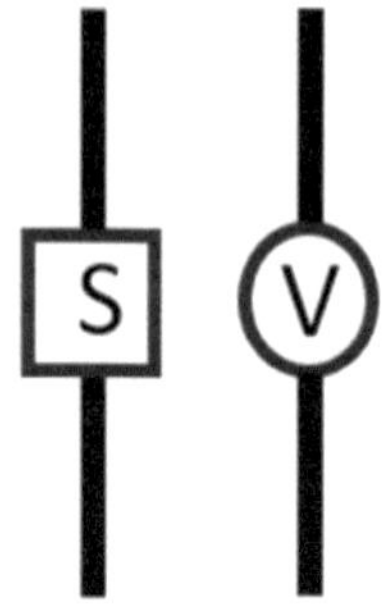


**Fig. 2.10** Switched parasitic dipole antenna (V represents the driven element; S represents the switch in the parasitic element)

MEMS switches, the antenna provides two switched beams at ± 25°. Parasitic antenna arrays involve one or few elements driven by the RF signal, while others are acting as parasitic directors or reflectors to focus the beam in a specific direction. In the proposed design slot antennas are used as radiating elements. The central slot is fed by a microstrip line, while both adjacent slots are loaded by coplanar open or short-circuited stubs. It has been shown that different radiation patterns can be generated depending on the loading stubs connected to the adjacent slots.

### *2.4.7 Other Pattern Diversity Methods*

In [42], the electrical length of the radiators to manipulate the radiation pattern to achieve pattern diversity is reported. For a radiator with a fixed physical length, the electrical length of the radiator is different at various frequencies. When the frequency is tuned, the main beam of the antenna shifts in different directions.

A pattern diversity antenna using the concept of the reconfigurable frequency-selective reflector is proposed in [43]. Here the antenna is at the center and surrounded by reflectors. In one switch state, the rectangular loops are in resonance and the waves will be reflected. In the other state, when loops are not in resonance, reflectors are transparent to the waves.

An annular slot antenna suitable for pattern diversity in wideband applications is presented in [44]. Pattern diversity is obtained by feeding the antenna simultaneously with two microstrip lines printed on the back of the substrate, which generates orthogonal radiation patterns. Two shorts are placed, one positioned between the feeding lines and the second one in its mirror position. Good isolation between ports is achieved when a single short is placed between the ports [45]. However, the insertion of a second short placed on this structure produces a more symmetrical current distribution, hence leading to symmetrical radiation patterns. The radiated fields appear rotated by 90° owing to the location of the shorts.

## 2.5 Pattern Diversity Antennas with Polarization Diversity

A pattern diversity antenna that includes polarization diversity is always advantageous as it can cover radiation beams from different directions in different orientations as well. Some of the pattern diversity antennas that include polarization diversity are discussed here.

### *2.5.1 Polarization Diversity*

Polarization in antennas is a combination of linear polarization (horizontal/vertical) and/or circular polarization (LHCP/RHCP). Linear polarized antennas have high losses in their path, and rarely suitable for the current designs. The LP antennas suffer from many losses such as absorption losses, change in polarization of incoming signal with the Faraday rotation effects, more sensitive to misalignment and require alignment at transmitter and receiver sides [46]. However, there are a number of advantages that circular polarization (CP) provides when it comes to signal propagation. One of the several prominent aspects of circular polarization includes decreased multipath losses and relatively lower sensitivity to the spatial arrangement of different antennas at the transmitter and receiver sides of the communication setup.

In polarization diversity, two antennas are positioned at the same place so that the antenna system can respond to two different orthogonal components of the incoming signal. The polarization of the received signal at the receiver is determined from the polarization of the base station signal and from the scattering parameters of the environment. The signals take multiple paths to reach the receiver which causes the fading phenomenon. If the transmitted signal is a vertically polarized signal, it can become horizontally polarized during transmission, generally referred to as Rayleigh fading. When a circularly polarized signal travels through obstacles, RHCP can change to LHCP. At the same time, the ground will have minimal effect on the signal polarization, which means the signal polarization remains the same after reflection from the ground. Polarization diversity antennas with multiple polarizations in an antenna structure can minimize the signal loss at the receiver. If the receiver has a pair of orthogonally polarized antennas, each antenna will receive different orthogonal components. Since these orthogonal components have a phase difference of 90° in between them, there is no coupling; thus, the correlation between them is very low.

***Need for Polarization Diversity***

- Polarization diversity allows antennas to be physically placed at the same location and embedded into a small space, lowering the cost.
- It is impossible to receive all the transmitted power using a single antenna. Antennas with circular polarization diversity can be used for random orientation of the signals.

- The use of polarization diversity can reduce the polarization mismatch and improve the signal-to-noise ratio even in the line-of-sight signal (LOS) channels [47]. The polarization mismatch due to the random orientation of the mobile handset can be reduced by implementing polarization diversity schemes.
- There is no single antenna that simultaneously receives both horizontally and vertically polarized signals. Thus, the polarization diversity scheme is required to receive two orthogonal horizontal and vertically polarized signals.

## 2.5.2 *Polarization and Pattern Diversity Antennas*

As multiple antennas get integrated into complex antenna systems, the placement of individual antennas becomes a daunting task. An insufficient separation between the antennas increases the spatial correlation reducing the achievable system performance. A close distance between the antennas can further degrade the performance due to increased mutual coupling [48]. The most effective schemes aimed to reduce both mutual coupling and spatial correlation in physically constrained antenna systems are polarization and pattern diversity. The pattern and polarization diversity techniques are very effective to solve multipath-fading effects in complex environments [49]. Comparing with classical antennas operated with a single radiation mode, the pattern diversity antennas, which are capable of radiating and receiving signals through different radiation modes simultaneously, have the advantage of performing higher effective gain while maintaining the same installation space [50]. Several antenna structures offering pattern and polarization diversities have been proposed in the literature.

### 2.5.2.1 Dual Circular Polarization with Pattern Diversity

A compact and geometrically simple circularly polarized MIMO antenna with polarization diversity is reported in [51]. A single-element antenna is first designed with the right-hand circular polarization (RHCP). The antenna comprises a simple coplanar waveguide (CPW)-fed monopole extension of the microstrip line and a modified coplanar ground plane. The vertical component of the CP is induced through a protruded stub in the right-side ground plane as shown in Fig. 2.11 and the horizontal components are excited along the width of the left-side ground plane. A high permittivity reflector is placed at a quarter wavelength away from the antenna to reduce the loading effects. The pattern diversity of the antenna can be realized when the antenna is in bent condition, in particular when worn on wrist, arm, or ankle. The effective pattern diversity between the two antennas of 120° is easily achieved without serious degradation of other performance figures.

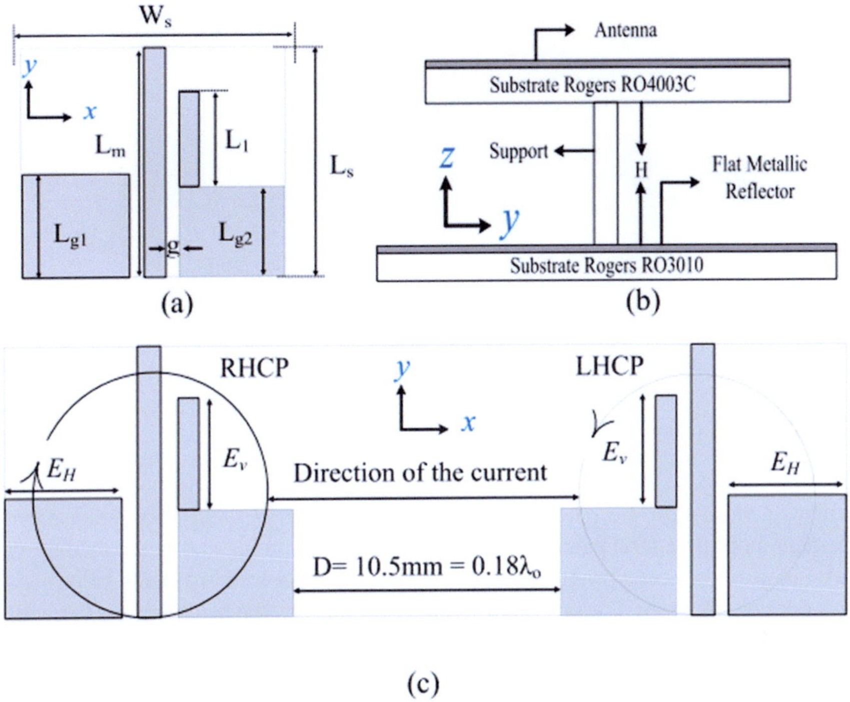


**Fig. 2.11** Circularly polarized MIMO Antenna with polarization/pattern diversity [51]: **a** top view **b** side view **c** MIMO configuration

#### 2.5.2.2 Dual Linear Polarization with Pattern Diversity

Polarization diversity is achieved due to two feed plates placed perpendicular to each other which produce cross-polarization. A UWB antenna design that is perpendicularly placed and shares the same monopole element is reported in [52]. A disk patch and a square patch are used as the monopole elements to realize antenna diversity. By changing the feeding signals applied at the two ports, different radiation characteristics can be obtained. When the antenna is fed at port 1 or 2 only, the antenna radiates at $\pm$ 45° direction. When the antenna is excited simultaneously at both the ports with in-phase and equal-magnitude signals, it radiates from both sides of the antenna at $\pm y$ direction and the antenna is horizontally polarized. When the antenna is excited simultaneously at both the ports with out-of-phase and equal-magnitude signals, it radiates from $+ x$ direction and the antenna is vertically polarized. By changing the input signals applied at the two ports, the radiation pattern and antenna polarization can be controlled.

Figure 2.12 presents a compact three-port antenna design suitable for use in mobile radio systems that require polarization and pattern diversity [53]. The antenna design

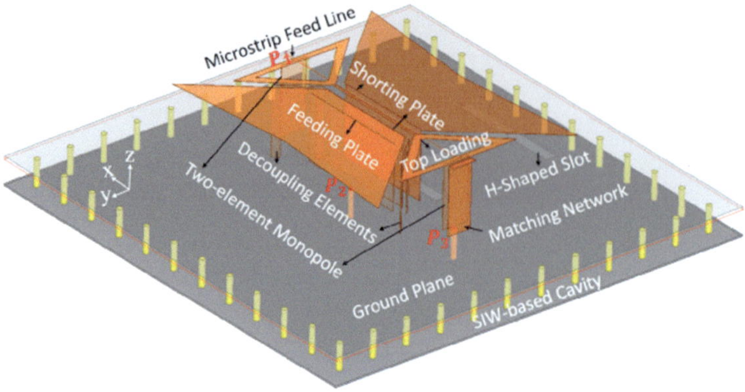


**Fig. 2.12** Three-port pattern and polarization diversity antenna [53]

utilizes a short bow–tie patch split in half with an air gap and a top-loaded two-element monopole. The patch is excited by aperture coupling and probe feeding via two independent ports, producing two orthogonally polarized broadside patterns for polarization diversity. The broadside pattern produced by the patch and the vertically polarized omnidirectional pattern generated by the monopole provide pattern diversity to the antenna module. Thus, the proposed antenna can exploit dual polarizations along with pattern diversity from two different antennas.

#### 2.5.2.3 Combined Linear and Circular Polarization with Pattern Diversity

A compact polarization and pattern reconfigurable metasurface antenna presented in [54] consists of an integrated structure of a microstrip patch and a square metasurface. The feeding network of the antenna enables the three polarization states of the antenna and a ring-shaped pixel structure around the metasurface realizes the pattern diversity of the antenna. Two orthogonal degenerate modes are excited by the probe feeding employed in the design, which realizes the antenna capable of switching between horizontal linear polarization, vertical linear polarization, and circular polarization. The designed antenna can produce the steerable beam directions of $\pm$ 40° with HP, $\pm$ 20° with VP, and $\pm$ 25° with CP in *y–z* plane by loading an annular pixel structure around the metasurface. Owning to the symmetry structure of the antenna, polarization, and pattern reconfigurable performances can also be realized in *x–z* plane.

A pattern diversity dual-polarized button antenna with two radiators is proposed in [55]. The first radiator is made by a crossed-bow–tie dipole printed on a rigid substrate generating an ultra-wideband, directional-radiation pattern, and a circularly polarized wave. The crossed-dipole radiator is fed by one port. An annular-ring-shaped radiator and a common ground plane are printed on a flexible felt substrate

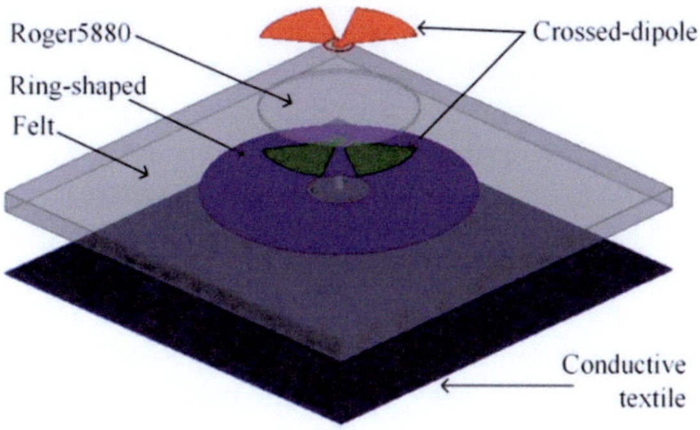


**Fig. 2.13** RHCP and linearly polarized pattern diversity antenna [55]

generating a triple band, on-/off-body radiation pattern, and linear polarization. The annular-ring radiator fed by the other port as shown in Fig. 2.13 adapt to versatile wireless applications of on-/off-body communications.

## 2.6 Pattern Diversity Antennas with Reconfigurability

The design of reconfigurable antennas has received much attention because it can provide some additional flexibility to pattern diversity antennas. These antennas also have the capability of reducing multipath effects making them an attractive candidate for wireless communication systems. Pattern reconfigurable antennas can select the best pattern in terms of SNR and avoid the interference of noise sources by redirecting the null position of its radiation pattern to the interferer. It can also provide larger coverage by redirecting the main beam.

Reconfigurability antenna designs use different methods to reconfigure the antenna, such as switching load [56], switching connection states by pin diodes [57], and switching feed network [58]. For these designs, the null or main beam position in the far-field patterns can be varied by reconfiguring the antenna. Several designs related to the reconfigurable pattern diversity antenna are proposed in the literature.

### *2.6.1 Reconfigurable Antennas*

The design of a single-feed square-ring patch antenna with switchable radiation patterns is presented in [59]. The antenna structure has four shorting walls in which two shorting walls are directly connected to the radiating patch and the connection states of the other two shorting walls are switched electrically by pin diodes. When the diodes are ON, the antenna is operated at the monopolar plate-patch mode that can radiate conical patterns. When the diodes are OFF, the antenna is operated in the

$TM_{11}$ mode which can radiate broadside patterns. The two modes can be designed at the same resonant frequency by adjusting the width of the shorting walls. The main beam positions of the radiation patterns are directed at the elevation angles of 50° and 0° when the diodes are ON and OFF, respectively. This characteristic allows the antenna to have the attractive feature of pattern diversity.

The antenna proposed in [60] operates at three different modes for achieving polarization and pattern diversity in an operating frequency band. The antenna utilizes three different radiating modes of a center-shorted microstrip patch to generate different polarizations and patterns. Reconfigurability is realized by using twelve PIN diodes distributed around the patch sides. The antenna can switch among two orthogonal polarizations with broadside pattern and one vertical polarization with omnidirectional pattern in the operating frequency band.

A novel reconfigurable microstrip antenna with radiation pattern selectivity and polarization diversity is presented in [61]. The antenna is fed through a four-way power divider made of three Wilkinson power dividers and interconnected with switches. By controlling the states of the switches, the antenna characteristics can be changed into two modes. When the four rectangular radiating patches are excited by four sources with equal amplitude and phase, a metamaterial antenna with a conical beam and linear polarization is achieved. When four sources with equal amplitude but 90 phase difference for each adjacent output is used for excitation, a wideband antenna with broadside beam and circular polarization is achieved.

### 2.6.2 *Reconfigurable Metamaterials*

A compact low-profile polarization and pattern reconfigurable metasurface antenna is proposed in [54]. The feeding network enables the antenna to switch between horizontal linear polarization, vertical linear polarization and circular polarization. The pattern reconfigurability of the proposed antenna has been realized by loading a ring-shaped pixel structure around the metasurface. The designed antenna can produce the steerable beam directions of ± 40° with HP, ± 20° with VP, and ± 25° with CP. Due to the symmetry of the antenna structure, polarization and pattern reconfigurable performances can also be realized in the orthogonal plane as well.

## 2.7 Conclusion

With antenna diversity, the obstacles in mobile channels that hinder efficient communication between a transmitter and receiver are attacked directly. Diversity techniques are powerful tools to improve the quality of wireless communication. Different diversity techniques are discussed followed by a detailed discussion of pattern diversity antennas in this chapter. Pattern diversity antennas are very effective in mitigating multipath fading due to reflection, diffraction, and scattering of signals in a dynamic and dense environment. Various techniques used for attaining pattern diversity along with designs from available literature are described in detail. This chapter could be useful for the industry as well as academics to understand various aspects of pattern diversity antenna designs for modern communication systems.

## References


1. James, J.R., Fujimoto, K.: Mobile Antenna Systems Handbook. Artech House, United Kingdom (2001)
2. Sasaoka, H.: Mobile Communications. Ohmsha Limited, Japan (2000)
3. Sizun, H.: Radio Wave Propagation for Telecommunication Applications. Springer, Germany (2005)
4. Goldsmith, A.: Wireless Communications. Cambridge University Press, Senegal (2005)
5. Saunders, S.R., Aragón-Zavala, A.: Antennas and Propagation for Wireless Communication Systems. Wiley, United Kingdom (2007)
6. Shankar, P.M.: Fading and Shadowing in Wireless Systems. Springer International Publishing, Switzerland (2018)
7. Lee, W.C.Y.: Wireless and Cellular Communications. McGraw-Hill Education, United Kingdom (2005)
8. Lee, W.C.Y.: Mobile Communication Engineering. John Wiley & Sons, New York (1982)
9. Vaughan, R.G., Andersen, J.B.: Antenna diversity in mobile communications. IEEE Trans. Veh. Technol. **36**(4), 149–172 (1987)
10. Arai, H.: Measurement of Mobile Antenna Systems. Artech House, United States (2013)
11. Vaughan, R.G.: Polarization diversity in mobile communications. IEEE Trans. Veh. Technol. **39**(3), 177–186 (1990)
12. Sánchez-Hernández, D.A.: Multiband Integrated Antennas for 4G Terminals. Artech House, United Kingdom (2008)
13. Kanaujia, B.K., Gupta, S.K., Kishor, J., Gangwar, D. (eds.): Printed Antennas: Theory and Design. Taylor & Francis Group, United Kingdom (2020)
14. Dietrich, C.B., Dietze, K., Nealy, J.R., Stutzman, W.L.: Spatial, polarization, and pattern diversity for wireless handheld terminals. IEEE Trans. Antennas Propag. **49**(9), 1271–1281 (2001)
15. Bartz, R.J.: Mobile Computing Deployment and Management: Real World Skills for CompTIA Mobility Certification and Beyond. Wiley (2015)
16. Mattheijssen, P., Herben, M.H.A.J., Dolmans, G., Leyten, L.: Antenna pattern diversity versus space diversity for use at handhelds. IEEE Trans. Veh. Technol. **53**(4), 1035–1042 (2004)
17. Rezazadeh, N., Shafai, L.: A pattern diversity antenna for ambient RF energy harvesting in multipath environments. In: 18th International Symposium on Antenna Technology and Applied Electromagnetics (ANTEM), 2018
18. Bruce, E.: Developments in short-wave directive antennas. Proc. IRE **19**(8), 1406–1433 (1931)
19. Bruce, E., Beck, A.C.: Experiments with directivity steering for fading reduction. Proc. IRE **23**(4), 357–371 (1935)
20. Friis, H.T., Feldman, C.B., Sharpless, W.M.: The determination of the direction of arrival of short radio waves. Proc. IRE **22**(1), 47–78 (1934)
21. Friis, H.T., Feldman, C.B.: A multiple unit steerable antenna for short-wave reception. Proc. IRE **25**(7), 841–917 (1937)
22. Gregory, D., Newsome, S.: Cooling radio station, Hoo Peninsula, Kent: An archaeological investigation of a short-wave receiving station. English Heritage Research Department, Portsmouth, UK, Technical Report 110-2010 (2010)
23. Vogelman, J.H., Ryerson, J.L., Bickelhaupt, M.H.: Tropospheric scatter system using angle diversity. Proc. IRE **47**(5), 688–696 (1959)

24. Gilbert, E.N.: Energy reception for mobile radio. Bell Sys. Tech. J. **44**(8), 1779–1803 (1965)
25. Noerpel, A.R.: Pattern diversity using a single horn reflector antenna. In: IEEE Global Telecommunications Conference and Exhibition. Communications for the Information Age, vol. 3, pp. 1457–1462 (1998)
26. Aubrey, T., White, P.: A comparison of switched pattern diversity antennas. In: Proceedings 43rd IEEE Vehicular Technology Conference, May 1993, pp. 89–92 (1993)
27. Vaughan, R.G.: Switched parasitic elements for antenna diversity. IEEE Trans. Antennas Propag. **47**(2), 399–405 (1999)
28. Dong, L., et al.: Multiple-input multiple-output wireless communication systems using antenna pattern diversity. In: Proceedings IEEE Global Telecommunications Conference (GLOBECOM), vol. 1, November 2002, pp. 997–1001 (2002)
29. Mattheijssen, P., et al.: Antenna-pattern diversity versus space diversity for use at handhelds. IEEE Trans. Veh. Technol. **53**(4), 1035–1042 (2004)
30. Sun, L., Huang, W., Sun, B., Sun, Q., Fan, J.: Two-port pattern diversity antenna for 3G and 4G MIMO indoor applications. IEEE Antennas Wirel. Propag. Lett. **13**, 1573–1576 (2014)
31. Yan, S., Vandenbosch, G.A.E.: Low-profile dual-band pattern diversity patch antenna based on composite right/left-handed transmission line. IEEE Trans. Antennas Propag. **65**(6), 2808–2815 (2017)
32. Xu, Z., Deng, C.: High-isolated MIMO antenna design based on pattern diversity for 5G mobile terminals. IEEE Antennas Wirel. Propag. Lett. **19**(3), 467–471 (2020)
33. Cui, L., Wu, W., Fang, D.-G.: Wideband circular patch antenna for pattern diversity application. IEEE Antennas Wirel. Propag. Lett. **14**, 1298–1301 (2015)
34. Yang, S.L.S., Luk, K-M.: Design of a wide-band L-probe patch antenna for pattern reconfiguration or diversity applications. IEEE Trans. Antennas Propag. **54**(2), 433–438 (2006)
35. Wei, K., Zhang, Z., Chen, W., Feng, Z.: A novel hybrid-fed patch antenna with pattern diversity. IEEE Antennas Wirel. Propag. Lett. **9**, 562–565 (2010)
36. Zhang, Y., Wei, K., Zhang, Z., Li, Y., Feng, Z.: A compact dual-mode metamaterial-based loop antenna for pattern diversity. IEEE Antennas Wirel. Propag. Lett. **14**, 394–397 (2015)
37. Krishnamoorthy, K., et al.: Low profile pattern diversity antenna using quarter-mode substrate integrated waveguide. Prog. In Electromagn. Res. Lett. **55**, 105 (2015)
38. Lai, A., Itoh, T., Caloz, C.: Composite right/left-handed transmission line metamaterials. IEEE Microwave Mag. **5**(3), 34–50 (2004)
39. Zhang, J., Yan, S., Vandenbosch, G.A.E.: Realization of dual-band pattern diversity with a CRLH-TL-inspired reconfigurable metamaterial. IEEE Trans. Antennas Propag. **66**(10), 5130–5138 (2018)
40. Thiel, D.V., Smith, S.: Switched Parasitic Antennas for Cellular Communications. Artech House (2002)
41. Petit, L., Dussopt, L., Laheurte, J.M.: MEMS-switched parasitic-antenna array for radiation pattern diversity. IEEE Trans. Antennas Propag. **54**(9), 2624–2631 (2006)
42. Ding, C.F., Zhang, X.Y., Xue, C., Sim, C.: Novel pattern-diversity-based decoupling method and its application to multielement mimo antenna. IEEE Trans. Antennas Propag. **66**(10), 4976–4985 (2018)
43. Ko, C.-H., Tarn, I.-Y., Chung, S.-J.: A compact dual-band pattern diversity antenna by dual-band reconfigurable frequency-selective reflectors with a minimum number of switches. IEEE Trans. Antennas Propag. **61**(2), 646–654 (2013)
44. Gallo, M., Antonino-Daviu, E., Ferrando-Bataller, M., Bozzetti, M., Molina-Garcia-Pardo, J.M., Juan-Llacer, L.: A broadband pattern diversity annular slot antenna. IEEE Trans. Antennas Propag. **60**(3), 1596–1600 (2012)
45. Antonino-Daviu, E., Gallo, M., Bernardo-Clemente, B., Ferrando-Bataller, M.: Ultra-wideband slot ring antenna for diversity applications. Electron. Lett. **46**(7), 478–480 (2010)
46. Ullah, U., Mabrouk, I.B., Koziel, S.: A compact circularly polarized antenna with directional pattern for wearable off-body communications. IEEE Antennas Wirel. Propag. Lett. **18**(12), 2523–2527 (2019)

47. Dietrich C.B. Jr., Dietze, K., Nealy, J.R., Stutzman, W.L.: Spatial, polarization, and pattern diversity for wireless handheld terminals. IEEE Trans. Antennas Propag. **49**(9), 1271–1281 (2001)
48. Clerckx, B., Craeye, C., Vanhoenacker-Janvier, D., Oestges, C.: Impact of antenna coupling on $2 \times 2$ MIMO communications. IEEE Trans. Veh. Technol. **56**(3), 1009–1018 (2007)
49. Behdad, N., Sarabandi, K.: Dual-band reconfigurable antenna with a very wide tunability range. IEEE Trans. Antennas Propag. **54**(2), 409–416 (2006)
50. Yang, S.L.S., Kishk, A.A., Lee, K.F., Luk, K.M., Lai, H.W.: The design of microstrip patch antenna with four polarizations. In: Proceedings IEEE Radio and Wireless Symposium, Jan 22–24, 2008
51. Ullah, U., Mabrouk, I.B., Koziel, S.: Enhanced-performance circularly polarized MIMO antenna with polarization/pattern diversity. IEEE Access **8**, 11887–11895 (2020)
52. Dong, Y., Itoh, T.: Planar ultra-wideband antennas in Ku- and K-band for pattern or polarization diversity applications. IEEE Trans. Antennas Propag. **60**(6), 2886–2895 (2012)
53. Choi, J., Dagefu, F.T.: A compact three-port antenna with enhanced inter-port isolation for polarization and pattern diversity. IEEE Access **11**, 111402–111409 (2023)
54. Li, W., Wang, Y.M., Hei, Y., Li, B., Shi, X.: A compact low-profile reconfigurable metasurface antenna with polarization and pattern diversities. IEEE Antennas Wirel. Propag. Lett. **20**(7), 1170–1174 (2021)
55. Le, T.T., Kim, Y.-D., Yun, T.-Y.: Wearable pattern-diversity dual-polarized button antenna for versatile on-/off-body communications. IEEE Access **10**, 98700–98711 (2022)
56. Chen, W.H., Feng, Z.H.: Planar reconfigurable pattern antenna by reactive-load switching. Microw. Opt. Technol. Lett. **47**, 506–507 (2005)
57. Nikolaou, S., Bairavasubramanian, R., Lugo, C., Carrasquillo, I., Thompson, D.C., Ponchak, G.E., Papapolymerou, J., Tentzeris, M.M.: Pattern and frequency reconfigurable annular slot antenna using pin diodes. IEEE Trans. Antennas Propag. **AP-54**, 439–448 (2007)
58. Yang, S.L.S., Luk, K.M.: Design of a wide-band L-probe patch antenna for pattern reconfiguration or diversity applications. IEEE Trans. Antennas Propag. **AP-54**, 433–438 (2006)
59. Chen, S.-H., Row, J.-S., Wong, K.-L.: Reconfigurable square-ring patch antenna with pattern diversity. IEEE Trans. Antennas Propag. **55**(2) (2007)
60. Nguyen-Trong, N., Mobashsher, A.T., Abbosh, A.M.: Reconfigurable shorted patch antenna with polarization and pattern diversity. In: 2018 Australian Microwave Symposium (AMS), 2018
61. Cao, W., Zhang, B., Liu, A., Yu, T., Guo, D., Pan, K.: A reconfigurable microstrip antenna with radiation pattern selectivity and polarization diversity. IEEE Antennas Wirel. Propag. Lett. **11**, 453–456 (2012)

# Chapter 3
# Electronic Beam-Scanning Antennas

## 3.1 Introduction

In present-day wireless communication systems, beam-scanning antennas play a very important role, as the instances of interferences are increasing in the multistandard wireless environment. In beam scanning, the antenna radiation patterns are scanned in a fixed sequence and the beam is pointed in the desired direction. Usually, it is a single main beam of the antenna that gets pointed in different directions; however, multibeam scanning antennas are also available and the beams are scanned when the required conditions are satisfied. A beam-scanning antenna can operate in one or two axes and scan the space in three different ways—in a finely quantified and continuous manner, with beam overlapping, or in a disjoint manner [1]. In continuous beam scanning, a single beam is angularly shifted in different directions across the entire coverage area. In beam overlapping, different beams are generated in the antenna system to scan the coverage area. In disjoint beam scanning, only required directions in the coverage area are scanned. Beam-scanning antennas offer a significant increase in the antenna coverage range and channel capacity. In mobile communication, a beam-scanning antenna reduces the multipath effect as the antenna receives signals from different directions [2]. It thus helps in improving the signal quality.

### *3.1.1 Need for Beam-Scanning Antennas*

The motivation for designing beam-scanning antennas is to improve the coverage of the overall antenna system. Such antennas also reduce the overall cost of the system as single antenna can provide a wide coverage rather than placing individual antennas for the same coverage. Care should be taken that such antennas are also highly power efficient as the extra circuitry that aids in beam scanning can consume relatively higher power. Another important factor for designing beam-scanning antennas is

 
S. K. Koul et al., *Antenna Systems for Modern Wireless Devices*, Signals and Communication Technology, https://doi.org/10.1007/978-981-97-3369-9_3

connectivity, especially connectivity in motion of the user. If a mobile device is traveling in a car or train, it should always be connected to the base station. Here the beam of the antenna should change based on the speed and position of the vehicle. The beam-scanning angle should also be wide enough to communicate with the base station. The increasing instances of electromagnetic waves in the environment have increased the significance of beam scanning and null scanning techniques [3]. For sensitive antenna systems that get affected by spurious EM interferences, narrow beam antennas with high gain improve the signal-to-interference ratio. For such antennas, beam scanning aids in continuous antenna coverage without compromising the signal strength. Multibeam scanning antennas have relatively higher diversity as they can include multipolarized and multiband beams for scanning. This helps in improving the scan coverage and capacity of the antenna system, as more signals are received at the receiver.

### *3.1.2 Applications of Beam-Scanning Antennas*

Beam-steering antennas are necessary for a variety of applications such as search, navigation and tracking, collision avoidance radars, imaging, drones, unmanned aerial vehicles (UAVs), airborne systems, and satellite systems [4] as illustrated in Fig. 3.1. High-gain beam-scanning antennas are going to be the key in the future wireless communication technologies like 5G communication. A beam-scanning antenna in a wireless power transfer system can be used to transfer the wireless power to different sensors and RF energy harvesting systems allows energy to be harvested from different directions. Some applications require the antenna to modify the radiation pattern so that better coverage or reception is achieved. This has promoted research into smart base station antennas for WLAN and cellular communication.

## 3.2 Different Techniques for Antenna Beam Scanning

Depending on the application, different types of beam-scanning methods are used. Based on the method of operation beam-scanning antennas can be classified into two broad categories—electronic and mechanical. Detailed description of scanning methods would be presented in subsequent chapters.

### *3.2.1 Electronic Scanning*

If an antenna uses electronic approaches to realize steering or scanning of the antenna beam in space, such an antenna is called an electronic scanning antenna system. Electronic scanning is a fast beam scanning and the beams are pointed accurately.

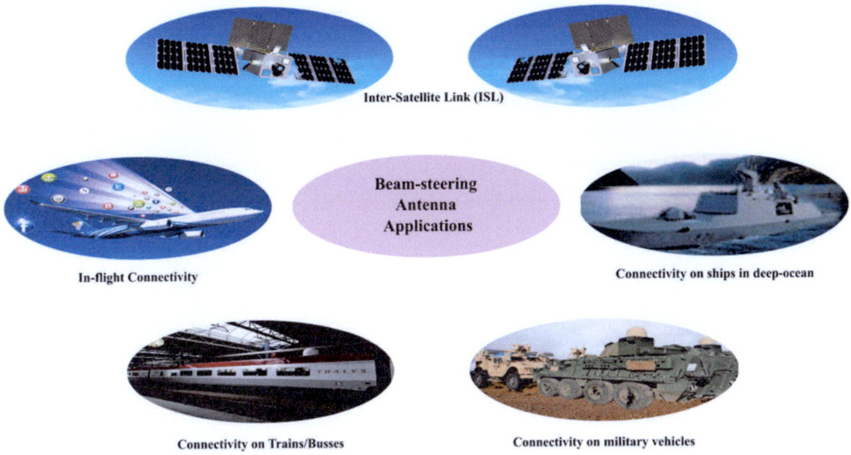


**Fig. 3.1** Application areas for the beam-steering antenna technologies [5]

It has a high beam resolution and fine beam control while scanning. It is also very compact and convenient to deploy. However, electronic beam scanning is complex due to the additional scanning control circuitry attached to the antenna making it expensive, and it suffers from higher scanning loss as well.

Electronic beam-scanning antennas generally consist of a phased array, an antenna array which steers the beam in different directions without moving the antennas. But they require phase shifters to control the phase of each element which in turn is limited by the bandwidth [6]. When an array is scanned with fixed units of phase shifters the main beam position changes with the applied progressive phase shift. The bandwidth of the phased-array antenna is limited by the phase shifters [7]. The scanning of the phased array is accomplished by using diode or ferrite phase shifters. Subsequently, such electronic scanning antennas can be categorized into phase scanning and frequency scanning antennas. These methods are generally dependent on supplying linear phase across the array elements.

***Phase Scanning***

In-phase scanning, the shape and direction of the beam depend on the relative phases and amplitudes of each antenna active in the array.

***Frequency Scanning***

In frequency scanning, the beam-scanning control is a function of the frequency of the transmitter. No phase shifters are present in frequency scanning array antennas. A change in the frequency of an electromagnetic signal propagating along a transmission line produces a change in phase. This phase difference introduces a shift in main beam position. The frequency scanning technique has two main drawbacks—they need wider bandwidth and the frequency scanning systems are very susceptible to jamming [8].

***Additional Techniques in Electronic Scanning***

Instead of using phase shifters to steer electromagnetic radiation, pin diodes, varactor diodes, RF MEMS switches can also be used to steer the beam, but the extra bias lines needed for their operation make the antenna design more complex. The electronic scanning can also be accomplished using liquid crystal and graphene materials. Due to their robust beam-scanning potential, leaky-wave antennas for beam scanning are also reported in the literature.

***Other Scanning Methods***

To overcome the problems encountered in frequency and phase scanning systems, a new scanning technique, amplitude scanning was proposed in [9]. In this technique, the magnitude of individual elements of the array is varied. It requires neither variable phase shifters, nor frequency scanning. It requires variable power dividers to change excitations of some of the array elements, and a single fixed phase shifter. Another type of scanning is optical beam scanning delivered by tunable laser or phased array [10], but optical scanning has a very limited range and is used mainly for sensor applications. A beam-scanning array by switched leaky-wave antenna is presented in [11] where the beam scanning is provided by wavelength sweeping and beam switching by Mach–Zehnder-type optical switch to extend scanning range.

### *3.2.2 Mechanical Scanning*

In mechanical beam scanning some part of the antenna has to be physically moved. Therefore, such scanning is used where rapid beam scanning is not required. The movable part can consist of

(a) The entire antenna system.
(b) The antenna part alone.
(c) The focusing aperture

Mechanical beam scanning is relatively cheap and more efficient than electronic beam scanning. It is often slow due to the large inertia of the moving parts. It is bulky and heavy due to the involvement of mechanical parts and also maintenance intensive. However, the mechanical beam scanning provides wide-coverage beam scanning without any scanning loss.

Conventional mechanical beam-scanning antennas consist of moving reflector antennas [12] that occupy massive size and volume. An alternate solution to reduce the cost and weight of the antennas is to use transmitarray [13] or reflectarray design [14]. The lens antenna designs are adjustable and offer flexibility to the antenna design [15]. However, its performance deteriorates for wide scan angles due to high scanning losses, and the lens makes the antenna bulky.

### 3.2.3 *Null Steering*

Like beam scanning, null steering antennas are also proposed in the literature. An interfering signal can be incident to the antenna from any direction in space. To counter such interferences, a null steering antenna is capable of placing a null in any desired position. In beam-scanning antennas, the part of the beam that carries maximum power is swept to the desired direction. On the other hand, in null scanning the part of the beam that carries minimum power is swept to the desired direction to negate the signals coming from those directions.

## 3.3 Frequency Beam Scanning

The electronic beam-scanning antennas can be divided into different categories depending on the method used by the antenna for beam scanning. Broadly it consists of frequency beam scanning—as variation in frequency causes the beam to change its direction and scan accordingly and phased-array antennas—where a variation in phase results in beam scanning. Other than these methods, one can use diodes as switching devices or use metasurfaces for achieving beam scanning. In this section, we will discuss how beam scanning is obtained by changing frequency.

Frequency beam-scanning antenna provides a simple and passive steering technology where the beam scanning is achieved by just sweeping through the available frequency channels. In this type of scanning, the shift in phase is achieved by frequency variation. Frequency scanning antennas change the direction of the main lobe whenever there is a change in the operating frequency. They are widely used in aircrafts, vehicular radars, and satellite communication systems [16], but a conventional frequency scanning antenna usually cannot scan through endfire direction.

In frequency scanning antennas (FSAs), the resonant frequency of the antenna provides the maximum radiation at $\theta = 0°$. When the frequency changes, maximum radiation deviates from the normal direction to $-\,\theta°$ or $+\,\theta°$. There are two main FSA designs available. A common one is a leaky-wave antenna (LWA). Another one is through proper design of the feed network.

### 3.3.1 *Leaky-Wave Antenna*

A waveguide is a structure that guides electromagnetic waves with minimal energy loss by restricting the transmission in one direction. When an electromagnetic wave-carrying waveguide possesses periodic slots at regular intervals it results in the radiation of a plane wave from them. This type of plane wave radiation is termed a leaky

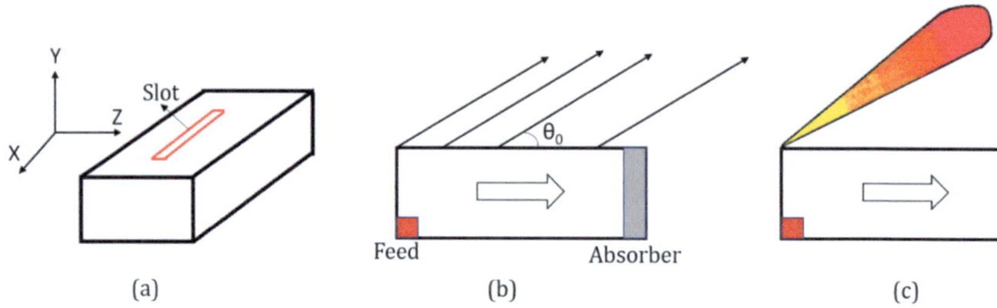


**Fig. 3.2** Illustration of a leaky-wave antenna **a** rectangular waveguide with a slot, **b** leaky-wave antenna **c** 1D unidirectional LWA radiating a scanned beam

wave. The leaky-wave antenna can be designed to radiate in a certain direction or scan through radiation between certain angles.

Leaky-wave antennas are distinguished by the electromagnetic field being stimulated by a wave incident on the interior or exterior of the waveguide. This results in currents propagating along the longitudinal direction of the antenna. During propagation, the wave moves along the waveguide and releases energy from the structure, leaving only a tiny field at the termination end of the traveling wave antenna [17]. Figure 3.2 shows an illustration of a leaky-wave antenna. An array of slots as shown in Fig. 3.2a is repeated along the $z$-direction to create a planar illumination. The narrow beam of the leaky-wave antenna is created at an angle of $\theta_0$. The radiation angle is generally given by

$$\theta = \cos^{-1}\frac{\beta}{k_0}, \tag{3.1}$$

where $\beta$ is the phase constant and $k_0$ is the free-space wavenumber [18]. The waves radiate through the slot into the free space, and at the other end, the remaining energy is absorbed by a matched load. 90% of the energy is radiated during this travel [19].

LWAs have been around since the 1940s, but the field has advanced significantly in recent years. Some of the leaky-wave antenna designs are discussed below.

1. Symmetric leaky-wave antenna

Due to their beam-scanning advantage, LWA offers wide range of applications in vehicular radars, surveillance, and point-to-point communication. The periodic leaky-wave antennas are more sophisticated type of LWA that have good scanning capabilities and provides the advantage of backfire to endfire scanning but suffers from degradation issue for broadside radiation. The instability of gain and efficiency during broadside scanning is the origin of this problem. The impacts of the open-stop band are countered by using a variety of strategies. One such LWA that makes use of symmetric microstrip patches along coupled lines is reported in [20]. Coupled lines with a quarter-wavelength on each side are joined to a symmetric, rectangular patch of half wavelength. Any connected line that performs the excitation also provides the feed. The suggested antenna's large working frequency spectrum and compatibility with applications requiring frequency scanning through the broadside made its deployment possible.

2. Surface-based leaky-wave antennas

Surface waves (SW) may be restricted to a surface or may exist at a material interface, mostly in the dielectrics used. Radar applications benefit from the usage of SW, which is crucial in electromagnetics. One such leaky-wave antenna that is activated by collimated surface waves was created in [21]. Two surface-wave launchers (SWL) and a high impedance surface made up the antenna, while a planar horn supplied the feed. A linear tapered microstrip is related to SWL's substrate-integrated waveguide. The antenna's performance demonstrated that it is easily capable of producing both a consistent radiation pattern with reasonably good gain values. Hence, they are useful in radars and satellite communications.

3. Circularly polarized leaky-wave antenna

Due to their reduced sensitivity to Faraday rotation, circularly polarized antennas have become very common in recent years. A novel compact high-gain broadband beam-scanning antenna is proposed based on $TE_{220}$ mode in SIW technology in [22]. The feed network of the antenna array is made simpler, and the processing tolerance is relaxed by the higher-order modes resulting in compact construction and higher gain. A microstrip-slot coupling structure is used to stimulate the antenna, and an electromagnetic wave is coupled out of the SIW cavity through a $2 \times 2$ slot array engraved into the top surface.

The four-element periodic array structure's beam-scanning range is increased by using spoof SPP slow-wave TL as a feeding mechanism. The linearly polarized wave of the antenna is converted into a circularly polarized wave using a polarization converter as shown in Fig. 3.3. Compared with LP beam-scanning antenna, CP antenna can reduce the influence of multipath reflection and rain fog interference.

## 3.4 Phased-Array Antenna

Phased-array technology, the idea of which was first conceived in the early 1900s, gained interest and development throughout the 1950s and 1960s. Over the years it has undergone a steady evolution from being mechanically controlling the phase shift at each element to controlling the phase shift electronically with more ease. Three general categories of phased arrays can be described as passive, active, and digital. A passive array is one in which the array is fed by a centralized high-power transmit amplifier, with a low-loss power-combining network and a phase shifter at every element. An active array is one where the transmit power amplifiers, low-noise amplifiers, and phase shifters are distributed throughout the array at the element level. A digital array is one where the radio frequency (RF) signals are converted to digital signals at the subarray or element level, with the beamforming being performed by a

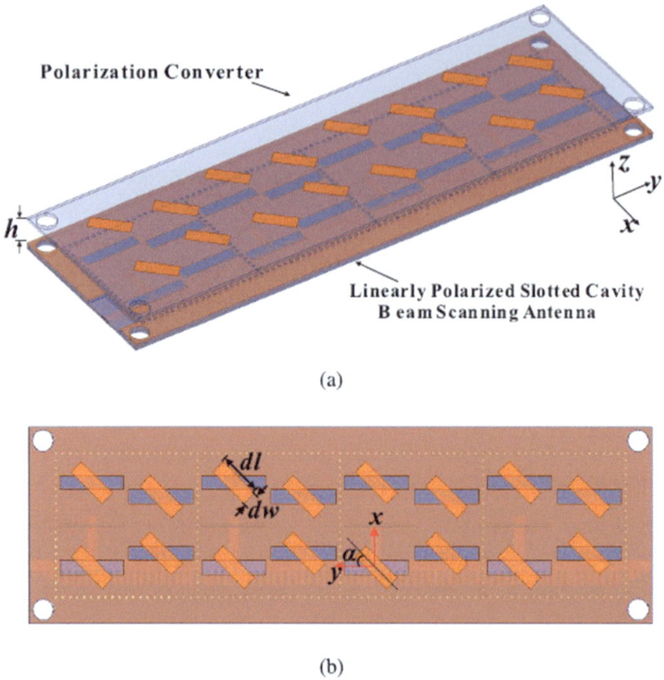


**Fig. 3.3** CP beam-steering antenna: **a** perspective view, **b** top view [22]

digital signal processor. As phase shifters are an important component in phased-array designs, a review of different phase shifter types is given below.

### *3.4.1 Phase Shifters*

Phase shifters are typically used in an antenna array for beam scanning. An antenna array consists of multiple antenna elements that are closely spaced. If same phase is applied to all the elements, the beam is formed at the broadside direction or 0° without any tilt. If different phases are applied to each element, the direction of the beam gets tilted. By controlling the phase of the elements, the direction of the beam can be controlled.

A phase shifter is an electronic circuit that adds a required phase shift to a signal without changing its amplitude or frequency. However, the amount of phase shift obtained changes linearly with frequency. Phase shifters can be designed to provide either fixed or variable phase shift. Fixed phase shifters provide a specific, predetermined amount of phase shift, while variable phase shifters can be adjusted to provide a range of phase shift values. The purpose of a phase shifter is to steer the direction of the antenna beam, without physically moving any part of the antenna. To understand how a change in phase results in beam scanning, one needs to understand how an antenna beam is steered in general. An antenna beam is steered when the radiations from each element combine constructively or destructively in a particular direction, producing a beam in the direction of constructive interference. Phase shifters are typically used in conjunction with an antenna array, where each element is fed with a phase-shifted signal. By using phase shifters alone, the beam of a single antenna cannot be steered. While phase shifters can be used to adjust the phase of the signal fed to a single antenna, this will only cause the phase of the radiated wave to change, but not the direction of the beam.

Different types of phase shifters are available in the literature. A passive phase shifter uses passive components such as transmission lines to achieve the desired phase shift. Active phase shifters use active components such as diodes and amplifiers. The choice of phase shifter depends on the design constraints and application requirements. A switched path transmission line phase shifter occurs when an additional length of transmission line is added to the signal path as shown in Fig. 3.4. The length corresponds to the required phase shift [23]. The switching is accomplished with pin diodes or RF MEMS capacitive membrane switches.

A reflection-type phase shifter consists of a non-reciprocal device like a coupler or circulator, and a variable length transmission line as shown in Fig. 3.5 [24]. The variable transmission line connects the device to a short and reflects the signal to the device and to the output port. The phase shift occurs when the length of the transmission line changes. Pin diode switches or RF MEMS capacitive membrane switches are used to change the line length by changing the location of the short.

A distributed phase shifter can be realized by capacitively loading a transmission line as shown in Fig. 3.6. A change in the shunt capacitance changes the line

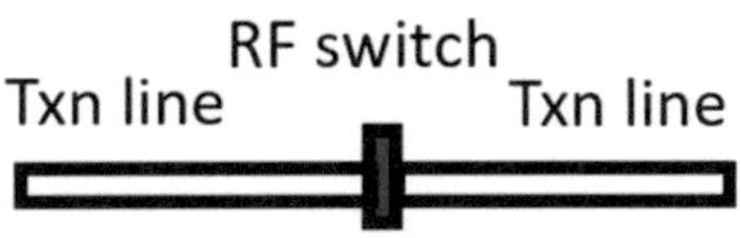


**Fig. 3.4** Transmission line phase shifter

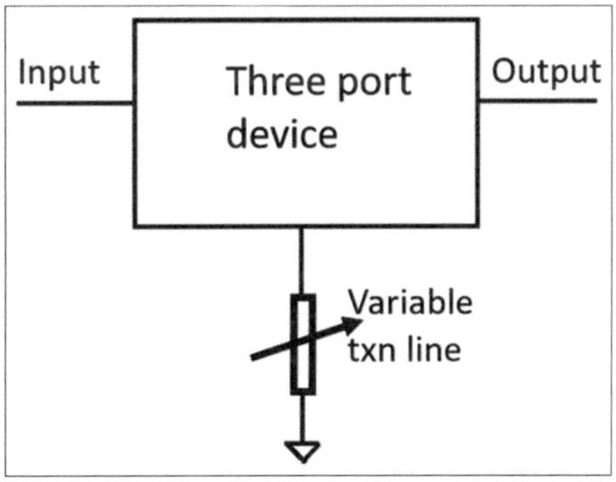


**Fig. 3.5** Reflection-type phase shifter

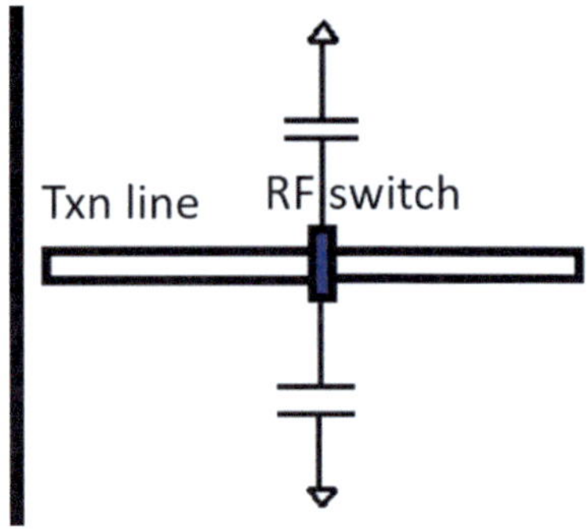


**Fig. 3.6** Capacitively loaded distributed transmission line phase shifter

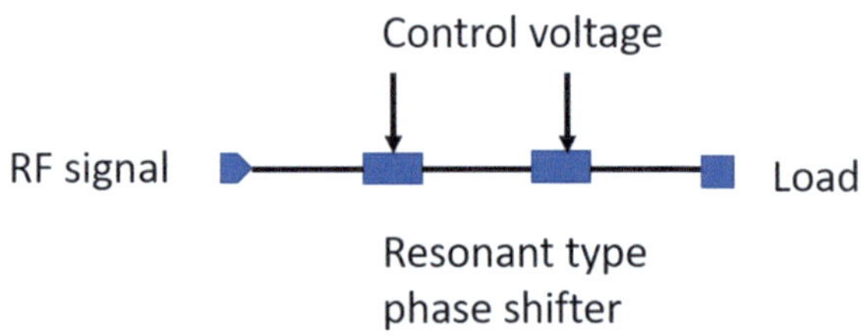


**Fig. 3.7** Resonant-type phase shifter

impedance and hence the phase shift of the line [25]. The capacitors are shorted to the ground.

Previously discussed beam-scanning techniques like frequency beam scanning or using multibeam antennas suffer from limitations in simultaneously achieving compact size and a wider scanning range. The discussed phase shifters in literature are also bulky in size. Novel resonant type of phase shifter topology that results in compact size, low cost, simple control circuitry with large phase-shifting ability is proposed in [26]. The phase characteristics are controlled by frequency and varactors as illustrated in Fig. 3.7. The beam can be steered to any direction within the $\pm$ 45° by changing the controlling voltage and working frequency. The tapered amplitude distribution applied onto the primary feedline guarantees low sidelobe levels.

In resonant-type phase shifter, the sidelobe level controlling strategies can be applied which is difficult to achieve in beam forming network-based arrays or frequency scanning antenna. The beam scanning can be done only at a single frequency and not over a wide frequency range making it less favorable in wireless communication systems. However, due to its simple control circuitry and low cost, it can be used in microwave imaging or low-cost radars.

### *3.4.2 Phased-Array Scanning*

The ubiquitous use of wireless communication and its resources have expanded the use of phased-array antenna systems as a widespread beam-scanning technique. It is

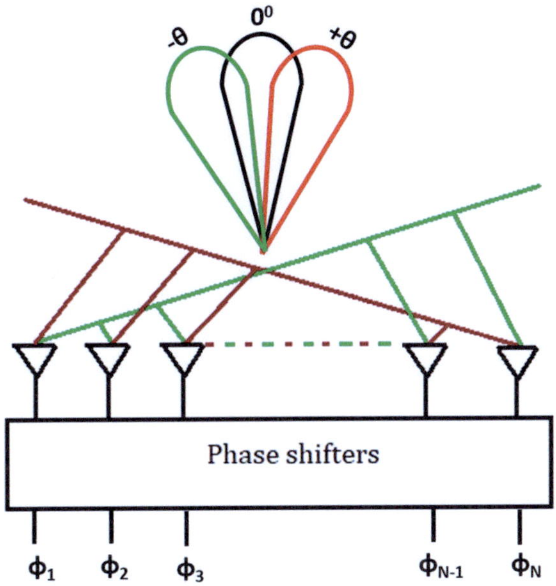


**Fig. 3.8** Illustration of phased-array scanning

extremely agile, flexible, and scans with high accuracy. Many wireless applications require wide bandwidth for operation, and this requires the use of true time delay-based phased-array antennas rather than a phase shifter to avoid beam squinting. The phased-array antenna scans by varying the phase of the antenna elements. An illustration of phased-array scanning is shown in Fig. 3.8. For an ideal wide-angle scanning phased array, the peak gain should be constant over its scanning range. However, the scanning range of a conventional phased-array antenna is usually limited due to considerable loss of antenna gain while scanning wide angles. A conventional phased-array antenna, while scanning from $-45°$ to $+45°$ experiences a gain fluctuation of 4–5 dB. This gain fluctuation is unavoidable due to limited element beamwidth and significant mutual coupling between the elements [27].

## 3.5 Electronic Scanning Using Discrete Devices

### *3.5.1 Pin Diodes*

In many electronic systems, including wireless LAN, cognitive radios, and fifth-generation cellular base stations, beam-switching antennas are becoming more and more prominent. Phased-array antennas have been used for these applications so far [28]. However, the design of a phase matrix for phased arrays is itself challenging. They occupy a large implementation area and are also power-hungry. Besides, full azimuthal coverage with minimum gain variation is difficult to achieve. To overcome

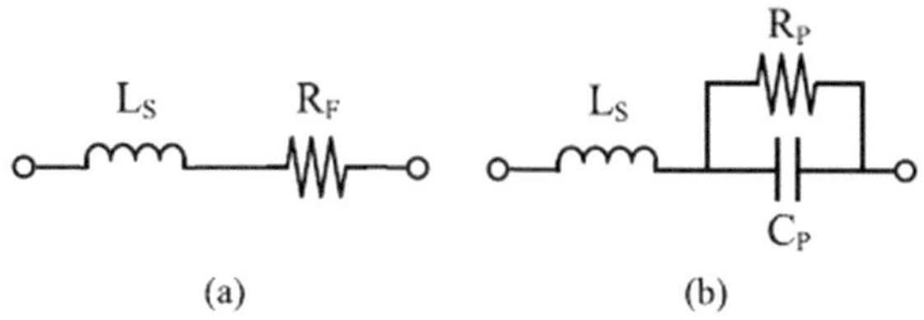


**Fig. 3.9** Equivalent circuits of a pin diode. **a** ON state, **b** OFF state

such issues in certain instances, we can use pin diodes to obtain beam switching. By carefully controlling the diodes, an antenna with several switchable patterns can be achieved. The pin diodes can be placed either in feed networks or in antenna aperture to switch between different beams. Higher the number of diodes used, higher could be the number of beams that could be switched. An equivalent circuit of pin diode is shown in Fig. 3.9.

The design concept in [29] is to combine several directional antennas to form a compact switched-beam antenna. The antenna is a four-element array with L-shaped slot as the radiation element. It presents a directional radiation pattern in the azimuth plane and its maximum radiation direction is toward the direction of the open end of the slot. The open ends of the four slot antennas are arranged toward 0, $\pi/2$, $\pi$, and $3\pi/2$, respectively. Four diodes are designed to control the antenna elements. When different antennas are switched on, the beam gets shifted to different directions. When the diagonal antenna pairs are on, the patterns will be nearly omnidirectional since the patterns of the antennas in a pair are almost complementary in the azimuth plane. Thus, by carefully controlling the diodes, a switched-beam antenna with eight directional patterns and two nearly omnidirectional patterns is achieved.

In [30], beam-switching ability is realized by electronically controlling the state of switches (pin diodes) on the side apertures of a square substrate-integrated waveguide (SIW). Four wide dipoles are loaded in front of the four apertures. When one of the beams is used, the corresponding aperture is opened to excite the above mounted dipole to radiate, the other three closed apertures block the waves, and the other three dipoles work as reflectors of the excited dipole. To generate one desired unidirectional radiation beam, the corresponding aperture is switched ON, which excites the corresponding dipole. Meanwhile, the other three apertures are switched OFF and perform as electric walls in the SIW. The antenna can generate four switchable beams at every 90° in the azimuth plane with identical patterns and well-controlled directional radiation patterns owing to the symmetrical antenna structure.

A single-layer planar coupled-slot antenna is presented in [31] for 360° azimuthal beam-scanning applications over a wide frequency band. The antenna uses eight sets of tightly coupled slots with each slot radiator covering a 45° sector in the endfire direction with maximum gain variation of 0.5 dB. Multiple beams are generated by exciting eight sets of radiators independently. The switching feed network and the radiators are designed in the single common ground plane resulting in a compact design. In [32], a circularly polarized reconfigurable antenna element is proposed for beam steering. The four states in the patch antenna (0°, 90°, 180°, and 270°) are excited by using pin diodes. The 0° and 180° phase states are realized by utilizing the symmetry of the CP patch antenna. As a result, the 180° phase shifter is integrated within the CP patch, which effectively reduces the dimension of the phase shift circuit. A 90° digital phase shifter is cascaded to the antenna element for the 90° and 270° phase states. Using reconfigurable feeding network to realize bit array is a common method in beam-steering applications. The eight-element array designed from the antenna element results in beam steering of – 49° to + 49° by appropriately controlling the ON and OFF states of the pin diodes.

A novel electronic phase-shifting structure, composed of fan-shaped open stubs with different sizes and PIN diodes, is proposed for beam scanning in [33]. The leaky-wave antenna uses interdigital slots as radiating elements, and the phase-shifting structure is placed between the adjacent radiating elements. By changing the switching states of these pin diodes, the phase difference between the adjacent radiating elements are controlled. The electronic phase-shifting structure consists of four fan-shaped open microstrip stubs of different sizes and six PIN diodes, is connected to the free edge of the half mode bent corrugated SIW. The beam scanning of 25° (34°–59°) at fixed frequency is achieved by changing the DC bias voltages of the six PIN diodes on each phase-shifting structure. As the top and bottom conductors of this antenna are DC isolated, it can be easily connected with other active components and devices.

### 3.5.2 *Varactor Diodes*

Frequency-dependent antennas have suffered a limitation because most modern wireless communication systems operate in a fixed-frequency band. Fixed-frequency beam-steering can be achieved by using various kinds of tunable bit components, such as binary switch, pin diode, and digital phase shifter, to switchover between different phase states but it is limited to several discrete radiation angles. Compared to binary switches, capacitively tuned components such as varactor diodes can achieve continuous beam-scanning. An equivalent circuit of varactor diode is shown in Fig. 3.10.

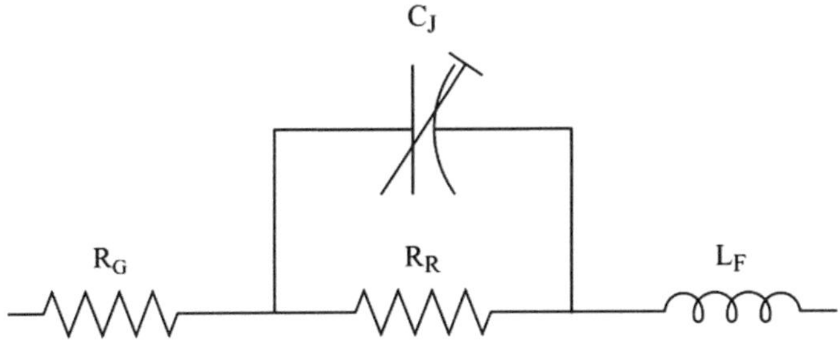


**Fig. 3.10** Equivalent circuit of a varactor diode

In [34], a CRLH transmission line (CRLH-TL) structure is loaded with varactor diodes to accomplish electrical beam steering. The LWA consists of square-shaped unit patches and a single-voltage bias circuit is introduced to control the tunable components, which consequently steers the beam direction. The unit cell includes three varactor diodes. The two series varactor diodes introduce the LH capacitor and the shunt varactor diode introduces the LH inductor, so both are electrically tunable. With single uniform bias voltage changing from 3.5 to 6.5 V, the main lobe scans from + 40° to – 17° continuously at 2.4 GHz. Another LWA that incorporates varactor diodes in its CRLH-TL is presented in [35]. The prototype antenna has nine-unit cells, where eighteen varactor diodes are employed for frequency reconfiguration and another twenty varactor diodes are employed for its beam reconfiguration. The varactor diodes for frequency reconfiguration are placed between the central rectangular pads and the main patches. Because these meander gaps introduce the left-handed capacitances, this location of the varactor diodes allows reconfiguration of that capacitance. The left-handed inductance is introduced when the shorting vias are turned on. Furthermore, this inductance can be tuned by varying the capacitance of the varactor diodes of beam reconfiguration. The presented design obtains a beam-scanning range of over 100° from 4.75 to 5.25 GHz. However, due to the limitation of commercially obtained varactor diodes, the measured beam-scanning range of the prototype is reduced.

Beam-steering LWA with wide-beam scanning angle at a fixed frequency is presented in [36]. The proposed LWA consists of an impedance modulation surface (IMS) and dc bias circuit. A three-strip structure is designed as the unit cell of the IMS. The varactor diodes are inserted in the unit cell and the impedance characteristics of every unit cell is controlled independently. Both average reactance and modulation periodicity of the IMS can be tuned for leaky-wave radiation with different beam angle. By tuning the DC voltages of varactor diodes, the beam angle can continuously scan in a wide range. The proposed LWA can be configured to work in single-beam mode or in dual-beam mode where each beam can be controlled independently. By controlling the dc voltages of varactor diodes, a scanning range of 118° can be achieved at 8 GHz. However, due to the limitation of adjustable range of capacitance values, the average reactance of the IMS cannot be tuned in a wide range, which in turn limits the reconfigurability of the beam angle.

For more flexible designs we can use a combination of pin diodes along with varactor diodes as presented in [37]. This antenna structure combines one patch antenna with two endfire radiation slot antennas. By loading varactors on the patch, the working frequency is continuously adjusted in the range of 2.24–2.58 GHz, and simultaneously the maximum beam direction could be reconfigured by controlling four PIN diodes located on two slots. The proposed antenna has discrete maximum radiation direction of – 78°, – 36°, 0°, + 36°, + 78°, and 180°, obtained by controlling the states of pin diode. The antenna has the merits of both wide-angle beam discrete scanning performance of ± 78° in the H-plane and switchable backward radiation feature within frequency tuning range.

### 3.5.3 RF MEMS Switches

RF micro-electro-mechanical systems (MEMS) switch technology has been introduced during the last 10–15 years as a prime candidate to replace the conventional GaAs FET and pin diode switches in RF and microwave communication systems, mainly due to their low insertion loss, very high isolation, linear characteristics, low power consumption, and low cost [38]. Hence, it is regarded as a promising technology for future reconfigurable microwave and antenna systems, particularly at millimeter-wave frequencies. It has also provided the way for the development of novel revolutionary RF circuits like low-cost reconfigurable and smart antenna system-on-chip that can be used in the next generation of broadband, wireless, intelligent communication and radar systems, and space-borne remote sensing.

MEMS switches have been used in different RF circuit applications: tunable antennas, tunable microwave filters, tunable phase shifters, and tunable matching networks. In recent years, researchers have incorporated RF switches in antenna configurations to attain more diverse functionalities like beam steering [39], dual-frequency band operation [40], and dual-polarized radiation [41]. Microwave and millimeter-wave technology that offers wide tunability is essential for today's cost-driven commercial and military industries. In order to meet the above requirements, micromachined tunable capacitors have been shown to have an adequate $Q$-factor when they are fabricated using alumina or polysilicon substrate [42] surface micromachining technology. Tunable capacitors are enabling components for high-frequency systems. There are two approaches to make such components. One is a chemical approach that improves properties of the materials, and the other is a physical approach that controls the gap or area of the dielectric layer for variable capacitance. MEMS switch precise, micrometer-level movements are ideal drives for the physical approach. A MEMS-based switching diaphragm could be used as a variable capacitor [43]. The tunability of this component is very impressive because a lossless, 2 $\mu$m air gap is used for the dielectric layer. However, the range of this variable capacitance is limited when the top membrane collapses onto the bottom plate.

One key factor of MEMS switches is input–output signal isolation in the down state. Traditional MEMS switches use $Si_3N_4$ as dielectric material, its lower dielectric constant (~7) limits its application at lower frequency. In order to further improve the performance of MEMS switches, a higher $C_{ON}/C_{OFF}$ ratio and thus larger down-state capacitance is required. Emerging barium strontium titanate (BST) thin film technology has been investigated for enhancing RF MEMS capacitive switches due to its high dielectric constant (> 200) in [44]. The switch provides better isolation than the nitride switches of the same type, but the isolation value drops considerably (< 15 dB) below 5 GHz. In [45], a highly compact low loss up to 40 GHz and linearly tuned capacitive cantilever MEMS switch using high quality BST as the dielectric layer is reported for the first time. It provides continuous (analog) tunability of the capacitor after the MEMS switch has been pulled down, due to the voltage controlled dielectric constant properties of the BST material. The proposed switch can be used to develop very compact digital capacitor banks with enhanced analog tuning for a variety of reconfigurable antenna systems and other networks.

RF switching is normally accomplished by using semiconductor devices such as diodes or FETs. However, these semiconductor devices have low $Q$ ($Q < 3$ @10 GHz) at high frequencies and therefore exhibit high insertion loss [46]. The use of MEMS switches has provided a reliable alternative to semiconductor devices [47]. MEMS switches have higher $Q$ ($Q > 10$ @10 GHz) compared to semiconductor devices. Although there are still problems to be solved such as MEMS switches reliability, packaging, and switching speed, when the $Q$-factor is a key parameter MEMS switches should be considered. Using RF MEMS switches, antennas with a variety of reconfigurability in terms of resonance frequency, impedance bandwidth, polarization, and radiation patterns have been recently introduced [48]. The goal of such reconfigurable apertures is to reduce the system complexity for operation over a wide frequency band or with diverse radiation characteristics. Low-loss RF MEMS switches are of great interest to enhance the antenna gain. RF MEMS technology has matured over the years, and it has now reached a level when RF MEMS switches for the first time have been successfully integrated into practical RF systems with proven long-term reliability [49].

Traditionally, MEMS switches are fabricated on silicon or GaAs substrates [50]. However, for antenna applications it is desirable to have switches fabricated on the same substrate where the antenna is printed. Doing so will eliminate the need for interconnections between antenna parts and switches, thus yielding higher performance and a fully integrated and lower cost system [51]. The antenna is normally fabricated on a printed circuit board material, such as RT-duroid or FR-4. The fabrication process of RF MEMS switch on the PCB requires some additional steps compared to silicon due to the rough PCB surface. Also low-temperature must be maintained during the entire fabrication process (below 200 °C) to ensure survivability of the PCB [52]. Researchers in [53] demonstrated that a reconfigurable scan-beam spiral antenna can be fabricated on traditional microwave substrates, using RF MEMS capacitive switches. Two antennas are fabricated respectively on quartz and PCB, and their performance is compared. The antenna printed on PCB is fed through a coaxial cable at its center point; the feed line goes through a via hole which is placed at the center of the antenna. For the antenna on quartz substrate, since it is difficult to create vias, we feed the antenna from the outer arm using a microstrip line. In both cases, the antenna beam is steered within a useful range of values while maintaining a good matching for different beam directions within the bandwidth of interest.

## 3.6 Material Engineering for Beam Scanning

In the following subsections, beam scanning using materials like liquid crystal and ferrites are discussed.

### 3.6.1 *Liquid Crystal*

Liquid crystals (LCs) have been quite successful in display and photonic applications, and therefore, they are well-known as optical components of Liquid Crystal Display (LCD) screens. Besides the optical region, characterizations of liquid crystals show that they also can be successfully used as a functional material in the microwave spectrum [54]. Liquid crystals consist of elongated rod-like molecules and exhibit different electrical characteristics depending on how an RF field is incident on the molecules owing to its anisotropy. The orientation of a liquid crystal bulk can be changed by external electric or magnetic fields or by surface anchoring [55]. Therefore, a desired effective permittivity can be obtained if it is possible to control the bulk orientation. Thus, liquid crystals can be used as tunable substrates. They possess mesophases between solid and liquid states, i.e., crystal solid, smectic, nematic, and isotropic liquid. LCs commonly used in the microwave and millimeter-wave bands are nematic LCs. The fundamental properties of nematic LCs versus temperature are shown in Fig. 3.11. LC exhibits an anisotropy in the crystalline as well as in the nematic phase, whereas only the nematic phase can be used for electrically tunable RF circuits/devices. The change of the rod-shaped molecules' direction can lead to the change of the LCs' relative dielectric constant. The LC molecules' axial direction can be deflected in an external electric field or magnetic field environment. Generally, when the applied electric field is consistent with the elongated or longer axis of the LC molecule, its measured dielectric constant is defined as $\varepsilon_{r\parallel}$; when the applied electric field is perpendicular to its long axis, the measured dielectric constant is defined as $\varepsilon_{r\perp}$, and the difference $\Delta\varepsilon$ is defined as the dielectric anisotropy [56].

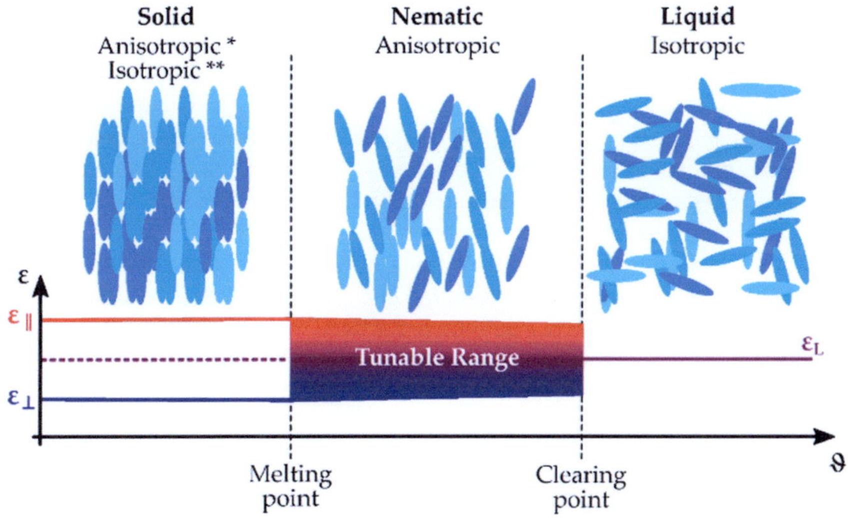


**Fig. 3.11** Nematic liquid crystal structure and permittivity for different orientations [56]

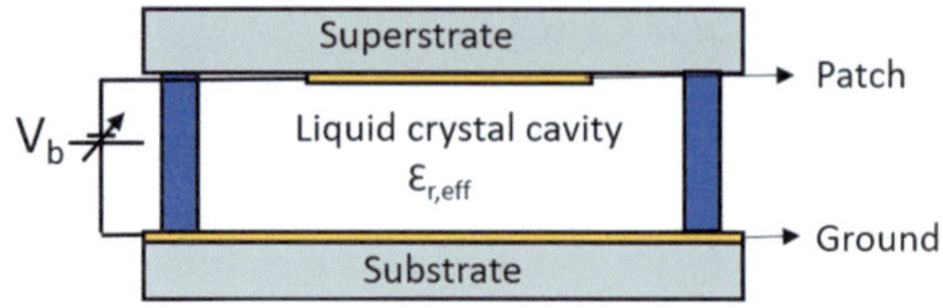


**Fig. 3.12** Cross section of the liquid crystal filled antenna element

A significant advantage of LCs is that their dielectric loss, and the loss tangent decreases with an increase in frequency, which means that the LCs are suitable tuning materials in millimeter wave and THz bands. Compared to other active components, such as a pin diode and a varactor diode, the loss tangent of the LC decreases as the frequency increases [57]. The microwave devices based on LC has the advantage of being lightweight and low cost [58].

In the application of tunable materials in reconfigurable antenna design, liquid crystals (LCs) can achieve continuous tuning of effective permittivity at low bias voltage. The DC voltage applied between the patch and the ground plane controls the crystal orientation within the cavity. In the absence of any bias voltage, the crystal molecules are perpendicular to the RF field between the patch and the ground plane. When a bias voltage is applied, molecules begin to turn leading to a variation in the effective permittivity of the liquid crystal bulk. If the voltage is increased further, molecules orient along the bias field lines and parallel to the RF field. The electronically controllable continuous permittivity variation can be used to obtain different reflection phase angle values. Bias voltages are applied via thin bias lines that are connected to the reflecting elements. An illustration is shown in Fig. 3.12. The biasing simplicity comes up as an advantage for the liquid crystal technology. Additionally, the biasing of the liquid crystal-based antenna element can be classified as low-power, since there is no need for any current flow through the element during the biasing.

An LC-based one-by-four dielectric resonator antenna, with improved beam-scanning capability, is proposed in [59]. The tunability of the structure, equivalently a desired differential phase shift, is achieved by controlling the orientation of the rod-shaped LC molecules with respect to the RF field of the inverted microstrip line. The inner sides of the front and back glasses are coated with an alignment layer that anchors the molecules parallel to the surface initially. In this case, the RF field is perpendicular to the LC molecules resulting in $\varepsilon_{r\perp} = 2.4$ and $\tan\delta_{\perp} = 0.006$. When a bias voltage is applied across the signal electrode and the ground electrode, the molecules tend to orient parallel to the applied bias field. For instance, after exceeding saturation voltage, i.e., 40 V, the LC molecules are aligned parallel to the RF field and therefore $\varepsilon_{r\parallel} = 3.2$ and $\tan\delta_{\parallel} = 0.002$. Thus, continuous tunability of the LC material is possible depending on the applied bias voltage. The antenna results in a beam-steering capability from $-30°$ to $30°$ due to LC tunability.

Researchers in [60] have presented an electrically steerable lens antenna with a single feed consisting of a semicircular-shaped parallel-plate waveguide (PPW) filled with LC. The radiating aperture along the lens consists of two permittivities defined by the LC material. Beam steering is achieved by using the LC's anisotropy by which the electromagnetic wave is guided in the area of the higher permittivity. Using an electrode network for LC biasing, the LC can be tuned, and therefore, the direction of the higher permittivity sector, and hence, the beam direction. For the first proof-of-concept, a simplified electrode biasing network, which is capable to reconfigure the beam in three predefined directions, $-30°$, $0°$, and $+30°$ is presented. With a more advanced electrode network and high-performance LCs, the proposed antenna could be potentially an efficient alternative to conventional phased arrays or beam-switching networks.

### *3.6.2 Ferrite Material*

The use of magnetized ferrite material in antenna apertures is one relatively undeveloped technique for providing electronic control although its deployment in discrete phase shifting components is well known. The Ferroscan idea consists of using ferrite, which partially fills an aperture, as a field displacement device [61]. For an incident field in a rectangular waveguide supporting dominant $TE_{10}$ mode, the displacement of the field occurs toward the more densely loaded portion of the aperture. Such devices ordinarily involve waveguides whose cross-sectional dimensions are less than 1/2 wavelength [62]. In ferrite-loaded radiating section, which is partially filled with ferrite material and the remaining volume occupied by a dielectric of relative dielectric constant $\varepsilon_r$, it is much wider than 1/2 wavelength and that the ferrite, which is operated in the remanent state, is magnetized in a stepwise fashion along its length by an external driver circuit. This can be used to generate a tilted phase front across the aperture and the outgoing wave is scanned.

With the availability of improved ferrites and more compact magnetizing circuits, several ferrite-loaded antenna configurations have been reported with varying degrees of beam-scanning capability. A ferrite-loaded microstrip antenna which has both low profile and a beam-steering capability with potential for conformal applications is described in [63]. The ferrite layer is placed behind the ground plane of a microstrip circular patch antenna and 15° beam scan is observed.

The characteristics of a Hertzian Dipole Antenna in various composite superstrate-substrate configuration are studied in [64]. If the geometry consists of layers which have constant permeabilities and permittivities, the radiation characteristics for a single dipole will be fixed by the choice of materials used for the layers and their thicknesses in terms of wavelength. However, if one of the layers is gyrotropic, the permittivity or permeability can be changed by adjusting the intensity or orientation of the magnetic biasing field or the operating frequency. Because of this, a variety of pattern shapes can be obtained for a particular antenna structure without modifying the hardware configuration. The top layer is a magnetically biased, fully saturated ferrite and the bottom grounded layer is an isotropic substrate with a relative permittivity $\varepsilon_r = 12.9$ and a relative permeability $\mu_r = 1.0$. The effect of changing the bias field intensity and direction, substrate thicknesses, and operating frequency have been investigated. It has been found that beam scanning or switching can be obtained by varying one of these parameters.

A magnetically scannable leaky-wave antenna employing a corrugated ferrite slab supported by a Teflon slab has been demonstrated in the millimeter-wave frequency range in [65]. The main beam direction, the beamwidth, and the radiation pattern of the leaky-wave antenna have been measured as a function of dc magnetic field in the frequency range 40–50 GHz. It is shown that the main beam direction can be steered continuously by altering the bias magnetic field and strongly depends on the bias magnetic field near the gyromagnetic resonance. The maximum scanning angle of the main beam direction is seen to be 41° associated with an average beamwidth of 3.4° with the bias magnetic field up to 1.4 T at 46.8 GHz. The maximum scanning rate is found to be 1°/0.02 T. To achieve a higher average scanning rate, ferrite materials with large saturation magnetization need to be developed.

Radiation from a ferrite-filled parallel-plate waveguide with a single slot has been studied in [66]. The slotted parallel-plate waveguide is filled with a transversely magnetized ferrite. The slot is installed on the upper plate in the ferrite-filled parallel-plate waveguide. It is found that the radiation region can be tuned without significantly changing the maximum radiation efficiency and the radiation bandwidth for variations of the applied dc magnetic field.

## 3.7 Conclusion

Beam scanning is a very vital requirement in antennas for wireless communication today. The chapter briefly covered various methods to achieve the beam scanning in antennas. The chapter also focused on electronic beam-scanning antennas. Based on the methodology used for scanning, the electronic beam scanning is classified into various categories. One type of electronic beam scanning is achieved by varying the phase or frequency causing the beam to change its direction. Leaky-wave antennas that are frequency scanned and phased-array antennas that scan the beam based on variation in phase at the individual antenna elements are discussed first. A brief discussion on phase shifters that can vary the phase of the antenna elements is included in the chapter. Different types of phased arrays like active, passive, and digital phased arrays are discussed followed by phased-array scanning. Beam scanning can also be performed using different discrete switches like pin diodes, varactor diodes and RF MEMS switches and its detailed discussion is provided in the chapter. Other than switches, the scanning is also possible using some exotic materials like liquid crystals and ferrites that can be used based on the applications and requirements.

## References


1. Lacomme, P.: Air and spaceborne radar systems: an introduction. United Kingdom, William Andrew (2001)
2. Gu, C., et al.: 3-d coverage beam-scanning antenna using feed array and active frequency-selective surface. IEEE Trans. Antennas Propag. **65**(11), 5862–5870 (2017)
3. Guney, K., Onay, M.: Amplitude-only pattern nulling of linear antenna arrays with the use of bees algorithm. Prog. Electromagn. Res. PIER **07**, 21–36 (2007)
4. Mailloux, R.J.: Phased Array Antenna Handbook, 2nd ed. Artech House, Inc. (2005)
5. Ahmed, F., Singh, K., Esselle, K.P.: State-of-the-art passive beam-steering antenna technologies: challenges and capabilities. IEEE Access **11**, 69101–69116 (2023)
6. Rocca, P., Oliveri, G., Mailloux, R.J., Massa, A.: Unconventional phased array architectures and design methodologies—a review. Proc. IEEE **104**(3), 544–560 (2016)
7. Hansen, R.C.: Phased Array Antennas, 2nd ed. Wiley (2009)
8. Li, X., Wan, Y., Liu, J., Jiang, D., Bai, T., Zhu, K., Zhuang, J., Wang, W.-Q.: Broadband electronically scanned reflectarray antenna with liquid crystals. IEEE Antennas Wirel. Propag. Lett. **20**(3), 396–400 (2021)
9. Sayidmarie, K.H., Saghurchy, M.N.: Array beam scanning by variation of elements amplitude-only excitations. In: 2011 4th IEEE International Symposium on Microwave, Antenna, Propagation and EMC Technologies for Wireless Communications, Beijing, China, pp. 749–753 (2011)
10. McManamon, P.F., et al.: A review of phased array steering for narrow-band electro optical systems. Proc. IEEE **97**(6), 1078–1096 (2009)
11. Van Acoleyen, K., et al.: Off-chip beam steering with a one-dimensional optical phased array on silicon-on–insulator. Opt. Lett. **34**(9), 1477–1479 (2009)
12. Yang, J., Qi, S.-S., Wu, W., Fang, D.-G.: A novel high-gain sum and difference conical beam-scanning reflector antenna. IEEE Access **8**, 103291–103300 (2020)
13. Ramazannia Tuloti, S.H., Rezaei, P., Tavakkol Hamedani, F.: High-efficient wideband transmitarray antenna. IEEE Ant. Wirel. Propag. Lett. **17**(5), 817–820 (2018)
14. Nguyen, B.D., Tran, S.V.: Beam-steering reflectarray based on two-bit aperture coupled reflectarray element. J. Electromagn. Waves Appl. **32**(1), 54–66 (2018)
15. Wang, H.-F., Wang, Z.-B., Wu, Z.-H., Zhang, Y.-R.: Beam-scanning lens antenna based on elliptical paraboloid phase distribution metasurfaces. IEEE Antennas Wirel. Propag. Lett. **18**(8), 1562–1566 (2019)
16. Schneider, D.A., Rosch, M., Tessmann, A., Zwick, T.: A low-loss W-band frequency-scanning antenna for wideband multichannel radar applications. IEEE Antennas Wireless Propag. Lett. **18**(4), 806–810 (2019)
17. Collin, R.E., Zucker, F.J.: Antenna Theory Part II, vol. 7. McGrawHill (1969)
18. Monticone, F., Alù, A.: Leaky-wave theory, techniques, and applications: from microwaves to visible frequencies. Proc. IEEE **103**(5), 793–821 (2015)
19. Fuscaldo, W.: Rigorous evaluation of losses in uniform leaky-wave antennas. IEEE Trans. Antennas Propag. **68**(2), 643–655 (2020)
20. Guo, T., Zhang, Q., Hou, X., Kandwal, A., Li, B.: Symmetry leaky-wave antenna without gain degradation at broadside. In: IEEE International Workshop on Antenna Technology (iWAT), pp. 1–3 (2018)
21. Ma, Z.L., Chan, C.H., Ng, K.B., Jiang, L.J.: A collimated surface-wave-excited highimpedance surface leaky-wave antenna. IEEE Antennas Wirel. Propag. Lett. **16**, 2082–2085 (2017)
22. Ma, W., Cao, W., Shi, S., Yang, X.: Compact high gain leaky-wave antennas based on substrate integrated waveguide $TE_{220}$ mode. IEEE Access **7**, 145060–145066 (2019)
23. Pillans, B., Eshelman, S., Malczewski, A., Ehmke, J., Goldsmith, C.: Ka-band RF MEMS phase shifters. IEEE Microw. Guided Wave Lett. **9**(12), 520–522 (1999)
24. Malczewski, A., Eshelman, S., Pillans, B., Ehmke, J., Goldsmith, C.L.: X-band RF MEMS phase shifters for phased array applications. IEEE Microw. Guided Wave Lett. **9**(12), 517–519 (1999)

25. Barker, N.S., Rebeiz, G.M.: Optimization of distributed MEMS phase shifters. In: 1999 IEEE MTT-S International Microwave Symposium Digest, pp. 299–302, Anaheim, CA, 1999
26. Zhao, H., Chu, H., Zhu, X., Guo, Y.: A compact full 90° cone-metric-space beam-scanning antenna array using resonant-type phase-shifters. IEEE Trans. Antennas Propag. **70**(6), 4404–4413 (2022)
27. Hansen, R.C.: Phased Array Antennas, 2nd ed. Wiley, Hoboken, NJ, USA (2009)
28. Lai, M.-I., Wu, T.-Y., Hsieh, J.-C., Wang, C.-H., Jeng, S.-K.: Compact switched-beam antenna employing a four-element slot antenna array for digital home applications. IEEE Trans. Antennas Propag. **56**(9), 2929–2936 (2008)
29. Ge, L., Li, M., Li, Y., Wong, H., Luk, K.-M.: Linearly polarized and circularly polarized wideband dipole antennas with reconfigurable beam direction. IEEE Trans. Antennas Propag. **66**(4), 1747–1755 (2018)
30. Kahar, M., Mandal, M.K.: A wideband tightly coupled slot antenna for 360° full azimuthal beam steering applications. IEEE Trans. Antennas Propag. **69**(6), 3538–3542 (2021)
31. Liu, P., Li, Y., Zhang, Z.: Circularly polarized 2 bit reconfigurable beam-steering antenna array. IEEE Trans. Antennas Propag. **68**(3), 2416–2421 (2020)
32. Lou, T., Yang, X.-X., Qiu, H., Luo, Q., Gao, S.: Low-cost electrical beam-scanning leaky-wave antenna based on bent corrugated substrate integrated waveguide. IEEE Antennas Wirel. Propag. Lett. **18**(2), 353–357 (2019)
33. Fu, J.H., et al.: An electrically controlled CRLH-inspired circularly polarized leaky-wave antenna. IEEE Antennas Wirel. Propag. Lett. **16**, 760–763 (2017)
34. Chen, S.L., Karmokar, D.K., Li, Z., Qin, P.Y., Ziolkowski, R.W., Guo, Y.J.: Continuous beam scanning at a fixed frequency with a composite right-/left-handed leaky-wave antenna operating over a wide frequency band. IEEE Trans. Antennas Propag. **67**(12), 7272–7284 (2019)
35. Wang, M., Ma, H.F., Zhang, H.C., Tang, W.X., Zhang, X.R., Cui, T.J.: Frequency-fixed beam-scanning leaky-wave antenna using electronically controllable corrugated microstrip line. IEEE Trans. Antennas Propag. **66**(9), 4449–4457 (2018)
36. You, C.J., Li, P.K., Dan Yang, D., Yao, C., Liu, S.H., Yin, G.: Frequency continuously tunable patch-slot antenna with wide-angle discrete scanning and optional backward radiation. IEEE Ant. Wirel. Propag. Lett. **20**(4), 607–611 (April 2021)
37. Brown, E.R.: RF-MEMS switches for reconfigurable integrated circuits. IEEE Trans. Microw. Theory Tech. **46**(11), 1868–1880 (1998)
38. Lim, S., Caloz, C., Itoh, T.: Metamaterial-based electronically controlled transmission-line structure as a novel leaky-wave antenna with tunable radiation angle and beamwidth. IEEE Trans. Microwave Theory Tech. **53**, 161–173 (2005)
39. Peroulis, D., Sarabandi, K., Katehi, L.P.B.: Design of reconfigurable slot antennas. IEEE Trans. Antennas Propag. **53**, 645–654 (2005)
40. Yang, F., Rahmat-Samii, Y.: Patch antenna with switchable slots (PASS): Reconfigurable design for wireless communications. Proc. IEEE Int. AP-S Int. **1**, 462–465 (2002)
41. Dec, A., Suyama, K.: RF micromachined varactors with wide tuning range. In: 1998 IEEE Radio Frequency Integrated Circuits (RFIC) Symposium. Digest of Papers (Cat. No.98CH36182), Baltimore, MD, USA, pp. 309–312 (1998)
42. Goldsmith, C., Lin, T.H., Powers, B., Wu, W.R., Norvell, B.: Micromechanical membrane switches for microwave applications. In: Proceedings 1995 IEEE MTT-S International Microwave Symposium Digest, Orlando, FL, pp. 91–94 (1995)
43. Liu, Y., Taylor, T.R., Speck, J., York, R.: High-isolation BST-MEMS switches. In Proceedings 2002 IEEE MTT-S International Microwave Symposium Digest, Seattle, WA, pp. 227–230 (2002)
44. Wang, G., Polley, T., Hunt, A., Papapolymerou, J.: A high performance tunable RF MEMS switch using barium strontium titanate (BST) dielectrics for reconfigurable antennas and phased arrays. IEEE Antennas Wirel. Propag. Lett. **4**, 217–220 (2005)
45. Rebeiz, G.M.: RF MEMS Theory, Design, and Technology. Wiley, New York (2003)
46. Papapolymerou, J., Lange, K.L., Goldsmith, C.L., Malczewski, A., Kleber, J.: Reconfigurable double-stub tuners using MEMS switches for intelligent RF front-ends. IEEE Trans. Microwave Theory Tech. **51**, 271–278 (2003)

47. Anagnostou, D.E., Zheng, G., Chryssomallis, M.T., Lyke, J.C., Ponchak, G.E., Papapolymerou, J., Christodoulou, C.G.: Design, fabrication, and measurements of an RF-MEMS-based self-similar reconfigurable antenna. IEEE Trans. Antennas Propag. **54**(2), 422–432 (2006)
48. Hallbjörner, P., Bergstrom, M., Boman, M., Lindberg, P., Öjefors, E., Rydberg, A.: Millimetre-wave switched beam antenna using multiple travelling-wave patch arrays. IEEE Proc. Microw. Antennas Propag. **152**(6), 551–555 (2005)
49. Goldsmith, C., Yao, Z.J., Shea, C., Eshelman, S., Denniston, D.: Performance of low-loss RF MEMS capacitive switches. IEEE Microwave Guided Wave Lett. **8**, 269–271 (1998)
50. Cetiner, B.A., Qian, J.Y., Chang, H.P., Bachman, M., Li, G.P., De Flaviis, F.: Monolithic integration of RF MEMS switches with a diversity antenna on PCB substrate. IEEE Trans. Microwave Theory Tech. **51**, 332–335 (2003)
51. Chang, H.P., Qian, J.Y., Cetiner, B.A., De Flaviis, F., Bachman, M., Li, G.P.: RF MEMS switches fabricated on microwave-laminate printed circuit boards. IEEE Electron Device Lett. **24**, 227–229 (2003)
52. Chang, C.H., Qian, J.Y., Cetiner, B.A., Xu, Q., Bachman, M., Kim, H.K., Ra, Y., De Flaviis, F., Li, G.P.: RF MEMS capacitive switches fabricated with HDICP CVD SiNx. Proc. IEEE MTT-S Int. **1**, 231–234 (2002)
53. Cheng, S., et al.: Switched beam antenna based on RF MEMS SPDT switch on quartz substrate. IEEE Antennas Wirel. Propag. Lett. **8**, 383–386 (2009)
54. Yang, D.-K., Wu, S.-T.: Liquid crystal physics. In: Fundamentals of Liquid Crystal Devices, pp. 1–37. Wiley, Hoboken, NJ, USA (2006)
55. Maune, H., Jost, M., Reese, R., Polat, E., Nickel, M., Jakoby, R.: Microwave liquid crystal technology. Crystals **8**(9), Art. no. 355 (2018)
56. Lapanik, V., et al.: New highly anisotropic liquid crystal materials for high frequency applications. Liq. Cryst. **45**(8), 1242–1249 (2018)
57. Perez-Palomino, G., et al.: Design and experimental validation of liquid crystal-based reconfigurable reflectarray elements with improved bandwidth in F-band. IEEE Trans. Antennas Propag. **61**(4), 1704–1713 (2013)
58. Karabey, O., et al.: Liquid crystal based phased array antenna with improved beam scanning capability. Electron. Lett. **50**(6), 426–428 (2014)
59. Bildik, S., et al.: Reconfigurable folded reflectarray antenna based upon liquid crystal technology. IEEE Trans. Antennas Propag. **63**(1), 122–132 (2014)
60. Li, X., et al.: Broadband electronically scanned reflectarray antenna with liquid crystals. IEEE Antennas Wirel. Propag. Lett. **20**(3), 396–400 (2021)
61. Lax, B., Button, K.J.: Microware Ferrites and Ferrimagnetics. McGraw-Hill, New York (1966)
62. Henderson, A., James, J.R., Fray, A.: Magnetised microstrip antenna with pattern control. Electron. Lett. **24**(1), 45–47 (1988)
63. Hsia, I.Y., Alexopoulos, N.G.: Radiation characteristics of Hertzian dipole antennas in a nonreciprocal superstrate-substrate structure. IEEE Trans. Antennas Propag. **40**(7), 782–790 (1992)
64. How, H., Fang, T.M., Guan, D.X., Vittoria, C.: Magnetic steerable ferrite patch antenna array. IEEE Trans. Magnet. **30**(6), 4551–4553 (1994)
65. Nishimura, K.: Radiation characteristics of a slotted parallel plate waveguide filled with a transversely magnetized ferrite. Electron. Commun. Jpn., pt. 2, **85**(7), 22–29 (2002)
66. Hwang, K.C., Eom, H.J.: Radiation from a ferrite-filled rectangular waveguide with multiple slits. IEEE Microw. Wirel. Compon. Lett. **15**(5), 345–347 (2005)

# Chapter 4
# Other Beam-Scanning Techniques

## 4.1 Introduction

Beam-steerable antennas capable of flexibly steering radiation beams are good candidates to perform communications with mobile users to maintain link budget and offer reliable connections. Due to their simple configuration, high gain, and low loss of transmitarray (TA) and reflectarray (RA) antennas, they have been widely investigated in the past decades, where both TA and RA antennas consist of a phase-shifting surface and a feed source. The methods to design beam-scanning reflectarrays can be normally categorized into three basic classes: electrical scan, frequency scan, and mechanical scan techniques. In electrical scan techniques, beam steering can be accomplished by utilizing discrete devices, like varactor diode, pin diode, and micro-electro-mechanical systems switches, or exotic materials, like ferroelectric material, liquid crystal, and graphene. Despite showing high-speed beam control, electrical control suffers from various shortcomings including design complexity, fabrication cost, and low efficiency caused by high loss. With regard to the frequency scanning method, the dispersion characteristics of reflectarray elements can be applied to realize beam scanning at terahertz frequencies for imaging application [1]. However, the applications of frequency scanning are still limited in modern wireless communication and radar systems.

## 4.2 Mechanical Beam-Scanning Antennas

Compared to conventional phased-array antennas, mechanically beam-steerable antennas do not use any electronic phase-shifting mechanisms. Thus, they benefit from lower losses, have reduced cost, and are often better suited for high-power applications. This, however, comes at the expense of reduced scanning speeds. Therefore, in applications that prioritize high-power handling, lower cost, and reduced

 
S. K. Koul et al., *Antenna Systems for Modern Wireless Devices*, Signals and Communication Technology, https://doi.org/10.1007/978-981-97-3369-9_4

complexity over high-speed beam scanning, mechanical beam-steering techniques may present good alternatives to conventional active or passive electronically steered arrays. The beams of passive TA and RA antennas can be mechanically steered by moving the feed source [2], using multiple feed sources [3] rotating the phase-shifting surface [4] or rotating the unit cells of the phase-shifting surface [5]. Among them, the movement of the feed source (equivalent to multiple feed sources) or rotating the phase-shifting surfaces introduces certain progressive phase gradients with respect to the feed source, thereby tilting the beams of the TA and RA antennas. The significant gain roll-off and remarkable sidelobe level degradations always occur when the feed source is moving outward to steer the beams of the TA and RA antennas whose phase-shifting surfaces are usually built according to the illumination of the center-located feed source (unifocal TA/RA antenna). To improve the gain roll-off, bifocal phase synthesis and phase-optimization-based methods have been proposed and leveraged to find the desired phase shifts to implement the phase-shifting surfaces. The bifocal phase synthesis can formulate explicit expressions to mathematically calculate the phase shift that each unit cell should offer. It is observed that the bifocal phase synthesis can indeed lower the gain roll-off but at the expense of reduced gains of scanning beams. The phase optimization-based method, by contrast, can define objective and cost functions, aiming at achieving a minor gain roll-off and comparable gains within the beam-scanning coverage with powerful algorithms [6]. It, however, is time-consuming as extensive iterative computations are needed to achieve the goals.

### *4.2.1 Steering the Feed Antenna*

One of the ways to steer the beam is to adjust the antenna position by physically moving it. Usually, the antennas are placed in some rotatable blocks that allow the antenna to scan when the block rotates. A fully metallic mechanically steerable slot array antenna in gap-waveguide technology is reported in [7] where the beam scanning is realized by adjusting the antenna feed positions. The lower rotatable block of the antenna rotates to obtain steering angles of up to 25° with gain levels of 21 dBi. The steering mechanism presented only requires the mechanical rotation of an internal part of the structure and such rotation movement can be implemented by relatively simple engines and it is a feasible option to attain beam-steerable slot arrays, especially compared to those of conventional positioning systems.

A conical beam-scanning reflector antenna, with the capability of tracking the beam-pointing angle is proposed in [8]. Rotational symmetric reflecting surface is illuminated by two primary omnidirectional horns as shown in Fig. 4.1. Two conical beams with different beam-pointing angles could be generated by offsetting the feed antenna. The inner horn is fed by a simple rectangular to coaxial waveguide transition. The outer horn is fed using four-way ridge waveguide to suppress the higher-order modes of the coaxial waveguide. The conical beam-scanning performance was realized by mechanically moving the total feed along the symmetry axis and taking full advantage of offset focus characteristic of the parabolic reflector. When the feed of a parabolic reflector antenna is displaced laterally from the focal point, the beam is shifted off-axis in the opposite direction to the feed displacement as shown in Fig. 4.2. Moving the feed horns along the symmetry axis allowed the sum conical beam to be scanned from 30.8° to 45.2°. This structure can make full use of the antenna aperture to obtain higher aperture efficiency and is easy to fabricate at K band.

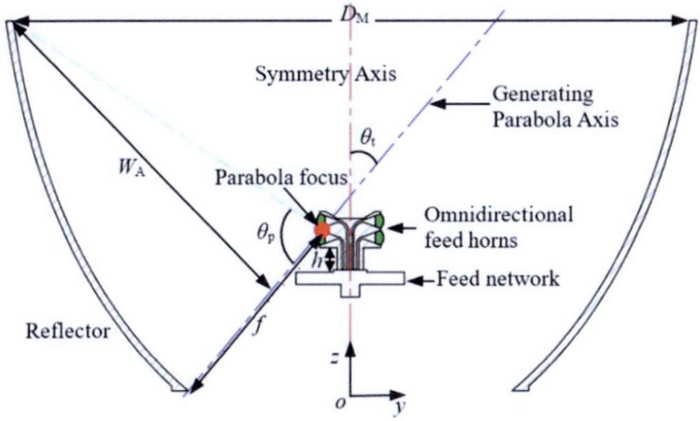


**Fig. 4.1** Cross-sectional view of the conical beam-scanning reflector antenna [8]

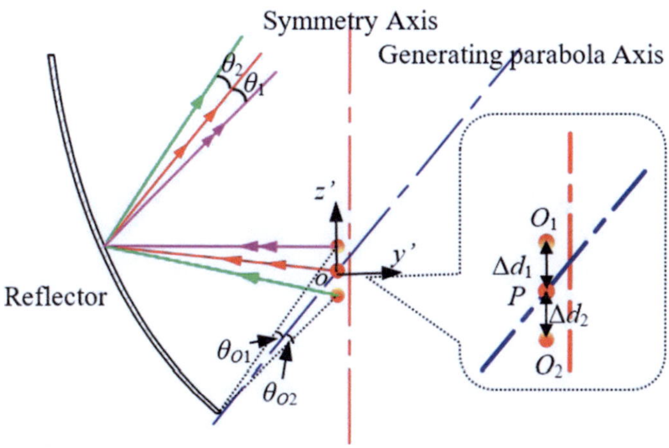


**Fig. 4.2** Displacement of the feed horns [8]

### *4.2.2 Steering the Focusing Aperture*

The reduction in volume and mass of small satellites (SmallSats) has enabled more efficient and robust missions in space [9], which has led to exponential growth in the number of SmallSats launched each year. A key requirement in advanced SmallSat applications is beam scanning [10]. However, traditional beam-scanning high-gain antennas are not practical for SmallSat applications given the power and space constraints. Therefore, current research is focusing on developing compact/deployable beam-scanning HGAs for space applications. Compared to the parabolic reflector and phased-array antennas, RAs provide highly directive and versatile beams while maintaining low mass, low fabrication cost, and low complexity [11]. Also, RAs have recently demonstrated the ability to achieve pattern reconfigurability [12]. However, current reconfigurable RA designs suffer from various limitations including lossy and complex DC biasing networks that drive reconfigurable elements [13], or inefficient mechanical systems [14]. Researchers in [15] presented a novel 1D beam-steering RA system that uses a simple and efficient scrolling mechanism to steer its beam along a single cut plane. It uses a flexible RA aperture that is wrapped around two columns, which can rotate, causing the aperture to roll. The aperture is first synthesized to steer its main beam toward the broadside direction. Then, the aperture is rolled, thereby defocusing the RA feed antenna causing the main beam to scan toward the direction of rotation. The rollable RA prototype achieves 42° of lateral beam scanning while maintaining a maximum realized gain of 25.1 dBi with a 1.5-dB variation. Our rollable RA uses a simple and low-profile mechanism, thereby eliminating the need for complicated, bulky, and inefficient feed networks or mechanical systems. This makes the design an ideal candidate for reconfigurable HGA applications in compact spaces, such as SmallSats or tactical equipment.

## 4.3 Mechanical Beam Scanning by Moving Antenna Parts

### *4.3.1 Rotating Prisms*

Instead of shifting the feed antenna for beam steering, we can steer the beam by moving some part of the antenna. Mechanically beam-steerable antenna systems based on the Risley-prism concept have attracted attention for their simplicity and good beam-scanning performance. The Risley-prism configuration [16] renders the antenna unique advantages, including low electronic complexity (no need for distributed RF device, active components, or bias network), low profile, and wide-angle beam-scan coverage. In the optical regime, a Risley-prism system typically consists of a pair of rotatable wedge-shaped prisms used in conjunction with a well-collimated beam of light (e.g., a laser beam) [17]. The beam can be steered when the prisms are rotated with respect to each other. In the microwave regime, different configurations of Risley-prism-based beam-steering systems have been proposed and implemented. A Risley-prism antenna (RPA), also known as the near-field meta-steering (NFMS) antenna [18], have three conventional configurations used at microwave frequencies where they are coaxially placed in the near field. Figure 4.3a shows the architecture of the first reported beam-steerable antenna that employed the Risley-prism concept. Here, the two wedge-shaped prisms used in optical systems are replaced by a pair of flat prisms, illuminated by a feed antenna. The spherical wave radiated by the feed antenna is collimated by the lens placed below the two flat prisms. The relative rotational displacement between Prism 1 and Prism 2 provides beam steering in both azimuth and elevation directions in the upper hemisphere of the outgoing aperture. This design configuration was demonstrated for beam-steering systems working at the Ka-band with a $\pm$ 75° scan range [19]. The configuration is relatively complex because it uses two flat prisms and a collimating lens, each of which is implemented with multiple metal and dielectric layers. The structure shown in Fig. 4.3b combines the collimating lens with the bottom flat prism. It not only collimates the spherical wave radiated by the feed antenna but also provides the corresponding phase gradient needed from the first prism for beam steering. The working principle of RPAs based on phase-shifting surfaces has been discussed using a phase analysis method in [20]. Essentially, the phase shifts of the two transmitarrays are accumulated to the wave as it travels through the system, and their summation becomes a new gradient phase at the exiting aperture. The value and the direction of this aperture phase gradient can be changed by orienting the two surfaces, and consequently, the far-field beam is scanned. Zhao et al. [21] presented a high power-capable beam-steering lens antenna to work at the X-band with a $\pm$ 20° beam-scan range. However, the structure had an overall profile of $17.15\lambda_0$ at 9.375 GHz due to the usage of the horn antenna, which must be placed at some distance from the input aperture of the collimating lens/first prism. This high overall thickness motivated the development of the low-profile Risley-prism-based beam-steerable antenna architecture, as shown in Fig. 4.3c. Here, the horn antenna is replaced by a planar antenna or an array to achieve a low-profile beam-steering system [22].

Although the antenna is low profile, it still requires the use of two flat prisms along with one antenna or planar array as the main feed. Each additional prism introduces ohmic and reflection losses, resulting in reductions of aperture efficiency and peak gain of the beam-steering system. Furthermore, each flat prism is generally implemented by multilayer printed circuit boards, which adds to the cost and weight of the antenna system. In [23], a compact and low-profile Risley-prism-based beam-steerable antenna system that uses only a single flat prism is presented. The proposed beam-steerable antenna consists of two parts: a holographic leaky-wave antenna and a phase-shifting surface (PSS) layer acting as a flat prism placed on top of it. The holographic leaky-wave antenna provides an electric field distribution with a linear phase shift gradient over the input aperture of the flat prism. Therefore, only a single flat prism with the same phase gradient is needed to perform the beam steering. In this design, the bottom holographic leaky-wave antenna is constructed using an array of subwavelength metallic patches with slits, placed on a grounded dielectric substrate. The flat prism is implemented using a low-pass PSS with hexagonal unit cells. 2D beam scanning in elevation and azimuth is achieved by rotating the flat prism and

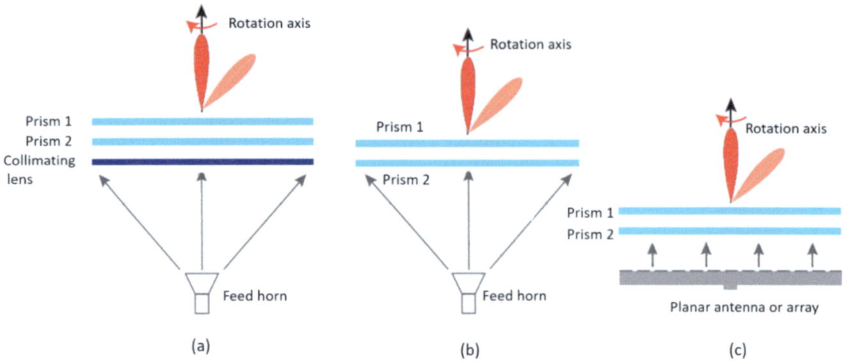


**Fig. 4.3** Different configurations of the Risley-prism-based beam-steerable antenna system: **a** system employing a feed horn, a collimating lens, and two flat prisms, **b** system where the collimating lens is integrated with the first prism, **c** system where the horn antenna is replaced with a planar antenna or array placed in close proximity to the two flat prisms

the leaky-wave antenna against each other. The main beam of the antenna can be scanned from the broadside direction to a maximum scan angle of $\pm 57°$ in 2D with a maximum scan loss of 3.6 dB. Since the feed antenna is also rotated, our proposed approach is only suitable for designing circularly polarized beam-steerable antennas. If the leaky-wave antenna is linearly polarized, beam scanning can be performed, but the polarization of the antenna will also change as a function of scan angle, which is generally not desirable. In circularly polarized systems, however, physically rotating the feed antenna does not adversely impact the polarization of the radiated wave. It has been demonstrated in [24] that the conventional RPA configuration can be simplified by replacing both the stationary feed and TA1 with a single rotatable gradient-phase feed. This new configuration offers the same 2D beam-steering capability but uses only two components. It can help lower the system profile and weight, simplify mechanics, and reduce transmission loss. The proposed RPA configuration reveals that the conventional three-component RPA configuration can be further simplified if a feed source radiating with a gradient phase is used. In this case, only one phase-shifting surface (TA2) is needed on top of the feed.

### *4.3.2 Rotating Other Antenna Sub-components*

A mechanical beam-steering array (MBSA) with a tunable height element is proposed in [25]. A phase range of 360° can be achieved by using a designed height-adjustable device. The $8 \times 8$ MBSA working at 10 GHz can realize beam steering in a large angle range and the height of each antenna element can be easily controlled by the circuit control board. Each antenna element is composed of a U-slot radiation patch and a stepper motor supported by the three-dimensional-printed structure. By controlling the height of each antenna element using a stepper motor, an arbitrary phase shift value in the range of 0°–360° can be achieved which can realize a beam scanning from − 40° to + 40°. In addition, steering high-gain beams can be generated by adjusting the height of each antenna element.

To enhance the application spectrum of beam-scanning antennas, a novel solution is developed using a combination of near-field phase transformation theory [22] using stepped dielectric (SD) and refraction of electromagnetic (EM) wave through a dielectric wedge (DW). An SD made of several dielectric section of different heights is placed above an array. The electric field, propagating along *z*-axis, at the input of the SD has uniform phase distribution represented by a parallel line. The electric field passes through each dielectric section of the SD. Each section has a different length and thus provides a different phase delay between two parallel planes marked at the input and output of the SD. The heights of sections in the SD can be adjusted to transform a uniform phase at the input plane into a linearly decreasing phase at the output plane to effectively tilt the beam by an angle $\theta°$. A single SD can tilt the array beam at a fixed offset angle away from the broadside direction. The beam direction can be dynamically changed using an optical beam-scanning concept of Risley prisms. In this optical-scanning method, two prisms are placed in front of an optical source where each prism tilts the optical beam away from the normal direction [26]. The prisms are placed in the far-field region of the optical source and while keeping the source fixed the prisms are rotated synchronously and independently to dynamically scan the optical beam in a large conical region. The azimuth and elevation angles of the beam depend on the orientation of the two prisms. Similar to Risley prisms, the proposed antenna uses a second beam-tilting structure made of DW, which is depicted in Fig. 4.4. The DW, unlike SD, is designed using ray-tracing method. The two opening angles of the DW are calculated by tracing rays from the input edge of the DW to the output edge. The beam can be moved to any intermediate elevation angle by changing the orientation angles of SD and DW.

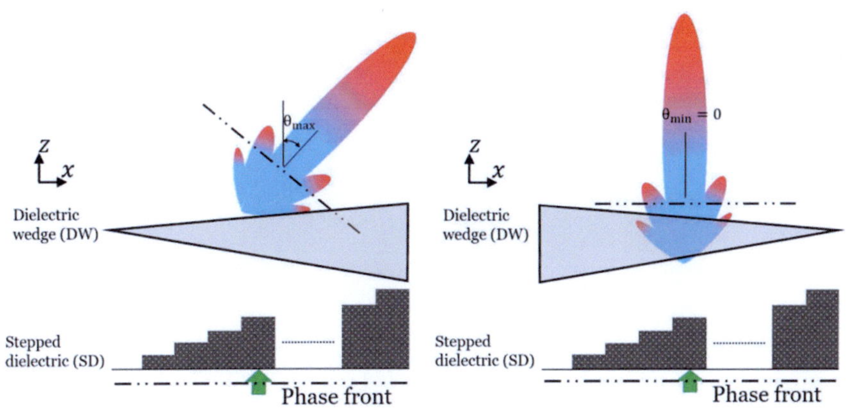


**Fig. 4.4** Pictorial depiction of stepped dielectric (SD) and dielectric wedge (DW) with beam directions [33]

The use of phase-transforming structures has advantage in terms of cost and power losses when compared with phase shifters in arrays. One type of free-standing phase-transforming structure is made from dielectric materials. It is well-known that the wavelength of electric field propagating inside a dielectric is less than that in the free space, which is expressed as $\lambda_g = \lambda_0/\sqrt{\varepsilon_r}$, where $\lambda_0$ is the free-space wavelength and $\varepsilon_r$ is the relative permittivity of the dielectric. A plane wave propagating through an infinitely extended dielectric material experiences a phase delay, which is directly proportional to the relative permittivity and length of the dielectric. The phase-delay through a dielectric can be varied either using different lengths of the same dielectric material or using different dielectric materials of a fixed length. The antenna presented in [27] focuses on using variation in the length of a fixed dielectric material to achieve the desired phase delay profile with an SD. It comprises of a resonant cavity antenna (RCA) and a pair of SD and DW as shown in Fig. 4.5

The SD is designed to realize near-field phase transformation and effectively tilts the RCA beam. The DW is optimally designed using optical ray tracing to tilt the beam further and to move the beam in the broadside direction when DW is co- and counter-aligned with the SD, respectively. The SD and DW are both 3D printed using a readily available polylactide (PLA) plastic and a moderate 3D printer. The SD and DW placed above the RCA are physically rotated, independently and synchronously, around the center of the aperture to scan the antenna beam in a 2D angular region. This method of beam scanning is continuous, unlike switched-beam-type antenna arrays. However, the beam-scanning speed of the proposed method is

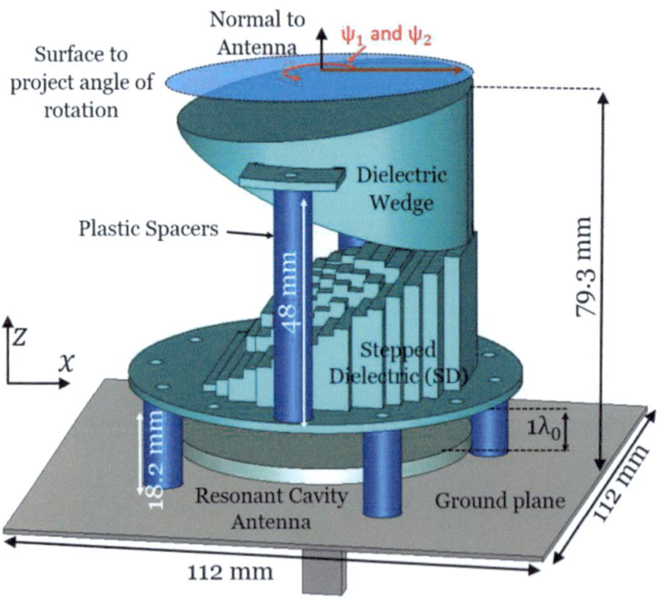


**Fig. 4.5** 3D model of the complete antenna system using SD and DW [27]

limited by the mechanical rotation of the dielectric structures and slower than the electronically switched arrays. The maximum height of the proposed design is only 8.4λ. Furthermore, the fabrication cost of the parts used in the proposed design is extremely low than that of the two stepped dielectrics developed using traditional subtractive manufacturing technique.

## 4.4 Beam Scanning with Lens

Lens antennas have been receiving considerable attention recently due to their advantages such as multibeam availability and potential for wide-angle beam scanning. 3D dielectric lenses of various shapes, like, sphere, ellipsoid, hemisphere, and cylinder are extensively studied for low cost and ease of fabrication. Among these lenses, hemispherical lenses have the advantage of easy integration due to their linear or planar feed arrays and are attractive for many applications, including high-data rate communication, automotive radar, and imaging systems. However, conventional hemispherical lens antennas often suffer from limited scanning range. To solve this problem, ray-tracing methods have been implemented in the literature [28].

### *4.4.1 Hemispherical Lens*

A closed-form relation between the scan angle and feed position for hemispherical lenses is studied in [29]. It demonstrated that the relation is linear in the small-angle limit, and the effective focal length determining the scan angle is presented. It is also demonstrated that the scan angle only depends on the geometrical configuration and that it is independent of the lens material. To determine the applicability of this scan angle relation to the design of focal plane arrays, it demonstrated that it can be used to determine the FPA spacing that results in − 3-dB overlap between switched beams. This linear relation between the scan angle $\gamma$ and the feed position $d$ for small angles holds for any thin lens of focal length $f_1$

$$\gamma \approx \frac{d}{f_1}, \tag{4.1}$$

where $\gamma$ is measured in radians. The scan angle can be estimated using

$$\tan \gamma = \frac{d}{L}. \tag{4.2}$$

The reason why the scan angle of the traditional hemispherical lens is hard to reach ± 45° can be explained from Fig. 4.6. The "array-feeding region" (AFR) stands for the region reserved for the linear PAA. Based on the CFM method or the theory that optical paths are reversible, an ideal plane wave could be generated by the array-feeding lens antenna only if the E-field power transmitted through the lens is perfectly focused to the AFR. However, as for the traditional hemispherical lens, the oblique incident power is mostly focused to the area which is beyond the AFR. Therefore, Liu et al. [30] aim to develop a modified hemispherical lens which is able to focus the incident waves into the AFR as shown in Fig. 4.7. The elliptical cylinder is chosen as the modification shape, considering the fabrication complexity. The permittivity of each layer of lens is fixed and has a dominant impact on the aperture efficiency of total hemispherical lens antenna. The incident plane wave with a 45° incident angle is completely transmitted through the modified lens and mainly focused on the desired AFR. Although the scanning loss is beyond the general limitation of 3 dB when scanning up to ± 60°, the main beam for 60° scanning angle is stable.

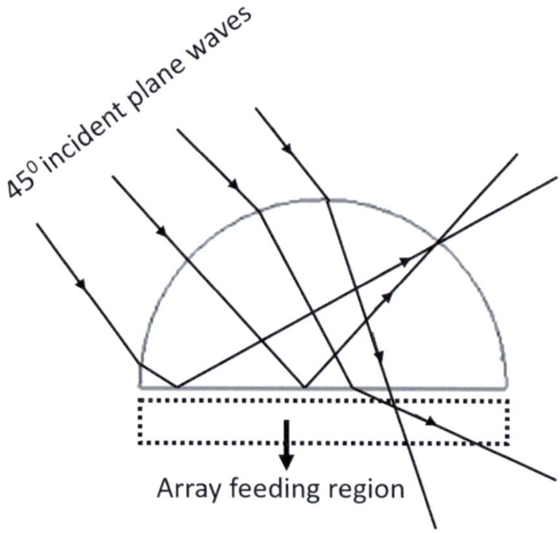


**Fig. 4.6** Array-feeding region of traditional hemispherical lens

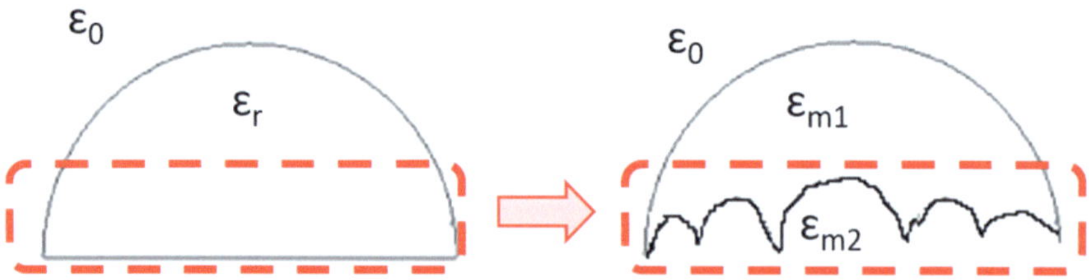


**Fig. 4.7** Schematic of the modification for the proposed lens

### 4.4.2 Dielectric Lens

Dielectric lenses are suitable candidates for MMW and THz operations owing to their space-fed nature and ease of obtaining high-gain performance with small feed and less conductor losses. Lens can transform the spherical wave front into the planar one in the far-field region. However, two shortcomings are inherent in the lens, i.e., reflection at the air–lens interfaces and lack of beam-scanning ability. Generally, a layer of antireflection (AR) coating is used to suppress the reflection at the interface. However, it makes the fabrication more complicated, and appropriate materials may be difficult to obtain [31]. In [1], discrete dielectric lenses are studied in which periodic antireflection (AR) structures are added to reduce impedance mismatch at the air–dielectric interfaces, leading to higher directivity. A multifrequency phase matching method is proposed to realize a beam-scanning lens in both MMW and THz regions. To demonstrate the effect of the AR structure, two dielectric lenses are designed at the operating frequency of 60 GHz. Figure 4.8 shows the mechanism of the lens antenna. The lenses are designed, which have square apertures with dielectric elements to provide the necessary phase compensations. Meanwhile, point sources are utilized to feed the lens in the simulations, corresponding to a focal to diameter ratio (*F*/*D*). The desired phase compensation at each unit cell is obtained according to Fermat's principle of equality of electrical path lengths of incoming electromagnetic waves.

$$\varphi_n = \frac{2\pi f}{c}\left\{\sqrt{F^2 + (x_n - x_0)^2} - F - x_n \sin\theta\right\} + \varphi_0, \tag{4.3}$$

where $F$ is the focal length, $x_0$ is the feed position along $x$-axis, $x_n$ is the $n$th element position, $\theta$ is the direction of the radiation beam, $f$ is the operating frequency, and $c$ is the velocity of light.

In general, the desired phase compensations are satisfied by choosing appropriate elements only at the center frequency. However, the desired phase compensations are not taken into consideration at other frequencies. To achieve the beam scanning, the desired phase compensation is regarded as a function of beam direction, and operating frequency. The electromagnetic waves are radiated in different directions at different frequencies, and a multifrequency matching method is employed to design the lens aperture. The designed phase compensation of the nth element can be rewritten as

$$\varphi_n(f, \theta) = \frac{2\pi f}{c} \times \left\{\sqrt{F^2 + (x_n - x_0)^2} - F - x_n \sin\theta\right\} + \varphi_0(f). \tag{4.4}$$

To obtain the capability of scanning, the desired phase compensation of each element in the lens must be met at the corresponding frequency. Different frequencies correspond to different radiation direction $\theta$.

The proposed scan solution in [32] consists of an axial symmetric dielectric lens that pivots with a simple two axis movement in front of a moderate directivity stationary feed. The main function of the lens is to tilt the beam with respect to

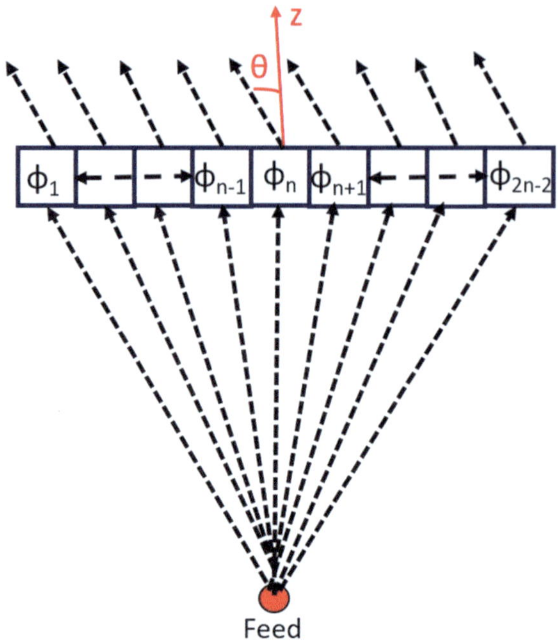


**Fig. 4.8** Phase correction mechanism for center-fed lens

the fixed feed axis, within a wide conic angular sector and to collimate the beam and push the gain up to more than 20 dBi. These two functions require appropriate shaping of the top and bottom lens surfaces. The feed is chosen to be stationary, to dispense expensive and fault prone millimeter-wave rotary joints which is opposed to the classical concept of scanning lenses where the lens is stationary and a cluster of feeds is distributed over the lens focal arc. In order to be able to tilt the lens without tilting the feed, it must be detached from the lens. For this a spherical air dome is opened at the base of the lens, centered with the feed phase center as shown in Fig. 4.9. The inner interface of the lens does not produce any refraction. In this way, the lens can tilt and steer the beam in elevation, while the feed remains stationary at the focal point. Figure 4.9 illustrates that the output rays always emerge parallel to the lens axis, and so the beam tilt is the same as the lens tilt. For any lens tilt, the lens can rotate about the feed axis to produce an azimuth beam scan. The subtended angle of the feed radiation must be restricted in order to avoid peripheral rays exceeding the lens collimating region. Based upon Geometrical Optics, the collimating region corresponds to

$$\theta < \theta_{\max} = a\cos\left(\frac{1}{\sqrt{\varepsilon_{\mathrm{r}}}}\right), \tag{4.5}$$

where $\varepsilon_{\mathrm{r}}$ is the lens material permittivity and $\theta$ is the angle between a ray path departing from the feed and the lens axis; $\theta_{\max}$ value corresponds to the first ray that hits the external cylinder wall of the lens. For larger angles, rays are refracted away

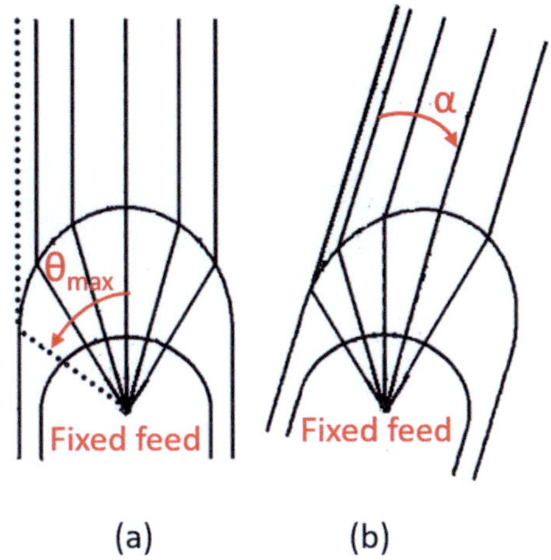


**Fig. 4.9** Ray tracing in polyethylene elliptical lens with spherical air cavity **a** zero tilt case, **b** 15° tilt case

from the main beam direction, contributing to sidelobes and to lowering gain. For a given feed, the impact of this effect is higher as the lens tilt increases, originating gain scan loss. The presented compact lens antenna works at 60 GHz with ± 45° scan capability.

### *4.4.3 Luneburg Lens*

Homogeneous lenses may allow the obtention of a wide scan antenna but it brings about a problem of retro-diffusion from the lenses to the sources. Luneburg lens avoids this problem because it presents a dielectric gradient-index lens whose relative permittivity varies radially according to the law given by Zimmerman [33]. Moreover, the inhomogeneous lens has infinite focus points making it difficult to achieve a very wide scan antenna. Practically, inhomogeneous lens can be manufactured either from materials with varying effective dielectric permittivity or by assembling a finite number of concentric homogeneous dielectric shells. This type of inhomogeneous spherical lens antenna has demonstrated off-axis performances by moving one source around the lens [34] and beam-shaping capability by associating several sources below the lens [35]. The permittivity distribution of the planar Luneburg lens is controlled by varying the hole density, dielectric thickness, or modifying the shape of the slab.

The antenna in [36] is based on a plate inhomogeneous Luneburg lens fed by several ridged waveguide sources to obtain a narrow beam (10°) for the azimuth plane and a wide beam (70°) for the elevation plane. First, a new mixed solution is chosen to manufacture the lens by using holes in a Teflon sheet ($\varepsilon_r = 2.04$) and Divinycell foam dielectric crowns ($\varepsilon_r = 1.45$ and $\varepsilon_r = 1.25$). Second, a ridged waveguide fed by a microstrip line illuminates the lens to realize the design of an electronically

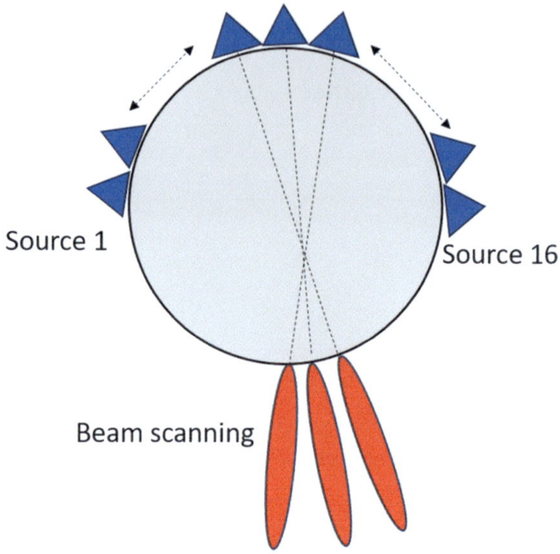


**Fig. 4.10** Beam-scanning angle depends on the feeding source position

beam-scanning antenna with a narrow beam (10°) over a wide scan angle. Third, the principle of the beam-scanning and beam-shaping antenna is matched by using one MMIC power amplifier behind each ridged source waveguide. A microstrip feeding line network is added to link up the different sources. Since the source antenna's width is small, numerous sources can be positioned closely together behind the Luneburg lens as shown in Fig. 4.10. The beam-scanning angle will depend on the excited source. Theoretically, if sixteen sources are placed behind the lens with a physical rotation angle of 10° between sources, the beam can be scanned over a wide area (± 75°) with a low decrease in gain. MMIC power amplifiers (used as switches) were used in the transmitter active prototype and a feeding line network was designed to link them. This active lens antenna allows to shape the beam (variable beamwidth) if several sources feed the lens simultaneously.

Luneburg lens antenna is a spherically symmetric lens antenna constituted by dielectric medium with gradient indices [37]. This lens can convert the spherical waves generated from the feed into quasi-plane waves, which leads to high-gain beam radiation. Luneburg lens antenna is especially attractive for beam-scanning or multibeam applications [38]. However, the curved surfaces of Luneburg lens are hard to fabricate [39]. Moreover, traditional Luneburg lens antennas have high profile and large weight, which obstruct their applications in moving platforms such as aircraft and missile-borne systems. By using transformation optics, the profile of Luneburg lens antenna can be greatly reduced, without sacrificing its beam-scanning ability. Metamaterial is a kind of effective medium constituted by artificial microstructures in subwavelength size. The effective permittivity and permeability of metamaterials can be controlled by adjusting the shapes or sizes of the microstructures. In this way, strong anisotropy permittivity or chiral materials can be designed and even the materials with negative permittivity and permeability can be realized [40]. The flat lens antenna designed in [41] is based on the theory proposed by Mateo-Segura et al. The flat lens with specific distribution of permittivity converts spherical signal generated from feed into plane wave, thus demonstrating high directivity. Ideal focusing performance can be achieved by increasing the number of the permittivity regions. Considering the effective medium theory of metamaterial and the fabrication process, nine permittivity regions are designed and optimized to maximize the efficiency of the lens antenna. The beam steering is achieved by mechanically changing the location of the feed horn. Non-resonant structure is used to construct the GIMs, so the metamaterial can work over broadband with negligible intrinsic loss. The antenna gain drops to about 2.5 dB when the beam scans to 40°. The prototype confirms the feasibility of using metamaterial technology for achieving high-efficiency and broadband beam-steering antenna.

### 4.4.4 Inhomogeneous Flat Lens

Researchers in [42] introduced a planar profile antenna configuration based on the switched-beam array antenna concept with an inhomogeneous gradient-index dielectric flat lens to steer and enhance the radiation in a specific direction, achieving a 2D beam scanning of high-gain radiation beams while maintaining a completely flat antenna profile that is suitable for medium-sized mobile devices.

(1) Circularly Distributed Parameters Flat Lens Concept

The inhomogeneous gradient-index circular flat lens in [43] consists of a set of six concentric rings of different permittivity ($\varepsilon_r$) materials, to produce the desired phase delays required to obtain a plane wave behind the lens, when the lens is illuminated from its central focus position. In the same way, when the feeding position is moved along *y*- or *x*-directions, the different permittivity values of the lens produce a linear phase slope that steers the beam, accordingly. Given the lens circular effective parameters distribution, the described behavior is independent of the axis in which the feeding antenna is moved along; the beam will be steered in the same manner.

(2) Cylindrically Distributed Parameters Flat Lens Concept

The cylindrically distributed parameters flat lens concept described in [44] consists of eleven rectangular sections of six different permittivity materials, to produce the desired phase delays required to obtain a plane wave. The lens is illuminated from its central focus position, in the same way as it has been described for the circular lens. Likewise, when the feeding position is moved along *y*-direction, the different permittivity values produce a linear phase slope, which steers the beam only along the gradient-index axis (i.e., along *y*-direction), accordingly. As a result of the lens cylindrical parameters distribution, if the position of the feeding element is moved along the constant-index axis (i.e., *x*-direction), the beam is maintained invariant, because the phase is not being corrected in this specific dimension, to finally obtain the desired fan-beam pattern.

### 4.4.5 Integrated Lens Antenna

A general ILA structure consists of a dielectric lens forming a narrow antenna beam by focusing the radiation and a feed array placed on the lens back-plane surface. Beam steering is provided by activation of one of the antenna elements in the array by using monolithic microwave integrated circuit (MMIC) switches that select the active antenna beam steered in the direction defined by the element displacement relative to the lens axis. Researchers in [45] proposed an ILA prototype for various applications in the E-band that allows switching the beam between 16 positions arranged in 2D angle space. Lenses of the prototypes have extended hemispherical shapes and are made from fused quartz ($\varepsilon_r = n^2 = 3.8$), which has low loss in the millimeter-wave band ($\tan\delta = 0.004$). The array has 16 planar aperture-coupled microstrip antenna elements and to provide electronic beam steering, the switching network includes five SP4T switches mounted on the same PCB together with the feed array. The ILA with the radius of 12.5 mm provides the coverage within the solid angle $\theta < 18°$ in any direction from the lens axis and an ILA with the radius of 7.5 mm provides higher coverage (within $\theta < 27°$), but with the lower directivity of 16 dBi.

Integrated lens antennas typically consist of an extended elliptical or extended hemispherical dielectric lens attached on top of the primary feed array. While offering a number of important practical advantages, ILAs often suffer from bulky dimensions, enhanced dielectric loss, and limited scanning capabilities. The significant scan loss in directivity observed for offset feeds is considered as a fundamental limitation of substrate lens antennas that restricts the use of ILAs in applications requiring a wide field of view. In the case of an array-fed multibeam ILAs, the scan loss originates from three main reasons, namely (1) degradation of the focusing capabilities of hemielliptic lenses in the case of incline illumination; (2) displacement of primary feeds from the focal arc; and (3) reduction of the projected area of the lens surface when illuminated by offset feeds [46]. Several solutions have been proposed in the literature to reduce the scan loss, including optimization of the lens size and shape as well as optimization of the feed position, orientation, and amplitude weighting [47]. These approaches often lead to complex lens shapes or sophisticated beamforming networks. Another available solution consists in using mechanical scanning [32]. This approach can provide a very low scan loss, but requires sophisticated mechanical positioning system that often suffers from poor reliability and may be too slow in some cases (e.g., automotive radars, real-time imaging). The scan loss problem can be significantly reduced using multishell spherical lenses [48]; however, certain fabrication and integration difficulties make such lenses less attractive.

## 4.5 Beam Scanning with Metasurface

Recently metamaterials such as negative refractive index transmission line (NRI-TL), composite right/left-handed transmission line (CRLH TL), and other structures have been used to design antennas with unique properties. Among them, antennas with capability of beam steering are also reported. The refraction of metamaterial could be positive, negative, or zero. This property provides a unique method of controlling the direction of emission. There are many types of metasurface elements including SRRs, square or circular patches, cross-shaped structures, etc. The parasitic structures mainly play the role of director or reflector. Normally, this kind of antenna adjusts the equivalent length of parasitic structure by welding PIN diodes [49], so as to steer its beam to the director and away from the reflector. Although this method is simple and easy to implement, the number of the beam's direction states is limited. The electromagnetic metasurface can achieve beam scanning easily by phase modulating. Some technologies such as active frequency-selective reflectors, phase gradient metasurfaces, and polarization grids are all used for beam steering. These antennas are capable of continuous beam scanning.

It has been shown in [50] that not only high antenna gain can be achieved, but when the position of the feeding source (slot) is shifted, the radiated beam from the metasurface can be steered in the opposite direction accordingly. The slot antenna is shown in Fig. 4.11. The radiated beam is steered from the center of 0° for $P = 0$, to the maximum steering angle of approximately 35° for the case of $P = 40$, and the beam is steered in the opposite direction to the movement of slot position. If the slot is moved in both directions, the beam-steering angle range of − 35° to 35° can be realized. When the position of slot is shifted, the antenna gains decrease accordingly which is a typical effect for phased-array antennas at higher steering angles. At higher frequencies, backlobes become larger, and a beam splitting effect in the forward direction starts to appear (the increase of sidelobe). The radiation is first coupled from the slot radiator to the nearest square ring, and then the neighboring rings are excited in sequence with different time delays. Effectively, different metasurface elements (square rings) are excited with different phase delays. For all cases, the magnitude variations are insignificant, while phase progresses along square ring elements. For the case of the slot located in the center of the antenna ($P = 0$), due to the symmetrical layout of square ring elements on both sides, the average phase delay, $\Delta\Phi$ from element #5 to element #1 is zero, and the resulting radiated beam is directed at 0°. When the slot position is shifted, the value of $\Delta\Phi$ gradually increases, thus the radiated beam is steered to an increasing angle accordingly. The following equation is used to calculate the theoretical phase delay by using parameters of the proposed metasurface:

$$\Delta\Phi = \frac{2\pi d \sin\theta}{\lambda_g}, \tag{4.6}$$

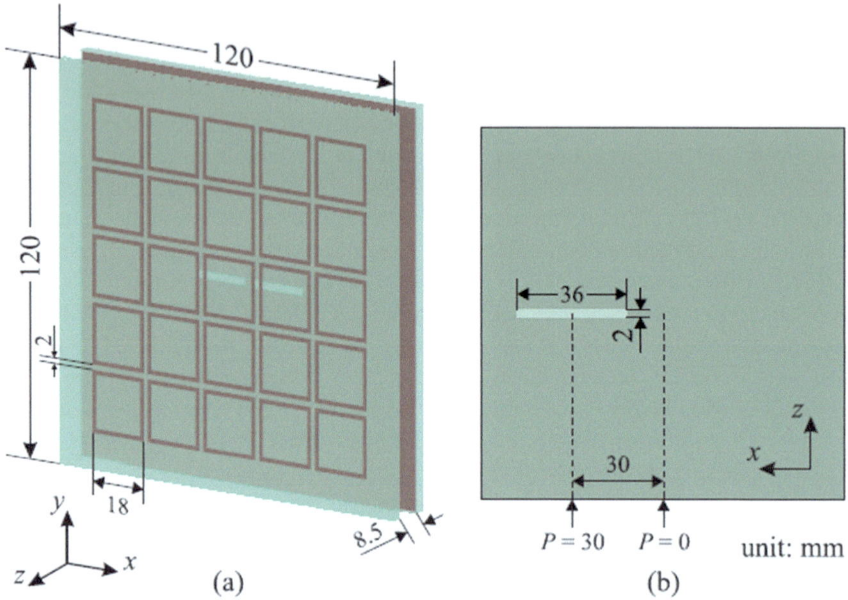


**Fig. 4.11** **a** Single-slot metasurface antenna consisting of a metasurface superstrate formed by closed square rings and **b** a conventional slot radiator as the feeding source [50]

where $d$ is the periodicity of metasurface elements, $\theta$ is the beam-steering angle, $\lambda_g$ is the guided wavelength on the metasurface which can be approximated as $\frac{\lambda_0}{\sqrt{\frac{\varepsilon_r+1}{2}}}$ where $\lambda_0$ is the free-space wavelength.

A novel compact single-patch antenna is designed in [51] to realize beam steering with complementary split-ring resonator (CSRR) structure as part of the ground plane. CSRR-loaded ground structure enables the property of wide-scale steering range and can greatly reduce the cost of a phased-array system to meet the requirements of the wireless communications. The refractive index of the loading part affects the beam direction. The beam of the double CSRR-loaded antenna can scan from − 51° to 48° via changing the parameters of CSRR structure. As the single CSRR-loaded antenna owns good beam steering on one direction, the double CSRR-loaded antenna achieves both positive and negative beam steering. The phase variation caused by the loading structure is given by

$$\theta = kd = \frac{2\pi}{\lambda_0} nD, \tag{4.7}$$

where $k$ is the equivalent wavenumber, $\lambda_0$ the wavelength of free space, $D$ the unit length in the propagation direction, and $n$ the effective refractive index. Different values of slot stub cause different refractive index, thus different phase shifts. Therefore, the CSRR loading part can be used as a phase shifter to achieve beam steering. Larger $n$ results in larger phase shift, thus larger beam-steering angle.

A novel anisotropic metasurface (AMS)-based integrated antenna system along with a polarization rotator (PR) to facilitate dual-polarized fan-beam patterns with independent beam-scanning ability is presented in [52]. The source antenna, a microstrip patch capacitively coupled with two metallic strips, is loaded with the PR by placing it at a height of $0.35\lambda_0$ from the radiating aperture. The polarization of the linearly polarized EM waves from the source antenna either remains preserved or gets rotated by 90° while transmitting through this PR. The proposed AMS lens is placed above the PR by integrating two cylindrical MS (CMS) lenses, CMSY and CMSX. This integration provides a unique phase profile to the AMS lens, thus introducing appropriate phase correction to the incident orthogonally polarized spherical EM waves along their respective polarization direction. The presented configuration of AMS lens results in the formation of the independent dual-polarized fan-beam radiation beams as shown in Fig. 4.12. The steering of the generated fan-beams along the direction of their respective polarization is facilitated using the proposed system by translating the AMS lens parallel to the source antenna. It is observed that the system with this kind of fan-beam pattern can easily detect any object located in the direction of its wide beamwidth. However, sometimes detection may be quite challenging if the object is placed along the orthogonal direction, where beamwidth of the corresponding fan-beam is somewhat less. Therefore, in order to effectively scan 2D space, this fan-beam is now being steered along the direction of its polarization, where the beamwidth is comparatively less. This 1D beam scanning in the elevation plane (the *YZ* plane) is obtained here by translation of the CMSY lens along the direction of polarization (which is *Y*-direction in this case) of fan-beam. This translation of the lens actually introduces a linear phase slope along the direction of its movement, resulting into the beam tilts in the *YZ* plane by an angle of $\theta_Y$ from the normal direction.

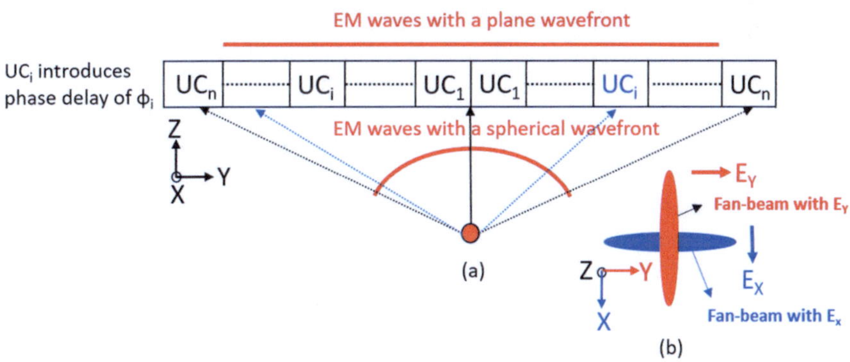


**Fig. 4.12** **a** Phase correction of $E_Y$ of spherical EM waves along the $Y$-direction by using CMSY lens, **b** two orthogonally oriented fan-beam patterns with the orthogonal polarizations

From simple array theory, the direction of the main beam ($\theta_Y$) can be estimated as

$$\theta_Y = \sin^{-1}\{\Delta\gamma_Y/(k \times L_{\mathrm{UA}})\}. \quad (4.8)$$

Here, $\Delta\gamma_Y$ a relative phase difference between the feed signals of the consecutive UCs of the lens is actually introduced by the movement of the lens. From the above relation, it is found that the continuous translation of the lens provides gradual variation in $\Delta\gamma_Y$, resulting into the fan-beam scanning (varying $\theta_Y$) along the direction of the lens movement.

## 4.6 Beam Scanning with FSS

A frequency-selective surface (FSS) is a periodic array of artificial structures that is designed to transmit, reflect, or absorb electromagnetic waves with frequency-selective features [53]. An FSS is mainly composed of two elements: resonant structures and dielectric mediums. The resonant structures are constituted by metal placed over a substrate, arranged in defined lattices and periodic patterns; meanwhile, the dielectric addresses mechanical stability, angular stability, phase control, absorption, and miniaturization of the structure. Depending on the periodic element design, an incoming plane wave will be either transmitted or reflected, completely or partially. Therefore, an FSS can pass or block the EM waves with certain frequencies in free space, so they are best identified as spatial filters. FSSs are widely used in many areas, such as spatial filters, absorbers, radar cross section reduction, and electromagnetic compatibility [54].

The frequency characterization of this type of surface depends mainly on the mutual interaction and the energy coupling of various decaying Bragg modes between adjacent elements [55]. When a plane wave strikes a surface with periodic structures, then every single element present in that planar structure should contain the same amount of field and current distribution. However, the phase of the incident wave and the phase shift in between unit cells, the phase of the reflected/transmitted waves might differ. The wave-propagation mechanism is such that, as the incident wave hits the filter, the electrons in the conductor start oscillating. At a particular frequency, a large section of this incident energy gets absorbed by the present electrons and therefore they re-radiate this energy in backward direction and cancel the initial field, leading to least transmission through the FSS, as shown in Fig. 4.13. If only a small portion of incident wave is absorbed, then no cancelation of incident wave takes place, and hence, strong transmission occurs, as shown in Fig. 4.13.

FSSs with different properties, such as high selectivity, multipolarization, tunability and angular stability, are the trend of future development. The single layered FSS structures have been proved inefficient due to unstable performance with the variation of EM wave incident angle. To overcome the limitations, multi-layered FSSs have been introduced, which offer additional flexibility of varying

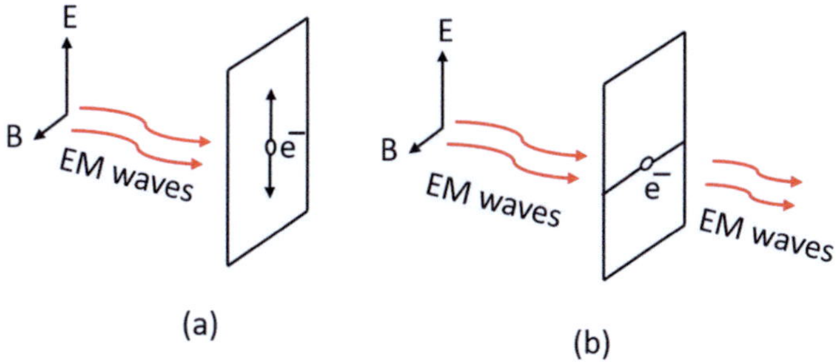


**Fig. 4.13** **a** Electrons undergo oscillation, therefore low transmission, **b** electrons undergo no oscillation therefore high transmission

parameters for desired performance [56]. Here we will have a look into how the FSS structures are being used for beam-scanning antenna applications. To make the passive FSS structures deflect the beam to different directions the usual practice is to use an active FSS or a tunable FSS that contains some switches that can steer the beam in various directions.

At present, tunable FSSs can be implemented in various ways, which can be roughly divided into the following categories:

(1) Tunable FSS Loaded with Active Components [57]

This is a well-known implementation of a tunable FSS. The active components could be varactor diodes, PIN diodes, Schottky diodes, etc. By adjusting the applied voltage of the active device, the resonance frequency of the FSS can be changed. This method can ensure a compact structure and a wide tunable range. However, the disadvantage is that active components need a feed network and an external power supply. Adding a bias network and isolating elements will also affect the original frequency response of the FSS, which may result in poor selectivity or a large insertion loss. In addition, the complex structure and too many active components also increase the manufacturing cost and restrict the practicability.

(2) Tunable FSS Based on an Active Dielectric

This method is realized by using materials with adjustable electrical properties, such as the liquid crystal [58]. By applying a changing voltage to the liquid crystal, the permittivity can be adjusted and tunability is achieved.

(3) Mechanically Tunable FSS

This kind of tunable FSS is generally novel and interesting, and tunability is achieved by changing the spring height via pressure to adjust the resonance frequency [59] or using pressure-driven fluidic control of mercury volumes [60]. Other examples included the use of a liquid metal [61], a rotatable square [62], and an origami structure [63]. Some designs also utilize MEMS metallic bridges [64] placed over slots to achieve tunability, whereas [65] used a MEMS rotating-dipole array to alter the resonance frequency.

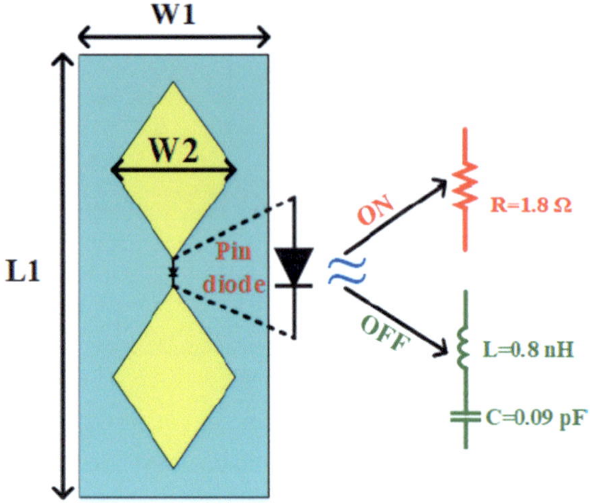


**Fig. 4.14** AFSS element with an equivalent circuit model of the diode's ON/OFF states [66]

A novel switched-beamforming antenna designed using active frequency-selective surfaces (AFSS) is presented in [66]. The antenna consists of two parts: a source with an omnidirectional radiation pattern and a novel active triangular frequency-selective surface (ATFSS) surrounding the source. Each unit cell in the ATFSS structure contains two diamond shaped patches connected with each other by a high-frequency pin-diode as shown in Fig. 4.14. The high-frequency pin-diode GMP4202-GM1 is used.

The basic principle for an AFSS is shown in Fig. 4.15, where the surface can control the wave propagation by switching the diode's ON/OFF states. The AFSS exhibits maximum reflection in the diode's ON state while maximum transmission in OFF state.

The ATFSS are arranged in an equilateral triangle to switch the beam in multiple directions using different combinations of the ATFSS diodes. Figure 4.16 shows the antenna and the ATFSS around the antenna.

The switched-beamforming antenna has the ability to sweep the radiations in the azimuth plane using three pin diode combinations on the ATFSS. By switching the pin diodes between ON/OFF state it can sweep the beam 120° for each condition in the azimuth direction. When the diodes in side 1 are OFF (combination A), the pattern is directed toward 240°, when the diodes in side 2 are OFF (combination B), the pattern is directed toward 120°, and when the diodes in side 3 are OFF (combination C), the pattern is oriented toward 0° Other beam-steering directions can be achieved by choosing different sectors ON/OFF state combinations. With this ATFSS, the proposed antenna provides beam-steering flexibility in the azimuth plane.

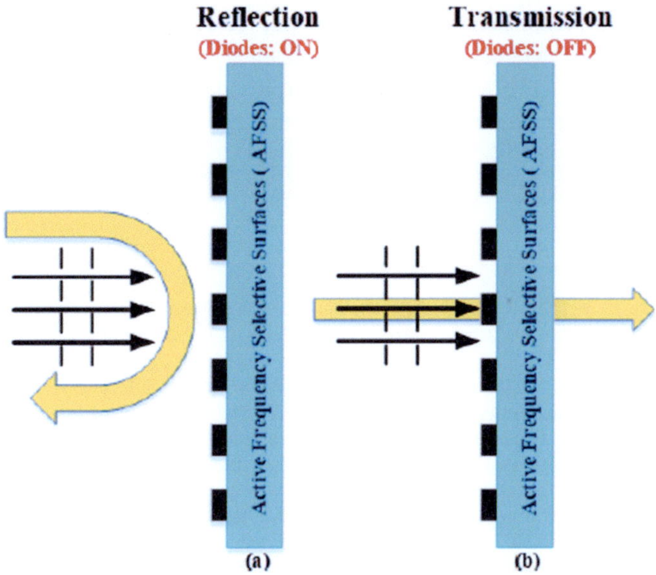


**Fig. 4.15** Basic principle of the AFSS in two diode states: **a** reflection **b** transmission [66]

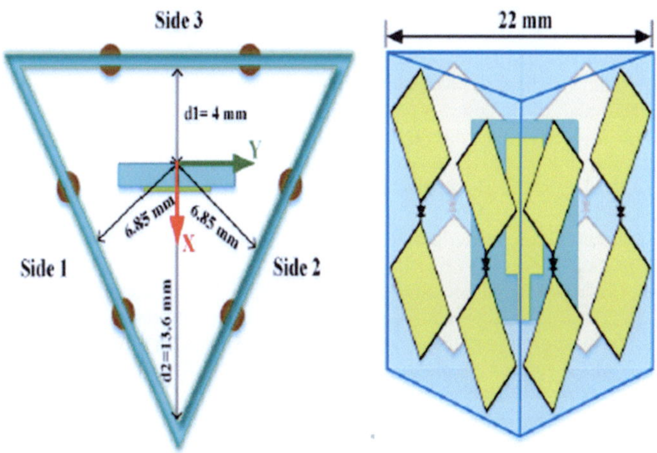


**Fig. 4.16** Simulated model of the proposed ATFSS antenna in top and side 3D view [66]

## 4.7 Conclusion

Beam-scanning methods other than electronic beam scanning are included in this chapter. Different methods of mechanical beam scanning of antennas are discussed first. Mechanical beam scanning can be done either by steering the antenna, steering the focusing aperture or by steering part of the antenna. Even though mechanical scanning is slower compared to electronic scanning, it has lower losses due to absence of active devices or switches. Other than mechanical scanning, beam scanning using lens, metasurface and FSS are also presented to provide an overall idea of the different types of scanning schemes available in the literature.

## References


1. Yi, H., Qu, S.-W., Ng, K.-B., Chan, C.H., Bai, X.: 3-D printed millimeter-wave and terahertz lenses with fixed and frequency scanned beam. IEEE Trans. Antennas Propag. **64**(2), 442–449 (2016)
2. Wu, G.-B., Qu, S.-W., Yang, S.: Wide-angle beam-scanning reflectarray with mechanical steering. IEEE Trans. Antennas Propag. **66**(1), 172–181 (2018)
3. Mei, P., Zhang, S., Pedersen, G.F.: A low-profile and beam-steerable transmitarray antenna: design, fabrication, and measurement [antenna applications corner]. IEEE Antennas Propag. Mag. **63**(5), 88–101 (2021)
4. Yang, J., Shen, Y., Wang, L., Meng, H., Dou, W., Hu, S.: 2-D scannable 40-GHz folded reflectarray fed by SIW slot antenna in single-layered PCB. IEEE Trans. Microw. Theory Techn. **66**(6), 3129–3135 (2018)
5. Yang, X., et al.: A broadband high-efficiency reconfigurable reflectarray antenna using mechanically rotational elements. IEEE Trans. Antennas Propag. **65**(8), 3959–3966 (2017)
6. Nayeri, P., Yang, F., Elsherbeni, A.Z.: Bifocal design and aperture phase optimizations of reflectarray antennas for wide-angle beam scanning performance. IEEE Trans. Antennas Propag. **61**(9), 4588–4597 (2013)
7. Sánchez-Escuderos, D., Herranz-Herruzo, J.I., Ferrando-Rocher, M., Valero-Nogueira, A.: True-time-delay mechanical phase shifter in gap waveguide technology for slotted waveguide arrays in Ka-band. IEEE Trans. Antennas Propag. **69**(5), 2727–2740 (2021)
8. Yang, J., Qi, S.-S., Wu, W., Fang, D.-G.: A novel high-gain sum and difference conical beam-scanning reflector antenna. IEEE Access **8**, 103291–103300 (2020)
9. Kopacz, J.R., Herschitz, R., Roney, J.: Small satellites an overview and assessment. Acta Astronaut. **170**, 93–105 (2020)
10. Gao, S., Rahmat-Samii, Y., Hodges, R.E., Yang, X.-X.: Advanced antennas for small satellites. Proc. IEEE **106**(3), 391–403 (2018)
11. Huang, J., Encinar, J.A.: Reflectarray Antennas, vol. 30. Wiley, Hoboken, NJ, USA (2007)
12. Nayeri, P., Yang, F., Elsherbeni, A.Z.: Beam-scanning reflectarray antennas: A technical overview and state of the art. IEEE Antennas Propag. Mag. **57**(4), 32–47 (2015)
13. Baracco, J.-M., Ratajczak, P., Brachat, P., Fargeas, J.-M., Toso, G.: Ka-band reconfigurable reflectarrays using varactor technology for space applications: a proposed design. IEEE Antennas Propag. Mag. **64**(1), 27–38 (2022)
14. Mirhamed, M., et al.: Mechanically reconfigurable, beam-scanning reflectarray and transmitarray antennas: a review. Appl. Sci. **11**(15), 6890 (2021)
15. Rubio, A.J., Kaddour, A.-S., Georgakopoulos, S.V.: A mechanically rollable reflectarray with beam-scanning capabilities. IEEE Open J. Antennas Propag. **3**, 1180–1190 (2022)
16. Li, A.: Double-Prism Multi-mode Scanning: Principles and Technology, vol. 216. Springer, Berlin, Germany (2018)
17. Yang, Y.: Analytic solution of free space optical beam steering using risley prisms. J. Lightw. Technol. **26**(21), 3576–3583 (2008)
18. Afzal, M.U., Esselle, K.P., Koli, M.N.Y.: A beam-steering solution with highly transmitting hybrid metasurfaces and circularly polarized high-gain radial-line slot array antennas. IEEE Trans. Antennas Propag. **70**(1), 365–377 (2022)
19. Gagnon, N., Petosa, A.: Using rotatable planar phase shifting surfaces to steer a high-gain beam. IEEE Trans. Antennas Propag. **61**(6), 3086–3092 (2013)

20. Wang, J., Rahmat-Samii, Y.: Phase method: a more precise beam steering model for phase-delay metasurface based Risley antenna. In: Proceedings URSI International Symposium on Electromagnetic Theory, pp. 1–4 (2019)
21. Zhao, X., et al.: All-metal beam steering lens antenna for high power microwave applications. IEEE Trans. Antennas Propag. **65**(12), 7340–7344 (2017)
22. Afzal, M.U., Esselle, K.P.: Steering the beam of medium-to high gain antennas using near-field phase transformation. IEEE Trans. Antennas Propag. **65**(4), 1680–1690 (2017)
23. Zhang, Z., Zhong, Y.C., Luyen, H., Booske, J.H., Behdad, N.: A low-profile, risley-prism-based, beam-steerable antenna employing a single flat prism. IEEE Trans. Antennas Propag. **70**(8), 6646–6658 (2022)
24. Wang, J., Rahmat-Samii, Y.: A simplified configuration of beam-steerable risley prism antennas: principles and validation. IEEE Antennas Wirel. Propag. Lett. **21**(11), 2288–2292 (2022)
25. Teng, M., Yu, S., Kou, N., Ding, Z., Zhang, Z.: Mechanical beam steering array antenna with tunable height. IEEE Antennas Wirel. Propag. Lett. **21**(11), 2293–2297 (2022)
26. Lu, Y., Zhou, Y., Hei, M., Fan, D.: Theoretical and experimental determination of steering mechanism for Risley prism systems. Appl. Opt. **52**(7), 13891398 (2013)
27. Afzal, M.U., Matekovits, L., Esselle, K.P., Lalbakhsh, A.: Beam-scanning antenna based on near-electric field phase transformation and refraction of electromagnetic wave through dielectric structures. IEEE Access **8**, 199242–199253 (2020)
28. Pavacic, A.P., del Rio, D.L., Mosig, J.R., Eleftheriades, G.V.: Three-dimensional ray-tracing to model internal reflections in off axis lens antennas. IEEE Trans. Antennas Propag. **54**(2), 604–612 (2006)
29. Frid, H.: Closed-form relation between the scan angle and feed position for extended hemispherical lenses based on ray tracing. IEEE Antennas Wireless Propag. Lett. **15**, 1963–1966 (2016)
30. Liu, K., Yang, S., Qu, S.-W., Chen, C., Chen, Y.: Phased hemispherical lens antenna for 1-d wide-angle beam scanning. IEEE Trans. Antennas Propag. **67**(12), 7617–7621 (2019)
31. Gatesman, A.J., Waldman, J., Ji, M., Musante, C., Yngvesson, S.: An anti-reflection coating for silicon optics at terahertz frequencies. IEEE Microw. Guided Wave Lett. **10**(7), 264–266 (2000)
32. Costa, J.R., Lima, E.B., Fernandes, C.A.: Compact beam steerable lens antenna for 60-GHz wireless communications. IEEE Trans. Antennas Propag. **57**(10), 2926–2933 (2009)
33. Zimmerman.: Luneburg lens and method of constructing same. U.S. patent 5 677 796, Oct. 14, 1997
34. Fuchs, B., Lafond, O., Rondineau, S., Himdi, M., Le Coq, L.: Offaxis performances of half maxwell fish-eye lens antennas at 77 GHz. IEEE Trans. Antennas Propag. **55**(2), 479–482 (2007)
35. Fuchs, B., Palud, S., Lafond, O., Le Coq, L., Himdi, M., Rondineau, S.: Antenna system having a radiating diagram reconfigurable from sectorial and directional radiating diagrams, and corresponding transmitter and/or receiver device. Patent no. WO2009013248 (2007)
36. Lafond, O., Himdi, M., Merlet, H., Lebars, P.: An active reconfigurable antenna at 60 GHz based on plate inhomogeneous lens and feeders. IEEE Trans. Antennas Propag. **61**(4), 1672–1678 (2013)
37. Luneberg, R.K.: Mathematical Theory of Optics, pp. 189–212. Brown University Press, Providence, RI, USA (1944)
38. Thornton, J.: Wide-scanning multi-layer hemisphere lens antenna for Ka band. IEE Proc. Microw. Antennas Propag. **153**(6), 573–578 (2006)
39. Rondineau, S., Himdi, M., Sorieux, J.: A sliced spherical Luneburg lens. IEEE Antennas Wireless Propag. Lett. **2**, 163–166 (2003)
40. Jian, M., Chen, Z.N., Zhang, Y., Hong, W., Xuan, X.: Metamaterial based thin planar lens antenna for spatial beamforming and multibeam massive MIMO. IEEE Trans. Antennas Propag. **65**(2), 464–472 (2017)

41. Jia, D., He, Y., Ding, N., Zhou, J., Du, B., Zhang, W.: Beam-steering flat lens antenna based on multilayer gradient index metamaterials. IEEE Antennas Wirel. Propag. Lett. **17**(8), 1510–1514 (2018)
42. Imbert, M., Papio, A., De Flaviis, F., Jofre, L., Romeu, J.: Design and performance evaluation of a switched-beam antenna array for 60 GHz WPAN applications. In: Proceedings 9th European Conference on Antennas and Propagation (EuCAP), pp. 1–5, April 2015
43. Imbert, M., Papió, A., Flaviis, F.D., Jofre, L., Romeu, J.: Design and performance evaluation of a dielectric flat lens antenna for millimeter wave applications. IEEE Antennas Wirel. Propag. Lett. **14**, 342–345 (2015)
44. George, J., Smulders, P.F.M., Herben, M.H.A.J.: Application of fan-beam antennas for 60 GHz indoor wireless communication. Electron. Lett. **37**(2), 73–74 (2001)
45. Artemenko, A., Mozharovskiy, A., Maltsev, A., Maslennikov, R., Sevastyanov, A., Ssorin, V.: Experimental characterization of E-band two-dimensional electronically beam-steerable integrated lens antennas. IEEE Antennas Wirel. Propag. Lett. **12**, 1188–1191 (2013)
46. Boriskin, A.V., Sauleau, R.: Numerical investigation into the design of shaped dielectric lens antennas with improved angular characteristics. Prog. Electromagn. Res. B **30**, 279–292 (2011)
47. Costa, J.R., Silveirinha, M.G., Fernandes, C.A.: Evaluation of a double-shell integrated scanning lens antenna. IEEE Antennas Wirel. Propag. Lett. **7**, 781–784 (2008)
48. Mosallaei, H., Rahmat-Samii, Y.: Nonuniform Luneburg and two shell lens antennas: radiation characteristics and design optimization. IEEE Trans. Antennas Propag. **49**(1), 60–69 (2001)
49. Towfiq, M.A., Bahceci, I., Blanch, S., Romeu, J., Jofre, L., Cetiner, B.A.: A reconfigurable antenna with beam steering and beamwidth variability for wireless communications. IEEE Trans. Antennas Propag. **66**(10), 5052–5063 (2018)
50. Hongnara, T., Chaimool, S., Akkaraekthalin, P., Zhao, Y.: Design of compact beam-steering antennas using a metasurface formed by uniform square rings. IEEE Access **6**, 9420–9429 (2018)
51. Cao, W., Xiang, Y., Zhang, B., Liu, A., Yu, T., Guo, D.: A low-cost compact patch antenna with beam steering based on CSRR-loaded ground. IEEE Antennas Wirel. Propag. Lett. **10**, 1520–1523 (2011)
52. Katare, K.K., Chandravanshi, S., Sharma, A., Biswas, A., Jaleel Akhtar, M.: Anisotropic metasurface-based beam-scanning dual-polarized fan-beam integrated antenna system. IEEE Trans. Antennas Propag. **67**(12), 7204–7215 (2019)
53. Ferreira, D., Caldeirinha, R.F.S., Cuiñas, I., Fernandes, T.R.: Square loop and slot frequency selective surfaces study for equivalent circuit model optimization. IEEE Trans. Antennas Propag. **63**(9), 3947–3955 (2015)
54. Zhao, R., Gong, B., Xiao, F., He, C., Zhu, W.: Circuit model analysis of switchable perfect absorption/reflection in an active frequency selective surface. IEEE Access **7**, 55518–55523 (2019)
55. Moallem, M., Sarabandi, K.: Miniaturized-element frequency selective surfaces for millimeter-wave to terahertz applications. IEEE Trans. Terahertz Sci. Technol. **2**(3), 333–339 (2012)
56. Abdelrahman, A.H., Elsherbeni, A.Z., Yang, F.: Transmission phase limit of multilayer frequency-selective surfaces for transmitarray designs. IEEE Trans. Antennas Propag. **62**, 690–697 (2014)
57. Mamedes, D.F., Gomes Neto, A., Silva, J.C.E., Bornemann, J.: Design of reconfigurable frequency-selective surfaces including the PIN diode threshold region. IET Microw. Antennas Propag. **12**(9), 1483–1486 (2018)
58. Hu, W., Dickie, R., Cahill, R., Gamble, H., Ismail, Y., Fusco, V., Linton, D., Grant, N., Rea, S.: Liquid crystal tunable mm wave frequency selective surface. IEEE Microw. Wirel. Compon. Lett. **17**(9), 667–669 (2007)
59. Azemi, S.N., Ghorbani, K., Rowe, W.S.T.: A reconfigurable FSS using a spring resonator element. IEEE Antennas Wirel. Propag. Lett. **12**, 781–784 (2013)
60. Lei, B.J., Zamora, A., Chun, T.F., Ohta, A.T., Shiroma, W.A.: A wideband, pressure-driven, liquid-tunable frequency selective surface. IEEE Microw. Wirel. Compon. Lett. **21**(9), 465–467 (2011)

61. Li, M., Behdad, N.: Fluidically tunable frequency selective/phase shifting surfaces for high-power microwave applications. IEEE Trans. Antennas Propag. **60**(6), 2748–2759 (2012)
62. Ferreira, D., Cuinas, I., Caldeirinha, R.F.S., Fernandes, T.R.: 3-D mechanically tunable square slot FSS. IEEE Trans. Antennas Propag. **65**(1), 242–250 (2017)
63. Fuchi, K., Tang, J., Crowgey, B., Diaz, A.R., Rothwell, E.J., Ouedraogo, R.O.: Origami tunable frequency selective surfaces. IEEE Antennas Wireless Propag. Lett. **11**, 473–475 (2012)
64. Safari, M., Shafai, C., Shafai, L.: X-band tunable frequency selective surface using MEMS capacitive loads. IEEE Trans. Antennas Propag. **63**(3), 1014–1021 (2015)
65. Zendejas, J.M., Gianvittorio, J.P., Rahmat-Samii, Y., Judy, J.W.: Magnetic MEMS reconfigurable frequency-selective surfaces. J. Microelectromech. Syst. **15**(3), 613–623 (2006)
66. Elzwawi, G.H., Elzwawi, H.H., Tahseen, M.M., Denidni, T.A.: Frequency selective surface-based switched-beamforming antenna. IEEE Access **6**, 48042–48050 (2018)

# Chapter 5
# Wide-Angle Beam-Scanning Antennas

## 5.1 Introduction

One of the key antenna requirements in many modern wireless systems for communications and sensing is wide-angle beam scanning (WABS). The traditional method to achieve wide-angle beam scanning is to employ mechanically controlled reflectors or lenses, which are bulky, heavy, and suffer from slow beam-scanning speed. An alternative technique is to employ electronically beam-scanning antennas using passive or active phased arrays [1]. Main disadvantages of phased arrays are high complexity, high-power consumption, and high cost due to a large number of radio frequency (RF) or microwave phase shifters and T/R modules required. The problems worsen for phased arrays at millimeter-wave, sub-THz, and THz frequencies, due to significant losses in phase shifters and feed networks at higher frequencies combined with lower efficiency of power amplifiers [2]. The digital beamforming approach is even more expensive and energy hungry due to the employment of large number of RF modules and digital devices [3].

For civilian applications, it is critical to develop low-cost wide-angle beam-scanning antenna technologies. The increasing demand in data traffic for wireless networks is expected to grow even faster with the upcoming beyond fifth generation (B5G) and sixth-generation (6G) communications. WABS antennas can deliver critical frequency reuse and significantly improved system capacity, which has been considered as a key enabler technology for B5G and 6G communications [4]. In satellite communication, electronic beam steering phased-array antennas (PAAs) are highly desired since they generally have lower profiles than lens or reflector antennas and do not occupy large space as the mechanically scanned arrays. Highly integrated PAAs with WABS ability to achieve global coverage are becoming very promising in SATCOM, especially for medium earth orbit (MEO) and low earth orbit (LEO) SATCOM [5]. Another important application of WABS PAAs is the sensing-based automotive radar and anti-collision radar. which is a key technology

 
S. K. Koul et al., *Antenna Systems for Modern Wireless Devices*, Signals and Communication Technology, https://doi.org/10.1007/978-981-97-3369-9_5

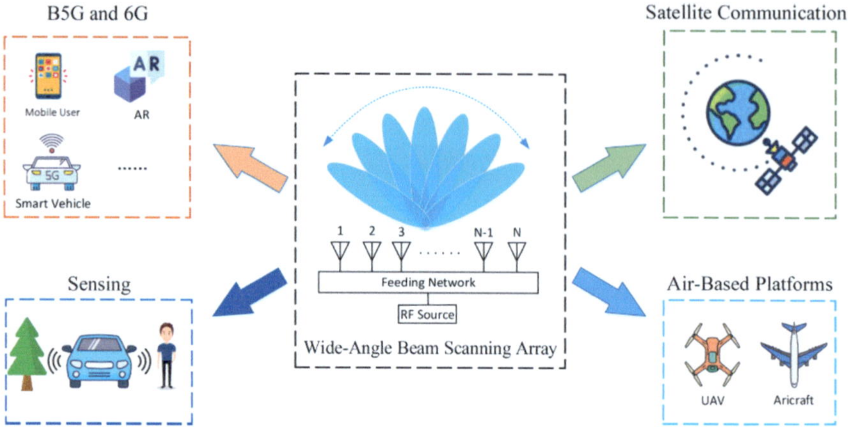


**Fig. 5.1** Several application scenarios of wide-angle beam-scanning (WABS) phased-array antennas (PAAs) [10]

to achieve autonomous driving [6]. Furthermore, WABS PAAs can also find applications in airborne platforms like unmanned aerial vehicles (UAVs) [7] and aircrafts [8], synthetic aperture radar [9], etc. This wide range of applications, also illustrated in Fig. 5.1, has largely driven the evolution and development of WABS PAA technologies.

To meet the market needs, a number of techniques have been developed in recent years to deliver low-cost wide-angle beam-scanning antenna solutions. The most widely known one beyond the antenna community is hybrid antenna arrays in which passive antenna beamforming, active analog beamforming, and digital beamforming are combined to lower the overall system cost with minimum performance degradation [11]. Although hybrid antenna arrays may have limited scanning range due to the use of subarrays, this drawback can be overcome by employing non-conventional subarrays such as lens antennas. Other pure antenna solutions include 1-bit reconfigurable reflectarrays, planar substrate-integrated waveguide-based leaky-wave beam-scanning antennas, small-director-array-based beam-steerable antennas, and active-frequency-selective-surface-based 3D beam-scanning antennas [12]. Some of the state-of-the-art wide-angle beam-scanning antennas will be discussed over the course of this chapter. This chapter also includes a discussion on scanning loss and some antenna designs with reduced scan loss in their scanning range. Finally, a detailed discussion and design evolution of a low scan loss antenna, with reduced scan loss across the beam-scanning range as well as across the wide bandwidth range is also included in this chapter.

## 5.2 Challenges in Wide-Angle Beam Scanning

Achieving WABS for PAAs has been one of the most challenging tasks. Specifically, good impedance matching and consistent gains should be maintained when the beam is scanned in a wide angular range. The wide angular range of PAAs is usually interpreted as being larger than 100°. In this section, challenges of a high-performance wide-angle beam-scanning phased-array antenna are introduced.

1. Beamwidth

Theoretically, gain pattern of an array is the product of the array factor and the element pattern when each element is assumed to have identical radiation pattern [13]. Therefore, for an antenna element with a traditional broadside radiation pattern, the array peak gain will decrease as the beam scans away from broadside due to the gain drop of the element pattern [14]. In order to mitigate significant gain drops at large scanning angles, the element pattern is expected to have a broad HPBW. This can be verified through a beam-scanning array reported in [15]. Its element antenna has HPBWs of 93° and 113° in the *E*-plane and H-plane, respectively. The array configuration with this element exhibited a 4.7-dB gain decrease with respect to the broadside gain when scanning to + 45° in the E-plane. The gain decreases only 2.3 dB when the beam is scanned to + 45° in the H-plane.

2. Mutual Coupling

The array elements are usually spaced at a relatively small distance (around 0.5λ) to avoid appearance of the grating lobes when the main beam is scanned. With a small element space, the radiation characteristics of one element in the array would be generally affected by the electromagnetic environment consisting of all the other elements. Mutual coupling among PAA elements will not only affect the beam shape of the element pattern, but also cause its impedance matching to deteriorate. Moreover, the impedance characteristics would be varied with a dependence on the beam-scanning angles. As a result, it would be a formidable task to match the impedance in a wide scanning angle range. Scan blindness is a common issue that occurs for WABS PAAs [16].

3. Grating Lobes

Grating lobe is the beam that has approximately the same gain as the main beam. It causes interference and significantly decreases the gain of the main beam. Although the grating lobe can be avoided by setting a relatively small element spacing, a large element spacing is inevitable in some specific applications. These include the size of the antenna element being larger than half a wavelength, the mutual coupling effect being severe when the spacing is smaller than half a wavelength, etc. With a large spacing, the grating lobe may appear in the visible region especially when the array beam is scanned to large angles. Thus, suppressing grating lobe is of great importance to increase array gain and its corresponding radiation efficiency. Theoretically, to avoid grating lobes in a uniformly linear or rectangular distributed planar array, the element spacing d should meet the relationship in (5.1)

$$d \leq \frac{\lambda}{1 + |\sin \theta_{max}|}, \tag{5.1}$$

where $\theta_{max}$ is the maximum scanning angle. If $\theta_{max}$ is anticipated to be 60°, $d$ should be maintained smaller than 0.53λ. Otherwise, grating lobes will appear in the visible region. From the above relation, the element spacing will be confined related to the wavelength of the highest operation frequency. To avoid the appearance of grating lobes in the whole operation band, elements should be very close at the low frequency, resulting in a strong mutual coupling and even structure intersection. Special techniques should thus be introduced to suppress grating lobes when a larger spacing is used [10].

4. Low Gain Fluctuation

Considering the realistic applications, the gain fluctuation of the phased array should be less than 3 dB during the scanning. According to the projection theory, for a two-dimensional phased array with a half wavelength element spacing, the directivity of the array factor will decline with the cos function of the scanning angle. Namely, the directivity of the array factor will decline by 3 dB when the array scans to ± 60°. For a wide working bandwidth, the electric aperture size of the element in the high frequency will be several times that at the low frequency. Based on the theory that a larger electric aperture results in a narrower beamwidth, the beamwidth at the high frequency is usually too narrow to support a low gain fluctuation during the wide-angle scanning. With the adverse conditions of the array factor directivity decline in wide scanning angle and a narrow element beamwidth at the high frequency, keeping a low gain fluctuation is difficult for a wide-angle scanning array, especially for wideband operation.

To realize a wideband and wide-angle scanning phased array for the satellite communication, grating lobe avoidance, active reflection coefficient suppression and gain fluctuation reduction are the main challenges. Obviously, enlarging the element spacing can reduce the mutual coupling between elements and be helpful in lowering the active reflection coefficient, but this may lead to grating lobe appearance. Besides, a wide beamwidth is beneficial for keeping a low gain fluctuation during the scanning, but also brings strong mutual coupling and results in high active reflection coefficient. Therefore, these challenges constrain each other and have attracted great research efforts along with significant advances for wide-angle beam-scanning antennas. In the following sections, we will review and discuss the latest developments.

## 5.3 Different Approaches for Wide-Angle Beam Scanning

In order to further extend the beam-scanning coverage, many related efforts have been carried out to break the bottleneck of limited scanning angles. Generally, there are two broad approaches to achieve wide-beam scanning.

The first approach involves obtaining a wide-angle impedance matching (WAIM) by compensating for mutual coupling between antenna elements. WAIM is a technique that views the degradation in gain as a result of an impedance mismatch between the array and free-space wave propagation. The WAIM can be accomplished by

1. ***Loading the Antenna Array with a Decoupling Surface***—Array-antenna decoupling surface (ADS) is used for reducing the mutual coupling between antenna elements in a large-scale array antenna [17]. An ADS is a thin surface that is composed of a plurality of electrically small metal reflection patches and is placed less than one-half the wavelength above the ground plane of an array antenna. An ADS is ingeniously designed to create partial reflective electromagnetic (EM) waves to cancel the coupled waves from the adjacent antenna elements. By appropriately adjusting the distance between the ADS and the array antenna and the sizes of the patches, the out of phase and the equal intensity conditions of the interference waves relative to that of the unwanted coupled waves can be achieved, leading to a high degree of cancelation of the unwanted mutual coupling between adjacent antenna elements.
2. ***Embedding Metallic* via *Walls***—Due to high-impedance properties of the embedded via wall structures in common ground plane between each antenna element, the mutual coupling path through surface current can be reduced [18]. To minimize leakage power caused by surface waves, low-loss coplanar waveguide ground (CPWG) feeding networks with via wall structures and island-shape pads are realized
3. ***Adding Parasitic Decoupling Walls***—Researchers in [19] presented parasitic decoupling walls as an effective structure to considerably increase the $E$-plane scanning angle of the phased array by reducing the magnitude of near-field mutual coupling.
4. ***Using Electromagnetic Bandgap Structures (EBG)***—A phased array in a triangular lattice, with a large element spacing with wide scan angle, is investigated in [20]. Scan blindness is observed for this triangular lattice array when the beam is scanned in the E-plane due to large element spacing, which provides a suitable propagation condition for the leaky surface waves in the array aperture. Therefore, compact electromagnetic bandgap (EBG) structures are loaded in the array element to suppress the leaky surface waves and then remove the $E$-plane scan blindness for the largest impedance bandwidth. For the entire operating frequency band, the gain drop is within a range of 2.5–5 dB as the beam scans from broadside to 60°.

Another approach involves constructing a broad beam antenna element. Many techniques to achieve this are described below:

1. ***Reconfiguring Narrow Beams***—A microstrip patch operating with different resonant modes will result in different radiation patterns. According to [21], when the resonant modes are $TM_{0,(2n+1)}$ and $TM_{(2m+1),0}$, the maximum radiation direction points to the broadside one, where m and n are both integers and their values start from 0. On the contrary, when the resonant modes are $TM_{0,(2n+2)}$ and

$TM_{(2m+2),0}$, the maximum radiation direction deviates from the broadside one. Based on the above intuitive phenomenon, mode reconfigurable elements with wide joint 3-dB beam coverage can be obtained with the reconfiguration technique. A phased array with such elements can scan a large range from broadside to near endfire direction, effectively. For example, the $TM_{11}$ mode and the $TM_{01}$ can obtain a joint radiation pattern covering all the half-space. This feature is conducive to achieve 2D wide-angle scanning. The multilayer microstrip patch element presented in [22] reconfigures the resonant modes $TM_{01}$ and $TM_{20}$, resulting in a wide joint 3-dB beam coverage. A $1 \times 4$ uniform linear phased array is designed where the main beam can scan from $-75°$ to $+75°$ with a gain fluctuation less than $\pm 1$ dB.
2. ***Combining Different Resonance Modes***—Researchers in [23] proposed a wide-angle *E*-plane scanning linear array which can scan from $-70°$ to $+70°$ with a gain fluctuation less than 2.7 dB. To reduce the gain fluctuation of a phased array at low elevation angle, a wide-beam antenna is used as the antenna element. The 3-dB beamwidth of the wide-beam antenna is 221° in the E-plane at 4.0 GHz. The WAIM is improved by the matching substrate that is arranged on the ground plane when the beam scans up to the low elevation angle. The excellent input impedance matching of the antenna element can positively affect the wide-angle scanning performance.
3. ***Adding an Artificial Magnetic Conductor (AMC) Ground***—Researchers in [24] proposed design for wide-angle scanning adding an artificial magnetic conductor (AMC) ground beneath an electric dipole. The main beam direction of the array can scan from $-89°$ to 90° in the *H*-plane with a gain fluctuation less than 3 dB and the scanning 3-dB beamwidth can cover a range from $-105°$ to 105°. An excellent wide-angle scanning performance from the broadside direction to the endfire direction can be obtained by the proposed method, which can be used to guide the design of wide-angle scanning arrays.
4. ***Loading a Parasitic Pixel Layer over a Patch Antenna***—The parasitic pixel surface is able to induce significant currents due to the strong coupling between the driven antenna and the parasitic surface [25]. For the purpose of controlling the radiation pattern of the antenna, the pixel surface is reconfigured to change the currents on the upper layer. Therefore, it is possible to achieve wide beam by connecting parts of the small patches on the pixel surface. A parasitic pixel layer-based antenna is used as the array element. The phased array is able to scan its main beam from $-75°$ to $+75°$ in both *XOZ*- and *YOZ*-planes, and it supports 3-dB beam coverages from $-85°$ to $+85°$ in the *XOZ*-plane and from $-85°$ to $+80°$ in the *YOZ*-plane.

The above approaches can be combined to build a phased-array antenna with wide scanning angle [26]. However, the mutual coupling between the antenna elements is still existent, resulting in noticeable gain drop and needs to be reduced. To further mitigate the gain variation, the coupling needs to be considered at the element design stage of the array. An array pattern can be decomposed into element pattern and array factor. Instead of focusing much on the array characteristics, Su [27] used a different strategy by focusing on the element pattern to widen the scanning range without an obvious gain drop. To calculate the optimal element pattern for wide-angle scanning a gain complementing technique is proposed. The idea is verified by using a cylindrical dielectric resonator antenna with metal ring and dielectric slabs that scanned from $-72°$ to $+72°$ with low gain fluctuation.

## 5.4 Wide-Angle Scanning Using Frequency Sweep

Phased-array antenna has been widely used in recent years because of its remarkable agility, flexible beam-scan capacity, high tracking accuracy, and adapting to the high-speed maneuver capacity of the carrier. However, phased-array antenna has a serious drawback which is the limited beam-scanning range of the antenna with high gain. For traditional phased-array antennas, the main beams can scan from $-45°$ to $+45°$ with a gain fluctuation of 4–5 dB [28], because the beamwidths of the elements are limited, and mutual coupling between elements is strong when the beam is scanned at low elevation (near endfire) angles.

In order to further extend the beam-scanning coverage, many related efforts have been carried out to break the bottlenecks of limited scanning angles. The electronically scanned PAAs showed attractive radiation and impedance performance with great flexibility. However, those architectures generally need to use many phase shifters to control the excitation phases of all the elements. As a consequence, the complexity as well as the system cost is greatly increased. To this end, frequency-based beam scanning is a promising choice for low-cost beam-scanning systems. They do not rely on any phase shifters for beam scanning. In contrast, they achieve beam scanning by changing the source frequency of the PAA. It is evident that the frequency scanning PAA has many significant advantages such as being simple in construction, easy for fabrication, and low-cost. This makes the frequency scanning PAA attractive for many modern applications including imaging systems and direction finding networks [29]. It is also acknowledged that owing to the dependence on the operating frequency, the freedom for beam scanning is much reduced in comparison with that with phase shifters. This may lead to performance degradation for frequency scanning PAAs compared to the electronically controlled PAAs. Moreover, the variation of frequency at different beam angles prohibits the PAA application in many systems such as most communications systems. Therefore, there is a trade-off between the system cost and the performance characteristics.

Figure 5.2 shows a systematic view of the leaky-wave antenna (LWA), one typical type of the frequency scanning PAA, for WABS. The LWA employs guiding structures to support wave propagation, and the electromagnetic energy is gradually radiated or leaked with the wave traveling along the waveguiding structures [30]. The guiding wave is represented by the long connected line and the radiators can be treated as an array of individual antenna elements, as seen in Fig. 5.2. Thus, the LWA is equivalent to a series feed antenna array.

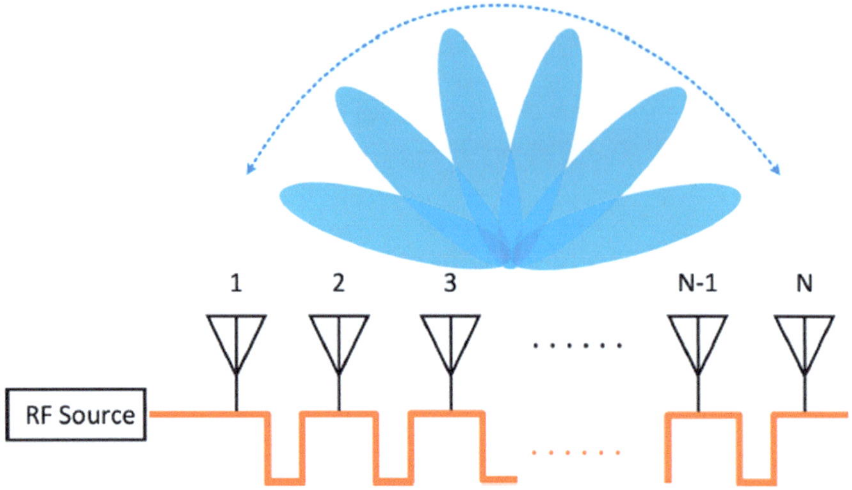


**Fig. 5.2** Schematic view of the leaky-wave antenna for WABS [10]

For one-dimensional (1D) LWA, the main beam angle θ can be obtained through the following relation [31]:

$$\theta(f) = \sin^{-1}\frac{\beta(f)}{k_0(f)}, \tag{5.2}$$

where $\beta$ is the phase constant in the waveguiding structure.

To facilitate WABS with LWAs, there are several critical challenges. The first is an open-stop band (OSB) issue. The OSB is caused by the reflection from each unit cell adding up in phase at the input port when the beam points toward the broadside direction [32]. It leads to deteriorated impedance matching and degraded gain. Therefore, the OSB effect prevents a continuous beam scanning through broadside for LWAs. To suppress the OSB, researchers have developed numerous techniques, including utilizing composite right/left-handed (CRLH) structures and employing impedance matching methods [33]. The second are those difficulties we have introduced and discussed for the electronically scanned PAA in the previous sections. For example, one needs to realize wide element pattern beamwidth to compensate for gain losses at large angles and maintain wide-angle impedance matching to avoid large gain variations.

A variety of techniques have been reported to overcome the above challenges and, hence, to facilitate LWAs with frequency-based WABS. A slotted SIW LWA with partially reflecting vias was reported in [34], and its configuration is shown in Fig. 5.3. Unlike the conventional SIW with two closely-spaced via walls (acting as PEC walls), the developed antenna has one via wall sparsely distributed for OSB suppression. The specified-shaped slot was used and optimized to improve the linear polarization purity. It is seen that the OSB was mitigated and a wide operating bandwidth was

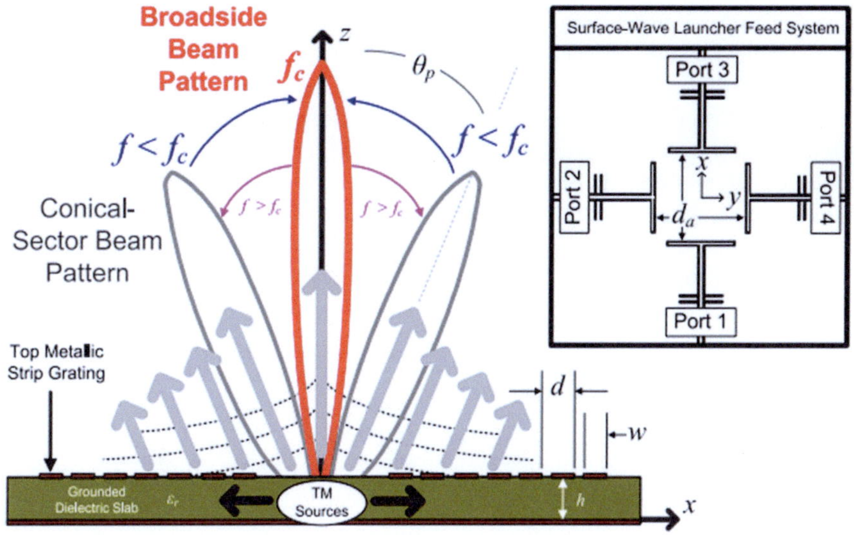


**Fig. 5.3** Planar leaky-wave antenna for broadside radiation

obtained. By sweeping the frequency from 7.625 to 11 GHz, the main beam was scanned from − 74° to + 45°, a total range of 119°.

As revealed in [35], the beam-scanning range can be further increased, being from − 76° to + 67°. This is accomplished by employing a combination structure of composite right/left-handed (CRLH) lines, magnetic dipole, and alumina ceramic block. These WABS LWAs can't point toward endfire directions. One main reason is the effect caused by the ground plane. In this regard, those ground-free structures are promising to further enhance the beam-scanning range from backward to forward and further to endfire directions. Various techniques including the Goubau line [36] and spoof surface plasmon polariton (SSPP) [37] are reported to achieve that goal. In [38], double layer glide-symmetric SSPP structures are reported that scans the beam from − 90° to + 44.3°.

The attained realized gain ranges from 8.03 to 12.17 dBi. Besides the elevational beam scanning as reported by the above WABS LWAs, the periodic coplanar-strip LWA reported in [39] realized an azimuthal beam-scanning range from − 75° to + 84°. The azimuthal directed beam was obtained by using open-ended slots.

## 5.5 Scan Loss

When the main beam is scanning along different angles, there is a gain reduction on the peak of the main lobe. This gain loss is called scan loss. The peak gain of one specifically beam-steered array pattern is proportional to the element gain at its main beam angle. As a consequence, the array gain will drop dramatically at large scanning angles if the element pattern has a narrow half-power beamwidth (HPBW). In the real application of phased-array antennas, the direction of the main lobe is not always at 0°. If the scan loss is too high, there would be a large amount of power loss resulting in lower SNR (signal noise ratio). Therefore, reducing the scan loss is an important design consideration. However, considering the permissible scan loss (for instance 3 dB or 4 dB), the scan range would be very limited. If the gain variation is not very large, then the scan loss can be reduced in a large angle range. This section describes some of the recent antenna designs with low scanning loss across their beam steering angle.

### *5.5.1 Low Scan Loss Antennas*

Beam-scanning antennas with low scan loss are very promising for WLAN applications. Ideally, a zero scan loss antenna system is desired in communication systems. A few design approaches are presented in this section.

A flat-focal-plane dual-metasurface lens (dual-metalens) has the potential to lower the scan loss and sidelobe level (SLL) of a metalens antenna. Metalens antennas are a class of lens antennas that fulfill and improve the performance of dielectric lenses using metasurfaces that effectively reduce the volume and complexity of the lens system [40]. To improve the scanning performance of a metalens antenna, feeds must be placed on a curved surface beneath the metalens [41]. However, the determination of the curved surface and installation of feeds on the surface are sophisticated and inconvenient.

Alternatively, multifocus or paraboloid-phase-distribution lens designs are proposed to reduce scan loss [42]. In optics, wide field-of-view imaging has already been facilitated in cameras and mobile phones by cascading lenses. In the microwave regime, Nguyen [43] demonstrates that two dielectric lenses can improve the scanning performance of the lens antenna. Reducing the comatic aberration of a metalens by designing a flat-focal-plane metalens can logically be an approach to improve the beam-scanning performance of the metalens antenna.

Another solution to the scan loss issue is proposed in [44], where the optimum feed position of a hemispherical and an elliptical integrated lens antenna is determined. The unwanted phase aberration can be minimized by locating the feed elements on the elliptical or spherical surface rather than on the planar surface. The concept solution is verified in [45] with a 100 mm diameter extended hemispherical ILA and manually moving WR-10 open-ended waveguide feed on the spherical back surface. The prototype achieved a very low scan loss up to $\pm 30°$.

## 5.6 Wideband Directional Cavity Antenna with Low Scan Loss for WLAN

The evolution of various wireless communication systems has given rise to design wideband antennas with high-gain and stable radiation patterns across the operating band for base stations to receive and transmit signals. The IEEE 802.11 ac WLAN has the bandwidth necessary to meet excess demand and high data rates [46]. Now, adequate angular coverage is necessary making beam scanning an important feature of the antenna systems. There are two scanning losses associated with antennas: First, the fall in gain when the beam is scanned from broadside to the scanned angle and second, the fall in gain when the frequency of the antenna is varied from the resonant frequency within the bandwidth. In this section, we propose a design that reduces these scanning losses to the lowest possible level. It is important in antenna designs to reduce these losses so that the signal strength is maintained throughout the desired range.

The impact on system performance due to the antenna tilt in Global System for Mobile Communications (GSM) is examined in [47]. It is seen that to enhance angular coverage on the ground, the base station antenna is tilted, mechanically. A power coupler comprising an electromechanical phase shifter that can adjust phase by the relative motion of components is claimed in [48]. This technique is widely being used commercially in many mobile systems. Even though it applies to wide frequency bands, this type of phase control by the phase shifter circuits results in higher losses and needs extra circuitry as well. Traditionally, the beam tilt of the base station antennas is controlled by ensuring that the antennas have sufficient tilt to reach the intended users. The antennas are tilted downwards physically along with required phase shifts [49]. But these are low-frequency GSM band antennas where such techniques provide expected beam tilts. The proposed beam-tilting technique is a much simpler solution for high-frequency antenna beam tilts. In this method, we shift the height of a slider inside an antenna cavity, while avoiding the use of any phase shifters throughout the design. This technique is much more robust than the standard antennas that are connected to feed networks and physically supporting networks.

In contrast to previous works, where researchers have proposed numerous wideband antennas with omnidirectional radiations [50, 51], the proposed design is a wideband unidirectional antenna with scanning loss of only 0.3 dB throughout the bandwidth. In previous works like [52], a wideband antenna with directional radiation, in just 1 GHz of antenna bandwidth, the antenna suffers from 1.1-dB scanning loss. The antenna design also lacks the beam-scanning ability. A wideband cavity antenna with a large ground plane where the gain varies from 0 to 6.8 dB within a 28% fractional bandwidth at 2.3 GHz is reported in [53]. The antenna has wide beamwidth making it poorly directional. The proposed antenna has a large ground plane and a unidirectional pattern that gives the opportunity to mount back-end electronics. When antennas are integrated with other back-end electronic devices, its performance is significantly affected by the ground plane structure [54]. This occurs due to the field coupling between the antenna and the device ground plane. In such instances, an electrically large ground plane would act as an isolating shield against the potential interference from the active electronic circuitry. The proximity to a conducting ground plane makes the bandwidth of a patch antenna deteriorate [55].

To increase the bandwidth of the antenna, we can increase the height or reduce the dielectric constant of the substrate. However as mentioned in [56], this approach can be used for a substrate thickness of less than 0.02λ only. [57] reports that when a rectangular plate is placed over a ground plane, as the height of the plate is varied from 0.125λ to 0.25λ there is a reduction of $Q$-factor by a factor of 2. In the proposed design, the height of the antenna is fixed at 0.1λ making it an electrically compact design, and at the same time the antenna is wideband in nature. Besides, the proposed design, having the benefit of being unidirectional, allows for easy integration with ceiling mounted back-end electronics without any influence on the performance of the antenna. The proposed antenna scans the beam from broadside to 20° with scanning loss of only 0.8 dB. So the proposed design is also able to maintain wide 20% 1-dB gain bandwidth throughout its scanning range. Phase shifters used in conventional array designs result in high insertion loss. The absence of any phase shifters or active devices in the proposed design consequently eliminates any insertion loss associated with the antenna.

### 5.6.1 Narrow-Band Cavity Antenna

Consider a conventional patch antenna. The bandwidth of a patch antenna can be increased by increasing the thickness of the substrate or by reducing the dielectric constant of the substrate [57]. To achieve finer characteristics than a conventional patch, a cavity antenna as shown in Fig. 5.4 is considered. The chosen substrate for the antenna design is polylactic acid (PLA) with $\varepsilon_r = 2.58$ and $\tan\delta = 0.015$. PLA is a user-friendly thermoplastic for 3D printing performed by the Fused Deposition Modelling (FDM) method. The height of the cavity antenna is 5 mm. Here, from the 5 mm PLA block, 4 mm of the substrate is removed to make the cavity. Now, the cavity consists of air as the substrate with a dielectric constant of 1. A 0.5 mm thickness of PLA substrate is used at the top and bottom to make the cavity antenna as shown in Fig. 5.4b. A 4 mm thickness was considered at the sides of the cavity walls to adequately support the antenna structure. This lateral support is necessary and can be obtained only through 3D printing as it will be attached firmly to the other sides as well which is only 0.5 mm in thickness. Using other dielectric substrate sections to build the cavity cannot ensure the stability obtained through the 3D printing method. Consequently, 3D printing using PLA was chosen even though the PLA substrate has a relatively high loss tangent, which could affect the radiation efficiency of the antenna.

The cavity antenna has a square structure with its edge lengths at 32 mm (0.6λ). The bottom of the cavity antenna body consists of a 59 mm ground plane. The input port is connected to the top layer of the antenna through a coaxial feed. The return

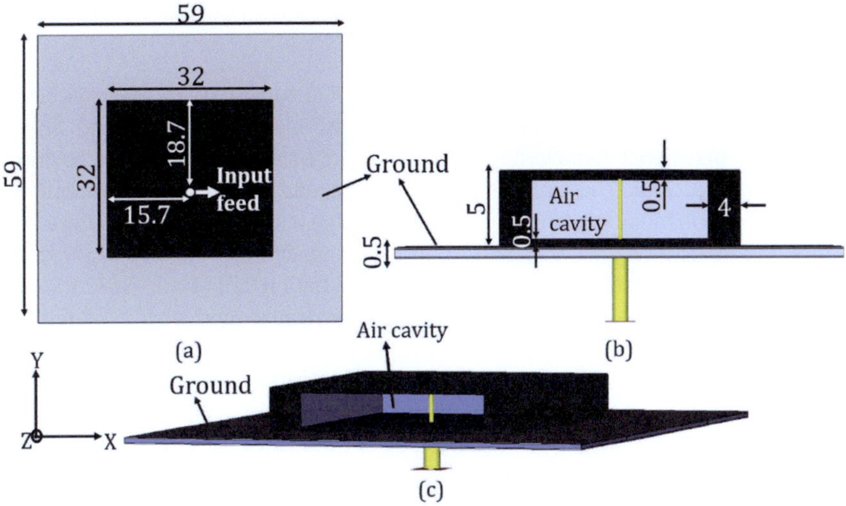


**Fig. 5.4** Schematic of narrow-band cavity antenna. **a** Top view. **b** Side view. **c** Isometric view (all dimensions are in mm) [58]

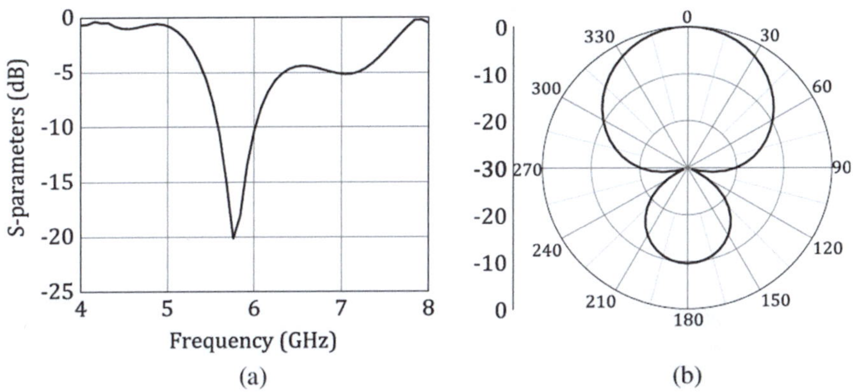


**Fig. 5.5** **a** Return loss of the narrow-band cavity antenna. **b** Radiation pattern of the narrow-band cavity antenna at 5.8 GHz [58]

loss of the cavity antenna is depicted in Fig. 5.5a with the bandwidth of the antenna from 5.58 to 6 GHz. The radiation pattern at 5.8 GHz is shown in Fig. 5.5b. The gain of the antenna at the resonant frequency of 5.8 GHz is 5.8 dBi.

### 5.6.2 *Wideband Cavity Antenna*

The narrow-band cavity antenna is further modified to realize a wideband cavity antenna. The schematic view of the wideband cavity antenna is shown in Fig. 5.6. To increase the impedance bandwidth, various slots are introduced at specific positions as shown in Fig. 5.6a. The narrow-band cavity antenna has a single opening in the cavity that aids in radiation. This is seen in Fig. 5.6b where the front part of the cavity is removed. To increase the bandwidth, slots were made in the metal and substrate, which increased the openings through which the radiation will occur. This increased energy loss reduces the quality factor of the cavity increasing the bandwidth of the antenna.

At first, the slots are introduced at the top metal part of the antenna with a slot width of 1 mm throughout. Figure 5.8a shows the return-loss variation of the antenna with the introduction of each circular slot. Slot 1 indicates the circle with a radius of 10.2 mm and a 1 mm slot toward the edge of the circle. Slot 2 is the circle at the center with a radius of 3 mm. Finally, two circles with a radius of 8 mm are introduced at two sides, referred to as slot 3. The slots introduced in the metal layer are sufficient for a wideband cavity antenna. However, it was observed that the slider that has to be placed inside the antenna cavity causes severe detuning of the antenna. To overcome this detuning effect, just like the slots in the metal layer, similar slots were created on the substrate as well. To ensure continuity in the substrate, some segments of the slots were closed strategically such that it does not affect the results considerably. The slots also increase the electrical length of the antenna. The surface current distributions of the cavity antenna without and with various slots are shown in Fig. 5.7a–d. The slots force the surface currents to meander, artificially increasing the electrical length of the antenna without changing the physical dimensions. This increased electrical length makes the cavity antenna

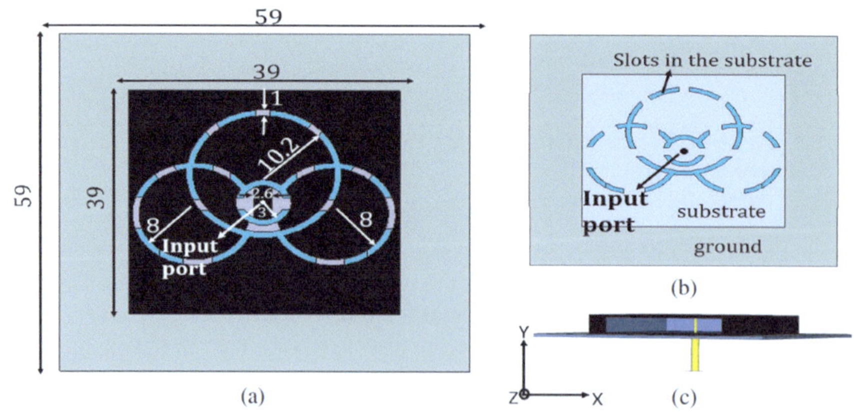


**Fig. 5.6** Schematic of wideband cavity antenna. **a** Top view. **b** Top view of the substrate. **c** Isometric view [58]

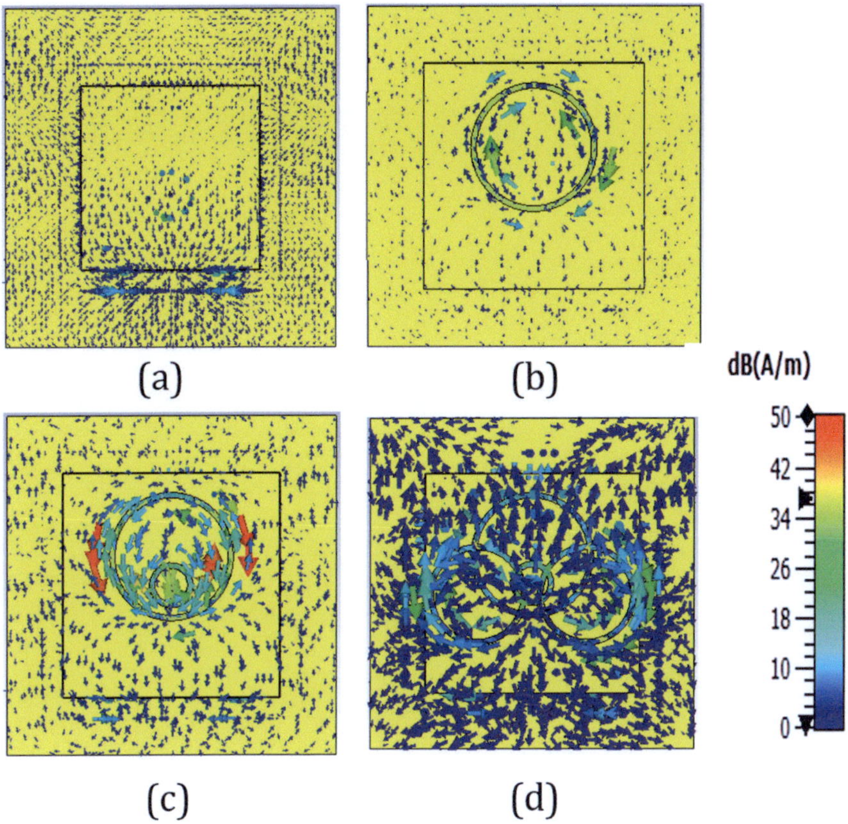


**Fig. 5.7** **a** Surface currents of antenna without any slots, **b** with slot 1, **c** with slot 2, **d** with slot 3 [58]

to also resonate in lower frequencies. By appropriately designing the slots, the cavity antenna starts to resonate in multiple frequencies which increases the bandwidth of the antenna. The final configuration of the substrate placed at the top part of the cavity is shown in Fig. 5.6b. This substrate is then covered with metal to complete the cavity.

The simulated impedance bandwidth for the antenna is 2.4 GHz (5.05–7.45 GHz) as shown in Fig. 5.8a. The gain of the wideband cavity antenna is shown in Fig. 5.8b. The simulated bandwidth has increased from 7.3% for narrow-band to 40% for wideband cavity antenna. The radiation pattern of the antenna is shown in Fig. 5.9. The half-power beamwidth (HPBW) is 96°, and the front-to-back ratio is 13 dB.

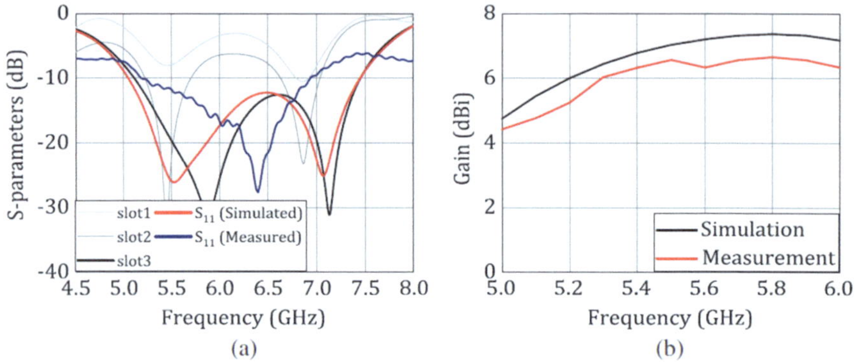


**Fig. 5.8** **a** Return-loss characteristics of wideband cavity antenna. **b** Gain of wideband cavity antenna [58]

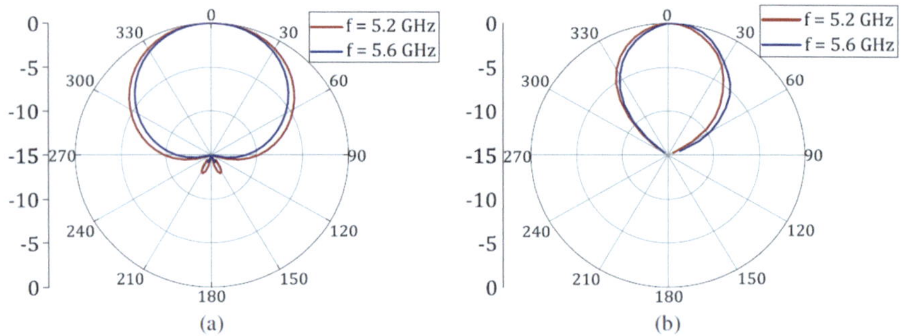


**Fig. 5.9** Radiation patterns at 5.2 and 5.6 GHz. **a** Simulation. **b** Measurement [58]

### *5.6.3 Wideband Cavity Antenna with Slider*

Now, the objective is to introduce a beam tilt in the designed wideband cavity antenna. The beam tilt must also ensure that the scanning loss is minimum at the scanned angle, and the tilt is stable for a wide frequency range. The component of the antenna responsible for beam tilting is the slider that is designed and placed within the cavity of the antenna as shown in Fig. 5.10. The slider is designed using a PLA substrate of thickness 0.5 mm. It is then placed in the cavity space between the wall at the rear end of the cavity and the coaxial input pin as shown in Fig. 5.10b. Initially, the slider shown in Fig. 5.10c is placed at the bottom of the cavity and then moved toward the top. For various heights of the slider from bottom to top, the beam tilt of the antenna increases from 0° to 20°. This tilt remains for frequencies from 5.2 GHz to 5.8 GHz or 10% of the bandwidth. Figure 5.10d depicts the design evolution for the slider design. It is observed that a slider with a rectangular pattern imparts maximum beam tilt, and a slider design with a circular pattern imparts beam tilt for a wide frequency range. So combining the patterns to form an elliptical pattern results in

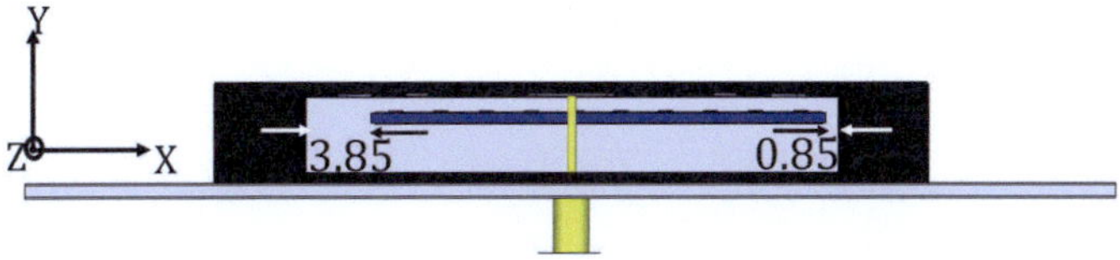


(a)

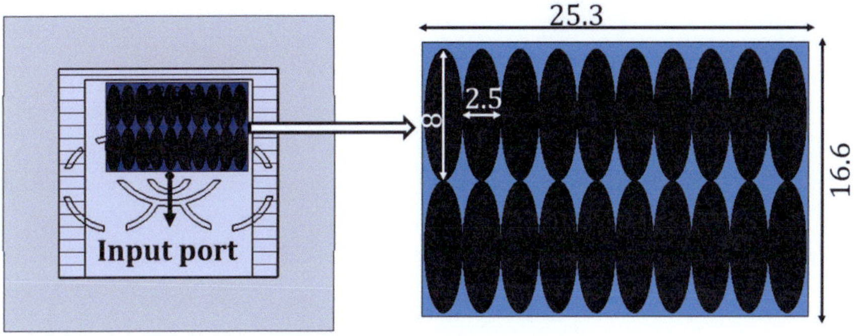


(b)

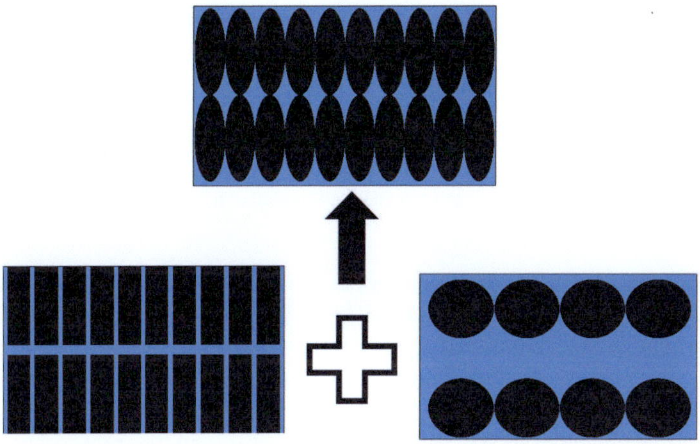

(c)

**Fig. 5.10** Schematic of wideband cavity antenna with slider. **a** Front view. **b** Top view. **c** Top view of the slider. **d** Design evolution for the slider [58]

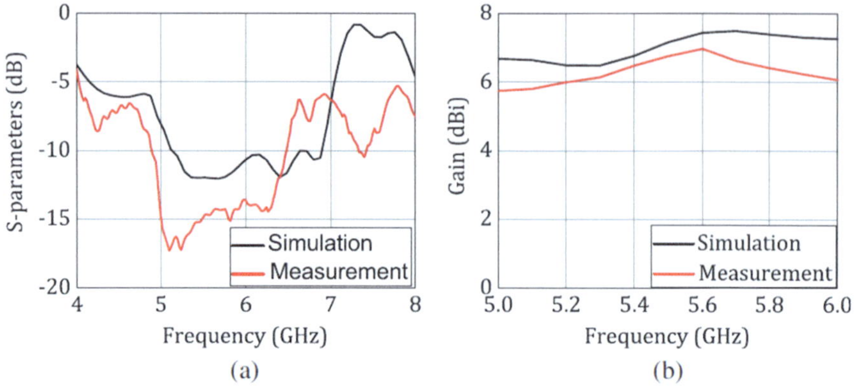


**Fig. 5.11** **a** Return-loss characteristics of wideband cavity antenna with slider, **b** gain of wideband cavity antenna with slider [58]

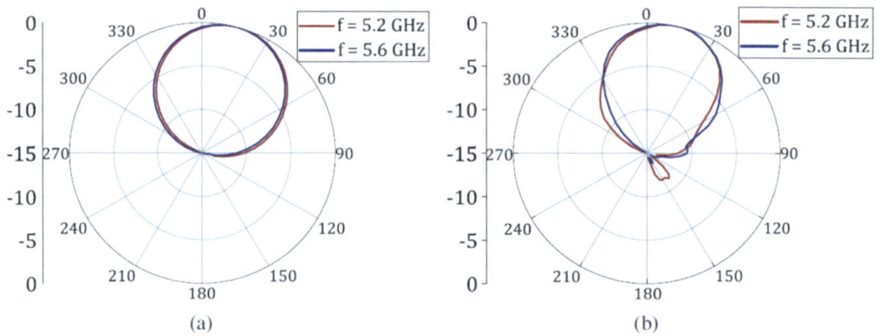


**Fig. 5.12** Radiation patterns at 5.2 and 5.6 GHz. **a** Simulated. **b** Measured [58]

maximum beam tilt for a wide frequency range. The simulated impedance bandwidth for the antenna is 1.72 GHz (5.18–6.9 GHz) as shown in Fig. 5.11a. The gain of the wideband cavity antenna after the placement of the slider is shown in Fig. 5.11b. The radiation patterns for the antenna for 5.2 and 5.6 GHz are shown in Fig. 5.12. A similar tilted radiation pattern is maintained for all frequencies in this range. The half-power beamwidth (HPBW) of the antenna is 72°, and the front-to-back ratio is 17.5 dB.

### *5.6.4 Theoretical Analysis of the Slider*

The slider inside the cavity influences the radiation pattern of the antenna resulting in a beam tilt. To better understand this phenomenon, the slider was modeled as a dielectric block with a high dielectric constant. Figure 5.13a shows the slider inside

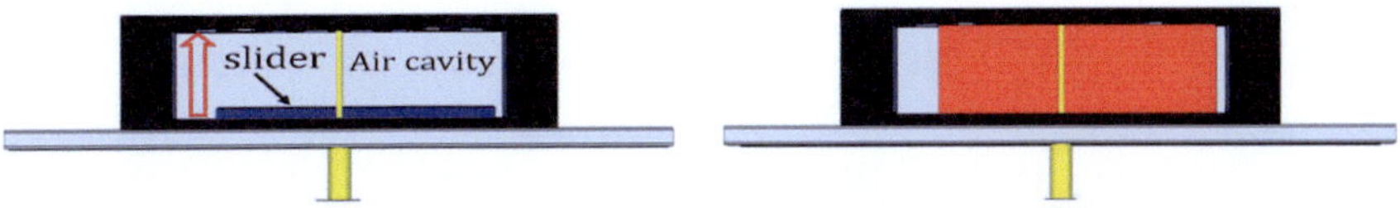


(a) (b)

**Fig. 5.13** Wideband cavity antenna with **a** slider, **b** varying dielectric block in the place of slider [58]

**Table 5.1** Antenna beam tilt with respect to dielectric constant

| $H$ | $\varepsilon$ | Beam tilt |
|---|---|---|
| 0.00 | 1.00 | 0° |
| 2.15 | 4.12 | 5° |
| 2.40 | 4.19 | 10° |
| 2.58 | 4.30 | 15° |
| 2.68 | 4.39 | 20° |

the cavity, and Fig. 5.13b shows the dielectric constant model inside the antenna cavity. This dielectric model has the same length and width as the slider, and the height is the same as the cavity height. As the slider moves inside the cavity air gap from bottom to top, the dielectric constant of the air cavity is affected. To verify this premise, the dielectric constant of the model was varied. It was observed that for dielectric constant variation from 1 to 4.39 the beam tilting of the antenna was comparable to the beam tilting that materialized while the slider moved from the bottom to the top of the cavity. The beam tilt of the antenna increases when the slider is moved from bottom to top inside the cavity. In the dielectric model of the slider, the beam tilt increases when the dielectric constant of the model is increased. The data are tabulated in Table 5.1.

The radiation patterns for both the sets are shown in Fig. 5.14 for beam tilting from 0° to 20°. As seen in Fig. 5.15a, b the $E$-field distribution along the mid-section of the antenna indicates the slider causing a field disturbance inside the cavity resulting in a deflected main beam. Figure 5.15c, d shows the electric field distributions inside the cavity without and with the slider. For a cavity antenna, there is a uniform distribution of the electric field at both edges of the cavity making the antenna to radiate in the broadside direction. When the slider is introduced, there is concentration of electric field toward the right edge of the antenna. As a consequence, the field intensity gets shifted which generates the tilted radiation beam.

### 5.6.5 *Fabrication and Measurement*

A prototype of the proposed antenna as shown in Fig. 5.16 was fabricated to validate the simulation results. The cavity structure for the antenna is 3D printed. The

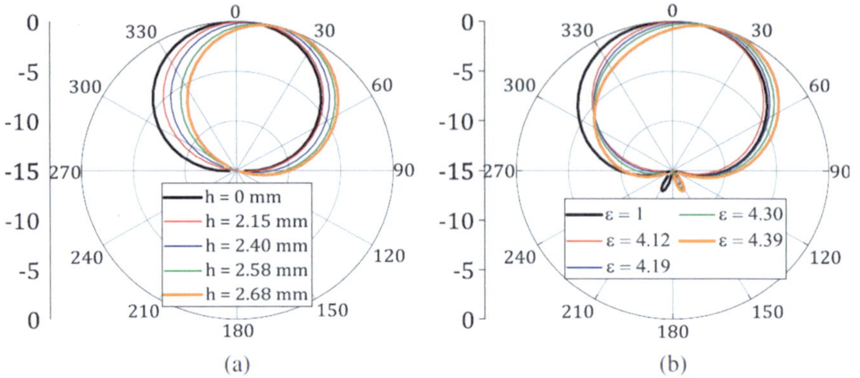


**Fig. 5.14** Radiation patterns at 5.2 GHz. **a** When height of the slider is varied across the cavity **b** when the dielectric constant of the block is varied [58]

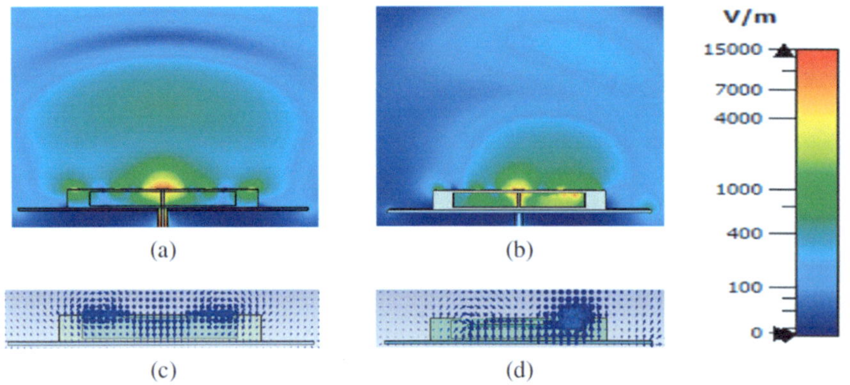


**Fig. 5.15** Electric field of the cavity antenna **a**, **c** without slider **b**, **d** with slider at 5.2 GHz [58]

3D printing was performed using a Raise 3D RxP2200® 3D printer a widely used commercial low-budget 3D printer. The lower part of the cavity antenna and ground plane is completed by pasting copper tape of 80 μm thickness. For the top part of the cavity, the copper was pasted, and the copper at the slots was removed by making proper incisions at slot positions. The slider design was fabricated by the photolithographic process. The slider is extended outside in order to assist in controlling the slider height from outside the cavity. 3D-printed blocks of required heights are then placed inside the cavity to keep the slider in position. The beam tilt is achieved from 0° to 20° by varying the height of the dielectric blocks that will keep the blocks at their required heights. Anritsu® MS2028C VNA was used to measure the reflection coefficient of the antenna. The far-field measurements were conducted inside an anechoic chamber where ETS-Lindgren® 3115 Model double-ridged waveguide horn antenna was used as the standard gain antenna for the measurement.

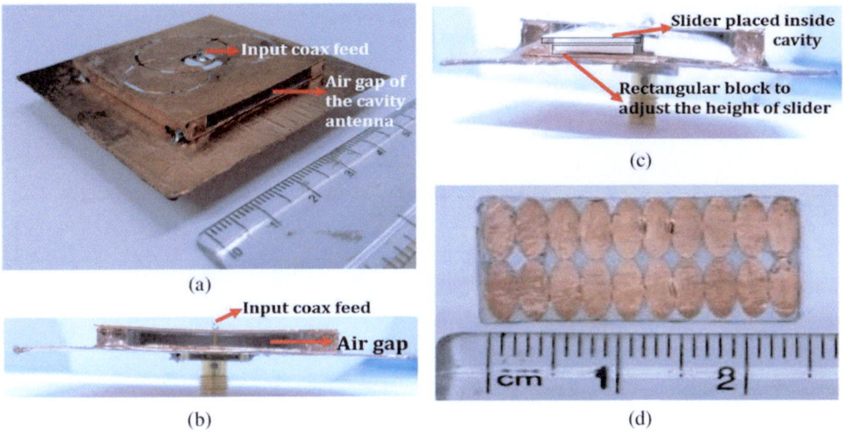


**Fig. 5.16** **a** Fabricated wideband cavity antenna. **b** Side view. **c** Side view of the antenna after placing the slider. **d** Slider to be placed inside the antenna cavity [58]

## *5.6.6 Two-Element Cavity Antenna Array*

Using the antenna element presented in Sect. 5.6.2, a linear two-element cavity antenna array is adopted as shown in Fig. 5.17. The array elements are arranged linearly on a ground plane of size 2.1λ × 1.2λ. The distance between the two elements is 1 mm, and the center-to-center distance between the elements is 0.8λ. The simulated impedance bandwidth for the antenna array is 2.43 GHz (5.07–7.5 GHz) as shown in Fig. 5.18a, and the gain of the antenna array is shown in Fig. 5.18b.

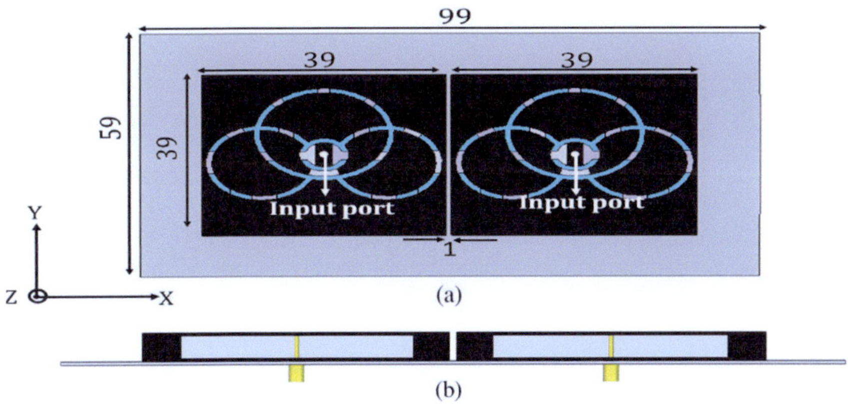


**Fig. 5.17** Schematic of two-element cavity antenna array. **a** Top view. **b** Front view [58]

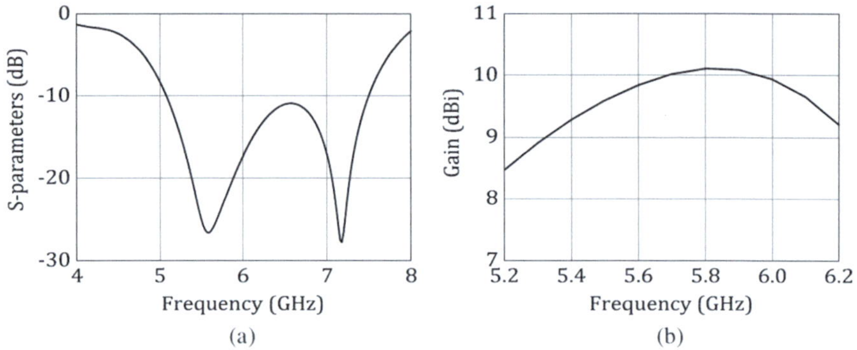


**Fig. 5.18** **a** Return-loss characteristics, **b** gain of wideband two-element antenna array [58]

### *5.6.7 Two-Element Cavity Antenna Array with Slider*

As shown in Fig. 5.19, in a two-element cavity array, the slider is placed inside the cavity of the antenna to the right. In this phase-shifter-less method of beam scanning, beam scanning occurs when the slider is moved from bottom to top inside the cavity. When the slider is placed in the right side cavity antenna and moved inside the cavity vertically the beam gets tilted toward + 20°. When the slider is placed in the left side cavity antenna and moved inside the cavity vertically the beam gets tilted toward − 20°. By using two sliders in both cavity antennas the beam can be scanned from − 20° to + 20° by keeping the slider at required heights inside the cavity.

The simulated impedance bandwidth for the antenna is 2.4 GHz (5.12–7.52 GHz) as shown in Fig. 5.20a, and the gain of the antenna is 9.6 dBi as shown in Fig. 5.20b. The gain of the antenna in Fig. 5.20b is at 20° scanned angle. It is clear from the graph

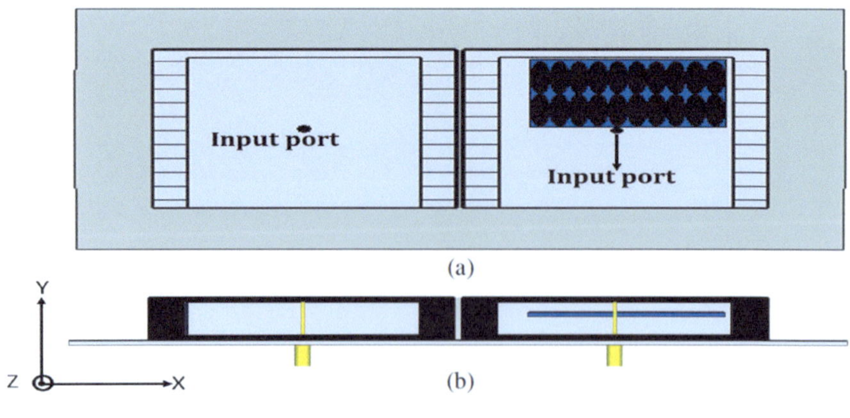


**Fig. 5.19** Schematic of two-element cavity antenna array with slider. **a** Top view. **b** Front view [58]

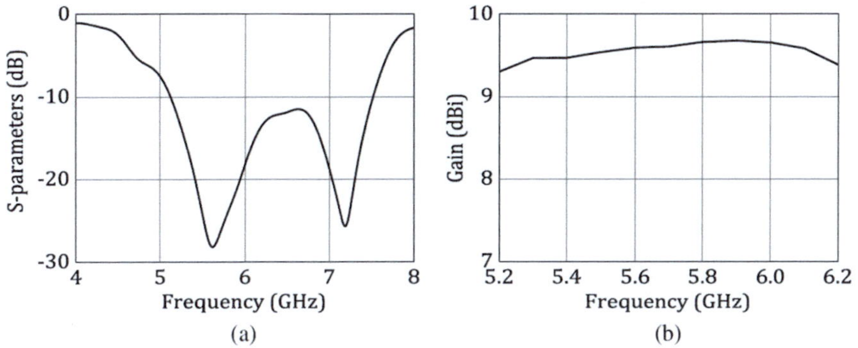


**Fig. 5.20** **a** Return-loss characteristics. **b** Gain of wideband two-element antenna array with slider [58]

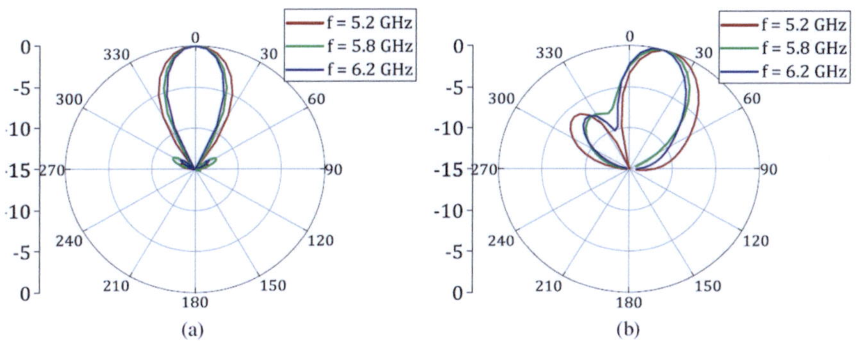


**Fig. 5.21** Simulated radiation patterns at 5.2 GHz, 5.8 GHz, 6.2 GHz. **a** Without slider. **b** With slider [58]

that for 20% fractional bandwidth the gain of the antenna varies from 9.6 to 9.3 dB limiting the scanning loss for wideband antenna within 0.3 dB. The radiation pattern for the two-element antenna array is shown in Fig. 5.21a. The half-power beamwidth (HPBW) of the antenna is 33°, and the front-to-back ratio is 19 dB. The radiation pattern for the two-element antenna array with slider is shown in Fig. 5.21b where the beam tilt for wide frequencies is visible. The half-power beamwidth (HPBW) of the antenna is 35°, and the front-to-back ratio of an antenna is 26 dB. The sidelobe levels can be further reduced by increasing the ground plane dimensions. The electric field distributions of the array with and without the slider are shown in Fig. 5.22. The field shown in Fig. 5.22b is at 20° beam scanning of the antenna array when the slider is at 3.25 mm from the bottom of the antenna cavity.

Comparisons of the bandwidth, scanning range, scanning loss along with the scanning technique of the proposed cavity antenna and some reported designs are described in Table 5.2. As observed from the table, the proposed design has only 0.3-dB scanning loss while scanning from broadside to 20° for a 20% fractional

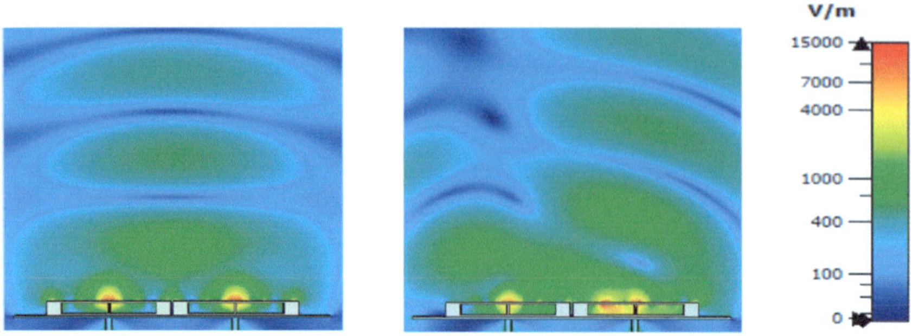


(a) (b)

**Fig. 5.22** Electric field distribution of two-element array **a** without slider **b** with slider [58]

bandwidth. The frequency range of the proposed antenna considered is the range that yields the beam tilt. For the proposed antenna the beam tilt does not vary for a 20% fractional bandwidth, and the beam is very stable at the scanned angle of 20°.

From Table 5.2 it is evident that the previous studies have achieved beam tilt only at the design frequency, and the scan angle varies for different frequencies within the bandwidth. The proposed design has a high radiation efficiency of 91% resulting from the presence of air cavity and absence of any phase shifters and active devices. Hence, the proposed antenna shows great potential for directional wideband antennas for WLAN applications.

Scanning loss reduction of a cavity antenna throughout the bandwidth as well as the scanned angle using a slider is presented. By shifting the slider from bottom to top inside the air cavity the beam is scanned from 0° to 20°. Considering the scanning range of the antenna, it can provide a total 3-dB coverage of 60°. The beam tilt of the antenna is sustained for 20% of antenna bandwidth with gain reduction of only 0.3 dB throughout the bandwidth. The bandwidth for which the beam tilt is sustained by the antenna is more than the reported bandwidth in the literature. Along with this high bandwidth, the gain variation while scanning the beam is also limited to only 0.8 dB. The proposed antenna thus ensures high signal strength for its 3-dB antenna beamwidth. Proper care must be taken to ensure that the appropriate slider height is maintained. The proposed design is validated for a single element antenna and for a two-element antenna array.

**Table 5.2** Comparison of the proposed antenna with other published works

| Ref. | [59] | [60] | [61] | [62] | [63] | Proposed antenna (PA) |
|---|---|---|---|---|---|---|
| Frequency range (GHz) | 9.05–9.65 | 11 | 8.86–9.84 | 8–12 | 16–18 | 5–6.2 |
| Fractional bandwidth (%) | 7 | – | 10.5 | 40 | 8.8 | 20 |
| Size ($\lambda$) | $\lambda \times 9.7\lambda \times \lambda$ | $6\lambda \times 6\lambda \times 0.5\lambda$ | $5\lambda \times 1.2\lambda \times 0.5\lambda$ | $3.23\lambda \times 3.1\lambda \times 0.27\lambda$ | $5.6\lambda \times 1.6\,\lambda \times 8\,\lambda$ | $2.1\,\lambda \times 1.2\lambda \times 0.1\lambda$ |
| Scan directions (°) | 36° | ± 51° | 14° | ± 37° | ± 26° | ± 20° |
| Gain (dBi) | 12.64 | 19.4 | 19.11 | 13.5 | 13.4 | 9.6 |
| Gain variation through scan angle (dB) | 1.3 | 3 | 0.8 | 2.1 | 1.5 | 0.8 |
| Scanning loss through bandwidth (dB) | Beam tilt only at 9.35 GHz | Beam tilt only at 11 GHz | Beam tilt only at 9.35 GHz | Beam scanningvaries within frequency range | Beam scanning only at 17 GHz | 0.3 |
| Beam-scanning technique | Rotating dielectric slab inside waveguide | Rotation of the metasurface pair | Rotating dielectricslabs inside waveguide | Physically rotating antennaelements | Insertion of tuning screws into waveguide | Varying the height of the slider inside the cavity |
| Radiation efficiency (%) | 61 | 89 | 79 | 51 | 90 | 91 |

## 5.7 Conclusion

Along with beam-scanning antennas, angle of scanning is also important as it determines how far the user can be without disruption in the signal. Wide-angle beam scanning and different techniques to achieve these are presented in this chapter. Even though wide-angle scanning was traditionally performed through mechanical scanning, electronic methods also need to be developed as mechanical scanning may not be feasible in every situation. Obtaining wide-angle scanning through varying the frequency is also discussed in this chapter. A brief discussion on scanning loss is also included. Detailed design, implementation and measurement of a wideband cavity antenna with low scanning loss is presented. Reduction in gain during scanning needs to be avoided in modern communication systems, and much more research is needed to make low scan loss systems more robust and reliable.

## References

1. Herd, J.S., Conway, M.D.: The evolution to modern phased array architectures. Proc. IEEE **104**(3), 519–529 (2016)
2. Asbeck, P.M., Rostomyan, N., Özen, M., Rabet, B., Jayamon, J.A.: Power amplifiers for mm-Wave 5G applications: technology comparisons and CMOS-SOI demonstration circuits. IEEE Trans. Microw. Theory Techn.**67**(7), 3099–3109 (2019)
3. Mailloux, R.J.: Phased Array Antenna Handbook (Antennas and Electromagnetics), 3rd edn. Artech House, Norwood, MA, USA (2017)
4. Feng, Y., et al.: A broadband wide-angle scanning linear array antenna with suppressed mutual coupling for 5G sub-6G applications. IEEE Antennas Wireless Propag. Lett. **21**(2), 366–370 (2022)
5. Aljuhani, A.H., Kanar, T., Zihir, S., Rebeiz, G.M.: A 256-element ku-band polarization agile SATCOM receive phased array with wide-angle scanning and high polarization purity. IEEE Trans. Microw. Theory Techn. **69**(5), 2609–2628 (2021)
6. Yi, Z., et al.: A wide-angle beam scanning antenna in E-plane for K-band radar sensor. IEEE Access **7**, 171684–171690 (2019)
7. Peng, F., Liao, C., Cheng, Y.F., Guan, L.X.: A multi-mode pattern reconfigurable antenna with wide beam coverage and its application to dual-polarized wide-angle scanning phased array. Int. J. RF Microw. Computer Aided Eng. **31**(9), Art. no. e22757 (2021)
8. Hu, J., Li, Y., Zhang, Z.: Vertically polarized 360◦ Azimuth scanning array. IEEE Antennas Wireless Propag. Lett. **21**(5), 898–902 (2022)
9. Gishkori, S., et al.: Adaptive subaperture integration for wide-angle synthetic aperture radar. IEEE Trans. Thz Sci. Technol. **12**(2), 118–129 (2022)
10. Li, M., Chen, S.-L., Liu, Y., Guo, Y.J.: Wide-angle beam scanning phased array antennas: a review. IEEE Open J. Antennas Propag. **4**, 695–712 (2023)
11. Jay Guo, Y., Ziolkowski, R.W.: Advanced Antenna Array Engineering for 6G and Beyond Wireless Communications. Wiley, Hoboken, NJ, USA (2021)
12. Gao, S., Guo, Y.J., Safavi-Naeini, S.A., Hong, W., Yang, X.-X.: Guest editorial low-cost wide-angle beam-scanning antennas. IEEE Trans. Antennas Propag. **70**(9), 7378–7383 (2022)
13. Hansen, R.C.: Phased Array Antennas. Wiley, Hoboken, NJ, USA (2009)
14. Mailloux, R.J.: Phased Array Antenna Handbook (Artech House Antennas and Propagation Library), 2nd edn. Artech House, Boston, MA, USA (2005)

15. Liu, S., et al.: A dual-band shared aperture antenna array in ku/Ka-bands for beam scanning applications. IEEE Access **7**, 78794–78802 (2019)
16. Cheng, Y.F., Ding, X., Gao, G.F., Shao, W., Liao, C.: Analysis and design of wide-scan phased array using polarization-conversion isolators. IEEE Antennas Wireless Propag. Lett. **18**(3), 512–516 (2019)
17. Wu, K.-L., Wei, C., Mei, X., Zhang, Z.-Y.: Array-antenna decoupling surface. IEEE Trans. Antennas Propag. **65**(12), 6728–6738 (2017)
18. Lee, J.-Y., Choi, J., Choi, D., Youn, Y., Hong, W.: Broadband and wide-angle scanning capability in low-coupled mm-wave phased-arrays incorporating ILA with HIS fabricated on FR-4 PCB. IEEE Trans. Veh. Technol. **70**(3), 2076–2088 (2021)
19. Arand, B.A., Bazrkar, A., Zahedi, A.: Design of a phased array in triangular grid with an efficient matching network and reduced mutual coupling for wide-angle scanning. IEEE Trans. Antennas Propag. **65**(6), 2983–2991 (2017)
20. Zou, W.-M., Qu, S.-W., Yang, S.: Wideband wide-scanning phased array in triangular lattice with electromagnetic bandgap structures. IEEE Antennas Wireless Propag. Lett. **18**(3), 422–426 (2019)
21. Balanis, C.A.: Antenna Theory, Analysis and Design, 2nd edn. Wiley, New York, NY, USA (1997)
22. Ding, X., Cheng, Y.-F., Shao, W., Wang, B.-Z.: A wide-angle scanning phased array with microstrip patch mode reconfiguration technique. IEEE Trans. Antennas Propag. **65**(9), 4548–4555 (2017)
23. Yang, G., Li, J., Zhou, S.-G., Qi, Y.: A wide-angle E-plane scanning linear array antenna with wide beam elements. IEEE Antennas Wireless Propag. Lett. **16**, 2923–2926 (2017)
24. Wang, R., Wang, B.-Z., Ding, X., Yang, X.-S.: Planar phased array with wide-angle scanning performance based on image theory. IEEE Trans. Antennas Propag. **63**(9), 3908–3917 (2015)
25. Cheng, Y.-F., Ding, X., Shao, W., Yu, M.-X., Wang, B.-Z.: 2-D planar wide-angle scanning-phased array based on wide-beam elements. IEEE Antennas Wireless Propag. Lett. **16**, 876–879 (2017)
26. Jin, F.-L., Ding, X., Cheng, Y.-F., Wang, B.-Z., Shao, W.: Impedance matching design of a low-profile wide-angle scanning phased array. IEEE Trans. Antennas Propag.**67**(10), 6401–6409 (2019)
27. Su, Z.-L., Leung, K.W., Lu, K.: A shaped-beam antenna for wide-angle scanning phased array. IEEE Trans. Antennas Propag. **70**(9), 7659-7669 (2022)
28. Balanis, C.A.: Antenna Theory. Wiley, New York, NY, USA (1982)
29. Li Z., et al.: IEEE Trans. Antennas Propag. **70**(1), 240–249 (2022)
30. Jackson, D.R., Caloz, C., Itoh, T.: Leaky-wave antennas. Proc. IEEE **100**(7), 2194–2206 (2012)
31. Guan, D.-F., Zhang, Q., You, P., Yang, Z.-B., Zhou, Y., Yong, S.-W.: Scanning rate enhancement of leaky-wave antennas using slow-wave substrate integrated waveguide structure. IEEE Trans. Antennas Propag. **66**(7), 3747–3751 (2018)
32. Chen, S.L., Karmokar, D.K., Li, Z., Qin, P.Y., Ziolkowski, R.W., Guo, Y.J.: Circular-polarized substrate-integrated-waveguide leaky-wave antenna with wide-angle and consistent-gain continuous beam scanning. IEEE Trans. Antennas Propag. **67**(7), 4418–4428 (2019)
33. Zhang, P.F., Zhu, L., Sun, S.: Microstrip-line EH1/EH2-mode leaky-wave antennas with backward-to-forward scanning. IEEE Antennas Wireless Propag. Lett. **19**(12), 2363–2367 (2020)
34. Comite, D., et al.: Design of a polarization-diverse planar leaky-wave antenna for broadside radiation. IEEE Access **7**, 28672–28683 (2019)
35. Li, Z., et al.: Investigation of leaky-wave antenna with stable wide beam-scanning characteristic. IEEE Trans. Antennas Propag. **70**(1), 240–249 (2022)
36. Rudramuni, K., et al.: Goubau-line leaky-wave antenna for wide angle beam scanning from backfire to Endfire. IEEE Antennas Wireless Propag. Lett. **17**(8), 1571–1574 (2018)
37. Yin, J.Y., et al.: Frequency-controlled broad-angle beam scanning of patch array fed by spoof surface plasmon polaritons. IEEE Trans. Antennas Propag. **64**(12), 5181–5189 (2016)

38. Zu, H.-R., Wu, B., Su, T.: Beam manipulation of antenna with large frequency-scanning angle based on field confinement of spoof surface plasmon Polaritons. IEEE Trans. Antennas Propag. **70**(4), 3022–3027 (2022)
39. Wang, H., Sun, S., Xue, X., Zhang, P., Xu, B.: A periodic coplanar strips leaky-wave antenna with horizontal wide-angle beam scanning and stable radiation. IEEE Trans. Antennas Propag. **70**(10), 9861–9866 (2022)
40. Milne, R.: Dipole array lens antenna. IEEE Trans. Antennas Propag. **AP-30**(4), 704–712 (1982)
41. Li, T., Chen, Z.N.: Compact wideband wide-angle polarization-free metasurface lens antenna array for multibeam base stations. IEEE Trans. Antennas Propag. **68**(3), 1378–1388 (2019)
42. Lima, E.B., Matos, S.A., Costa, J.R., Fernandes, C.A., Fonseca, N.J.G.: Circular polarization wide-angle beam steering at Ka-band by in-plane translation of a plate lens antenna. IEEE Trans. Antennas Propag. **63**(12), 5443–5455 (2015)
43. Nguyen, N.T., Boriskin, A.V., Le Coq, L., Sauleau, R.: Improvement of the scanning performance of the extended hemispherical integrated lens antenna using a double lens focusing system. IEEE Trans. Antennas Propag. **64**(8), 3698–3702 (2016)
44. van der Vorst, M.J.M., de Maagt, P.J.I., Herben, M.H.A.J.: Scan optimized integrated lens antennas. In: Proceedings of 27th European Microwave Conference, Jerusalem, Israel, vol. 1, pp. 605–610 (1997)
45. Karttunen, A., Ala-Laurinaho, J., Sauleau, R., R¨ais¨anen, A.V.: Extended hemispherical integrated lens antenna with feeds on a spherical surface. In: Proceedings of 7th European Conference on Antennas Propagation, pp. 2539–2543 (2013)
46. Karmakar, R., Chattopadhyay, S., Chakraborty, S.: Impact of IEEE 802.11n/ac phy/mac high throughput enhancements on transport and application protocols—a survey. IEEE Commun Surv Tutorials **19**(4), 2050–2091 (2017)
47. He, Y., Tian, W., Zhang, L.: A novel dual-broadband dual-polarized electrical down tilt base station antenna for 2G/3G applications. IEEE Access **5**, 15241–15249 (2017)
48. Izzat, N.M.K.M., Zimmerman, M.L., Linehan, K.E.: Antenna, base station and power coupler. U.S. Patent 6,922,169 B2, July 26, 2005
49. Luk, K.-M., Chen, Z.N.: Antennas for base stations in Wireless Communications, 3rd edn.. McGraw Hill (2009)
50. Zhang, H., Yin, Y.-Z.: Single-layer single-feed wideband omnidirectional microstrip antenna with rotating square patches. Progress Electromagnet. Res. Lett. **93**, 27–34 (2020)
51. Roy, S., Baishnab, K.L., Chakraborty, U.: Beam focusing compact wideband antenna loaded with mu-negative metamaterial for wireless lan application. Progress Electromagnet. Res. Lett. **83**, 33–44 (2018)
52. Ghaemi, K., Behdad, N.: A low-profile, wideband antenna with vertically polarized directional radiation. IEEE Antennas Wirel. Propag. Lett. **15**, 1093–1096 (2016)
53. Martinis, M., Bernard, L., Mahdjoubi, K., Sauleau, R., Collardey, S.: Wideband antenna in cavity based on metasurfaces. IEEE Antennas Wirel. Propag. Lett. **15**, 1053–1056 (2016)
54. Sanchez Hernandez, D.: Multiband Integrated Antennas for 4G Terminals. Artech House Inc (2008)
55. Tayli, D., Gustafsson, M.: Physical bounds for antennas above a ground plane. IEEE Antennas Wirel. Propag. Lett. **15**, 1281–1284 (2016)
56. Garg, R., Bhartia, P., Bahl, I., Ittipiboon, A.: Microstrip Antenna Design Handbook, 1st edn. Artech House Inc., USA (2001)
57. Ludvig-Osipov, A., Hannula, J.-M., Naccachian, P., Jonsson, B.L.G.: Physical limitations of phased array antennas. IEEE Trans. Antennas Propag. **69**(9), 5512–5523 (2021)
58. Swapna, S.P., Karthikeya, G.S., Koul, S.K., Basu, A.: Wideband directional cavity antenna with low scanning loss for WLAN. Progress Electromagnet. Res. C **118**, 231–245 (2022)
59. Ghasemi, A., Laurin, J.-J.: Beam steering in narrow-wall slotted ridge waveguide antenna using a rotating dielectric slab. IEEE Antennas Wirel. Propag. Lett. **17**(10), 1773–1777 (2018)
60. Afzal, M.U., Esselle, K.P.: Steering the beam of medium-to-high gain antennas using near field phase transformation. IEEE Trans. Antennas Propag. **65**(4), 1680–1690 (2017)

61. Ghasemi, A., Laurin, J.-J.: A continuous beam steering slotted waveguide antenna using rotating dielectric slabs. IEEE Trans. Antennas Propag. **67**(10), 6362–6370 (2019)
62. Yao, Y.-L., Zhang, F.-S., Zhang, F.: A new approach to design circularly polarized beam-steering antenna arrays without phase shift circuits. IEEE Trans. Antennas Propag. **66**(5), 2354–2364 (2018)
63. Sanchez-Olivares, P., Masa-Campos, J.L., Muriel-Barrado, A.T., Villena-Medina, R., Fernandez-Romero, G.M.: Mechanically reconfigurable linear array antenna fed by a tunable corporate waveguide network with tuning screws. IEEE Antennas Wirel. Propag. Lett. **17**(8), 1430–1434 (2018)

# Chapter 6
# Antenna Gain Enhancement Techniques

## 6.1 Introduction

This chapter discusses the concept of antenna gain, gain enhancement, and various gain enhancement techniques for WLAN applications. We often think about gain in terms of an amplifier whose gain is the ratio of power output to the power input. But an amplifier is connected to a power supply which uses the energy from the supply to amplify the signal input. However, the antenna is a passive device. So in an antenna, gain is about focusing the radiated power in the required directions and reducing the power radiated in the undesirable directions. To achieve this, various structures are introduced in the antenna design some of which will be discussed in this chapter.

### *6.1.1 Antenna Gain*

The gain of an antenna indicates how strongly an antenna radiates in any direction in free space. Antenna gain is usually given in dBi where i stands for isotropic. This means that the intensity of signal radiated by an antenna is compared to an isotropic antenna. Isotropic antenna is a theoretical antenna that radiates equally in all directions. The formula for calculating antenna gain is given below [1]:

$$\text{Gain} = 4\pi \frac{\text{radiation intensity}}{\text{total input (accepted) power}} \text{ (in linear scale)}, \tag{6.1}$$

$$\text{Gain (dBi)} = 10 \ \log \text{ (Gain) (in log scale)}. \tag{6.2}$$

 
S. K. Koul et al., *Antenna Systems for Modern Wireless Devices*, Signals and Communication Technology, https://doi.org/10.1007/978-981-97-3369-9_6

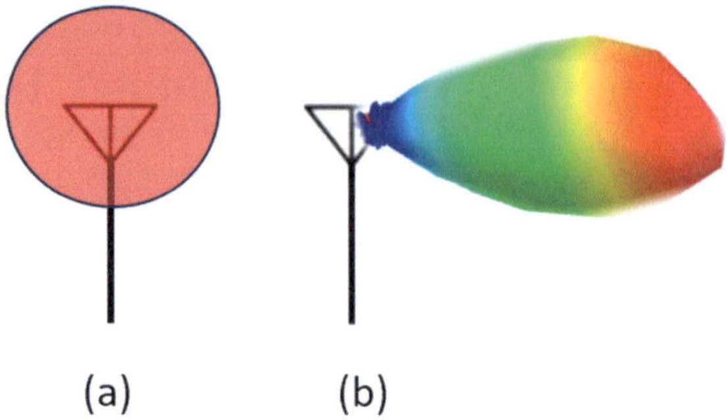


**Fig. 6.1** **a** Isotropic and **b** directional radiations

***Antenna Gain Enhancement***

The term antenna gain defines the degree to which an antenna concentrates radiated power in a given direction or absorbs incident power from that direction, compared with a reference antenna. Figure 6.1a shows an isotropic source which radiates in all directions equally. The 3D polar plot resembles an ideal sphere. On the other hand, Fig. 6.1b shows a source that has directional radiation. The purpose of gain enhancement techniques is to reduce the beamwidth of the radiation patterns in a specific radiation.

By improving the gain, the ability of antennas to communicate over large distances can be enhanced as well due to the FRIIS transmission link principle. High-gain antennas provide precise way of targeting radio signals and are therefore essential for long-range wireless networks. A common example is the microwave point-to-point link in cellular base stations. They even amplify weak signals used in satellite communication, as observed in commercial dish antennas used in the TV reception.

### *6.1.2 Methods to Enhance Antenna Gain*

Increasing the antenna aperture increases the antenna gain, but it is not necessarily a practically viable solution. So, instead of increasing the antenna size, we can increase the number of antenna elements to form an antenna array. Increasing the number of elements makes the radiation pattern to be more directive, which increases the antenna gain. Since electromagnetic waves have a phase in addition to an amplitude, they must be combined coherently (all at the same phase) or the sum of the signals will be much less than the maximum possible. Hence, the individual elements as well as the combination of the signals through the feed network are important.

In the recent years, researchers have come up with various methods to enhance the antenna gain without increasing the number of antenna elements. One such method is the use of metamaterial structures. Different types of metamaterial structures are used with the antenna to enhance the gain, some of which will be discussed in this chapter. A Fabry–Perot cavity antenna that uses a partially reflecting surface (PRS) can be used to enhance the gain through multiple reflections within the cavity. In addition to this, other methods like reflectors, directors, shorting pins, and substrate engineering to enhance the gain would be discussed in this chapter. High-gain planar antennas are usually preferred in modern communications systems, whether installed on terrestrial locations or on satellites. In modern wireless communication systems, the frequency of operation varies from microwave to millimeter-wave spectra such as 28, 38, and 60 GHz [2]. However, based on Friis's formula, these frequencies experience high free-space path loss, thereby providing short-range communications compared to the commercial 4G bands. To combat this free-space path loss, high-gain directional antennas are utilized at both ends of the communication system [3].

## 6.2 Gain Enhancement Using Metamaterials

Metamaterials (MTMs) are generally defined as artificially engineered materials with distinctive effective medium properties that are not naturally obtainable. Over the last two decades, these materials attracted considerable attention due to their extraordinary electromagnetic characteristics such as negative permittivity $\varepsilon$, permeability $\mu$, and index of refraction. These media are arranged as periodic unit cells whose average size is significantly smaller than the guided wavelength $\lambda g$, specifically around $\lambda g/4$. The first demonstration of the irregular electromagnetic response of the MTM, which exhibits simultaneously negative $\varepsilon$ and $\mu$, was performed in the theoretical study by Veselago in 1968. After almost three decades, Pendry et al. proposed the effective medium theory to enable the construction of such materials [4]. In 2000, the first MTM structure consisting of split-ring resonators (SRR) and metallic cut wires was experimentally verified by Smith et al. [5].

### *6.2.1 Metamaterial Classification*

The two fundamental quantities, the electric $\varepsilon$ and magnetic $\mu$, also known as the constitutive parameters, are used to describe the electromagnetic wave nature and its behavior in a medium. These parameters and the boundary conditions demonstrate the response of the medium to an incoming electromagnetic wave. The electromagnetic characteristics of MTMs can be expressed by the Lorentz–Drude model with two frequency-dependent macroscopic parameters, $\varepsilon_r(\omega)$ and $\mu_r(\omega)$, as follows [6]:

$$\varepsilon_{\mathrm{r}}(\omega) = 1 - \frac{\omega_{\mathrm{pe}}^2}{\omega^2 - \omega_{0,\mathrm{e}}^2 + i\omega\gamma_{\mathrm{e}}}, \tag{6.3}$$

$$\mu_{\mathrm{r}}(\omega) = 1 - \frac{\omega_{\mathrm{pm}}^2}{\omega^2 - \omega_{0,\mathrm{m}}^2 + i\omega\gamma_{\mathrm{m}}}, \tag{6.4}$$

where $\omega_{0,e}$ and $\omega_{0,m}$ are the transverse resonance frequency, $\omega_{pe}(\omega_{pm})$ is the electric (magnetic) coupling strength, and $\gamma_e$ and $\gamma_m$ are the damping factors (material losses).

The classification of MTMs relies on the signs of the ε and μ of a homogeneous material. The material properties can be characterized over a wide frequency range using proper values in (6.3) and (6.4). According to Veselago, when both $\varepsilon$ and $\mu$ are simultaneously negative, some unusual phenomena occur, such as reversed Snell's law and Doppler shift [7]. The material classifications based on $\varepsilon$ and $\mu$ with the electromagnetic wave interacting with four possible materials are depicted in Fig. 6.2. The materials with both $\varepsilon$ and $\mu$ which are positive are known as double-positive (DPS) materials, which include the majority of conventional dielectric materials. This type is shown in the top-right quarter of Fig. 6.2. There are three classes of MTMs depending on the signs of their $\varepsilon$ and $\mu$: ENG, MNG, and DNG.

The origin of Fig. 6.2 is classified as zero-index material (ZIM), where $\varepsilon$ and $\mu$ are at zero. The materials with $\varepsilon < 0$ and $\mu < 0$ are called left-handed materials (LHMs) or MTM. LHMs support backward-wave propagation, where the phase velocity and group velocity are antiparallel. The electromagnetic wave interacts with the four materials, as shown in Fig. 6.2. The principle of Snell's law is used to explain the

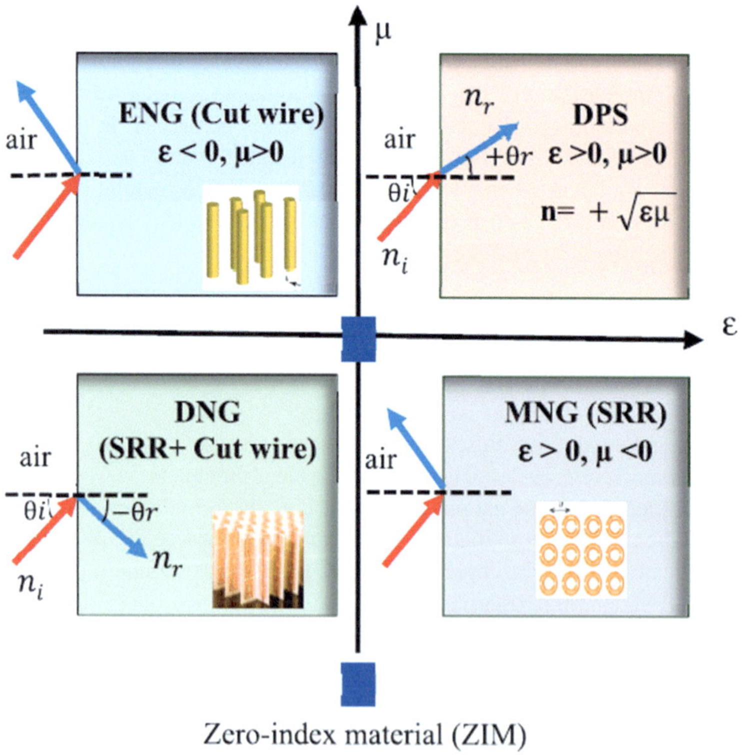


**Fig. 6.2** Classifications of the materials based on permittivity and permeability with the electromagnetic wave interacting with four possible materials [7]

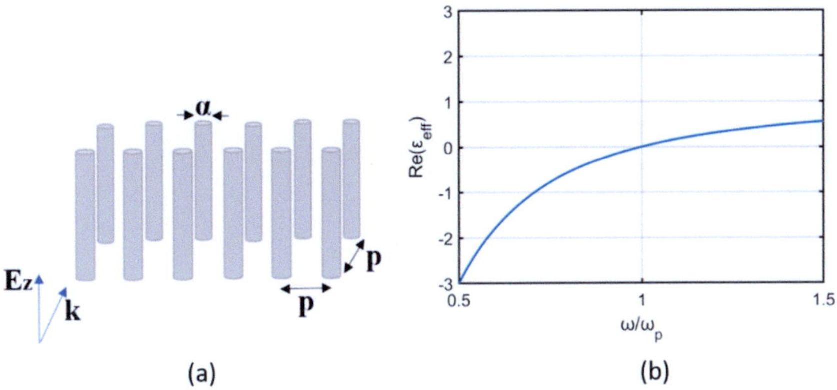


**Fig. 6.3** Periodic arrangement of thin conductive wires for the realization of negative permittivity. Real part of the effective permittivity of periodically arranged infinitely long thin wires [8]

propagation in materials, which is expressed as follows.

$$\frac{\sin\theta_{\mathrm{i}}}{\sin\theta_{\mathrm{r}}}=\frac{n_{\mathrm{r}}}{n_{\mathrm{i}}}, \tag{6.5}$$

where $n_{\mathrm{i}}(\theta_{\mathrm{i}})$ and $n_{\mathrm{r}}(\theta_{\mathrm{r}})$ are the refractive indices (angles) of the incident and refracted rays. A positive $\theta_{\mathrm{r}}$ can be obtained when both $n_{\mathrm{i}}$ and $n_{\mathrm{r}}$ are positive. Thus, the electromagnetic wave is normally refracted for double-positive (DPS) materials [top-right quarter of Fig. 6.2]. On the other hand, $\theta_{\mathrm{r}}$ is indeed negative when $n_{\mathrm{i}}$ and $n_{\mathrm{r}}$ are positive and negative, respectively, as depicted in the bottom-left quarter of Fig. 6.2. In other words, the refracted wave bends negatively at the interface. The electromagnetic wave does not propagate in the MNG and ENG materials. The epsilon negative (ENG) material is presented in the top-left quarter, which depicts a negative $\varepsilon$ at some frequency regimes and positive $\mu$ at all frequencies. This type of material can be achieved by periodic wire structures [8] as illustrated in Fig. 6.3.

The mu-negative (MNG) material is presented in the bottom-right quarter. This material is realized by means of a periodic structure of split-ring resonators (SRRs) [5] as shown in Fig. 6.4.

The bottom-left quarter is the most interesting one, in which $\varepsilon$ and $\mu$ are simultaneously negative, known as double-negative materials (DNG) as illustrated in Fig. 6.5. Such materials have to be designed and not found exclusively in the nature.

### *6.2.2 Metamaterial Integration for Gain Enhancement*

Gain is proportional to the aperture of the antenna [1]; therefore, high-profile antennas and array antennas are the most common methods to obtain higher gain. On the other hand, compact antennas are in high demand due to their small physical size. However,

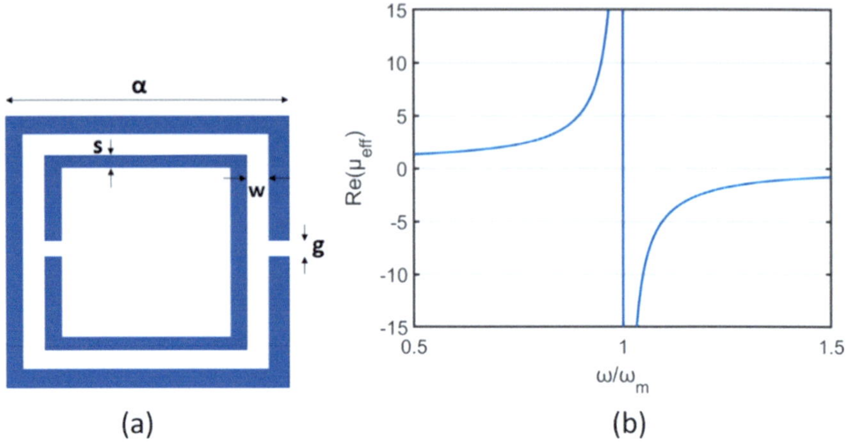


**Fig. 6.4** **a** Geometry of a square split-ring resonator unit cell. Blue parts depict metallization. **b** Real part of the SRR's effective permeability [8]

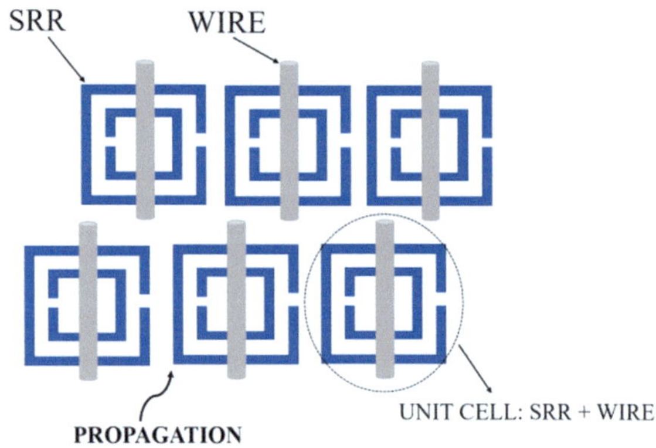


**Fig. 6.5** DNG medium consisting of unit cells that are synthesized by wires (ENG) placed in front of SRRs (MNG) [8]

gain of such antennas prohibits their use in many applications where high gain is needed. In this regard, MTMs are strong candidates to realize high-gain compact antennas with required characteristics. MTMs have been proposed during the last two decades as a low-cost approach for gain improvement without a significant increase of the antenna profile. These artificial materials are placed close to the radiating elements or etched in the same substrate to manipulate the radiation pattern. Gain enhancement mechanisms based on different metamaterials are illustrated in Fig. 6.6 and a detailed description is given below.

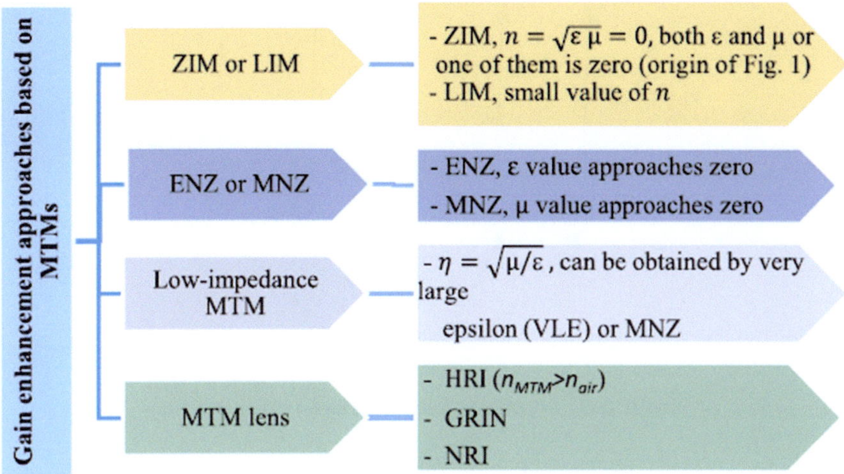


**Fig. 6.6** Antenna gain enhancement approaches based on MTMs [7]

#### 6.2.2.1 Zero-Index Materials (ZIMs) and Low-Index Materials (LIMs) (or Near Zero-Index Materials (NZIMs))

Pendry et al. [9] showed both theoretically and experimentally that a plasma frequency characterizes the periodic thin wires, $\varepsilon = 1 - \omega_p/\omega$, where $\omega_p(\omega)$ is the plasma frequency (operating frequency). The refractive index, $n = \sqrt{\varepsilon\mu}$, approaches zero when the operating frequency approaches the plasma frequency. The phase variation of the electromagnetic waves entering these types of structures is extremely small. This feature can be exploited in various applications such as wave tunneling through small cross sections with irregular shapes, cloaking, and lenses. The phase of the wave "freezes" inside a zero-index material suggesting that the phase front of the wave exiting the structure always conforms with the material interface. The gain enhancement capitalizes on Snell's law of refraction. Here, when the rays of incident angle $\theta_i$ travel from zero-index medium ($n_i = 0$) to air with a high index of refraction ($n_i = 1$), the refracted rays will spread in a direction normal to the interface [10]. Therefore, the phase change in the radiated wave front of the electromagnetic wave approaches zero or equals zero, thereby inducing a gain enhancement. This is illustrated in Fig. 6.7, in which the electromagnetic rays pass from ZIM to air in a direction very close to the normal. These characteristics (ZIMs, LIMs, or NZIMs) can be translated into a gain improvement when embedded within antennas.

A ZIM that enhances the antenna gain is presented in [11]. The anisotropic ZIM lens is placed in front of the Vivaldi antenna as shown in Fig. 6.8a. Different from isotropic ZIM, the impedance of anisotropic ZIM along the wave-propagation direction can be designed to match with that of background. Hence, high antenna efficiency can be achieved using the anisotropic ZIM. In the presented anisotropic ZIM lens, the $x$-component of the permittivity either approaches to zero ($\varepsilon_x = 10^{-4}$) or is small

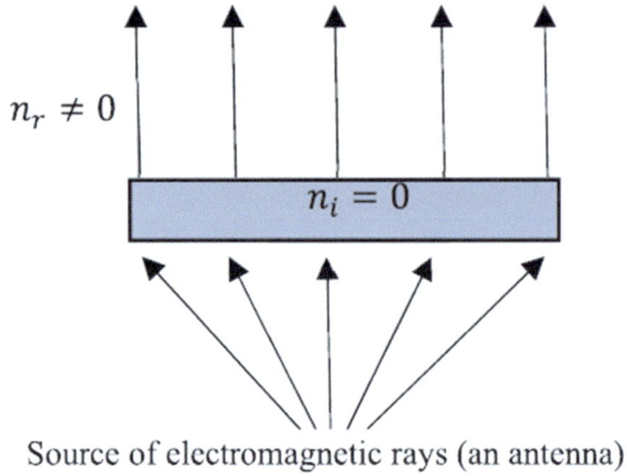


**Fig. 6.7** Control system for propagation with ZIM layer

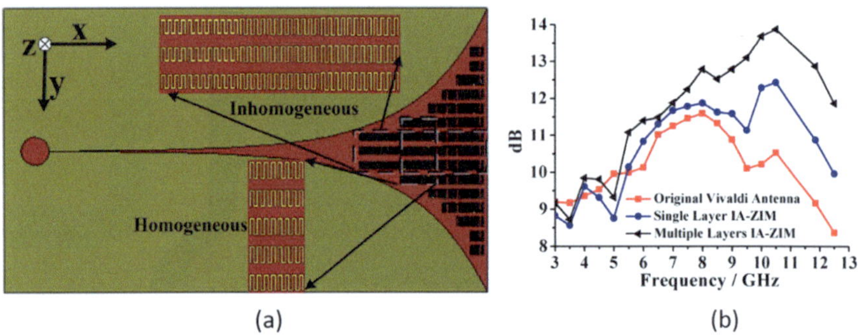


**Fig. 6.8** Designed and fabricated Vivaldi antennas with (**a**) inhomogeneous anisotropic ZIM, (**b**) gain enhancement due to ZIM [11]

($\varepsilon_x = 0.2$), while the $y$-component is equal to 1 ($\varepsilon_y = 1$). The gain enhancement of the antenna is shown in Fig. 6.8b. In the region of approximate anisotropic ZIM from 9.5 to 12.5 GHz, the gain of the Vivaldi antenna increases by around 2 dB for a single layer and 4 dB for multiple layers.

#### 6.2.2.2 Epsilon-Near-Zero/Mu-Near-Zero (ENZ/MNZ)

The refractive index $n$ is computed based on $n = \sqrt{\varepsilon\mu}$; if the MTM structure has zero or low $\varepsilon$ or $\mu$, the resultant refractive index also has zero or low value. This makes that the phase velocity of propagating electromagnetic waves tends to infinity and the resulting wavelength also tends to infinity. This property has numerous applications in antennas. Conventional antennas like dipole antenna have a wide beamwidth with a spherical wave front. An ENZ slab with an appropriate shape can be used to convert these spherical wave fronts to nearly planar wave fronts. Thus, the resulting radiation pattern and the mechanism of gain enhancement are based on Snell's law of refraction, as explained in the previous section.

Researchers in [12] presented an ENZ metamaterial structure to enhance the gain of a Vivaldi antenna. Two configurations as shown in Fig. 6.9a, b are reported. It is seen in Fig. 6.9c that both configurations showed gain enhancements with higher performance for the second configuration that shows an improvement in antenna gain around 5 dB in the operating frequency range.

High-gain endfire antenna with a wide 1-dB gain bandwidth using an array of Mu-near-zero (MNZ) unit cells is presented in [13]. The antenna consists of rounded bow–tie radiators, and to increase the antenna gain, seven slabs are integrated vertically in front of the bow–tie antenna as shown in Fig. 6.10. Each slab consists of a 2 × 7 array of double-sided split-ring resonators (SRRs). The proposed SRRs act as MNZ media, which is analogous to a metalens. Integrating an array of MNZ unit cells in front of the bow–tie antenna provides the measured antenna gain of 10.25–13.85 dBi over 5.65–7.85 GHz. There is a 5–6.5 dBi gain enhancement compared with the conventional bow–tie antenna with a protrusion dimension of $0.6\lambda_0$ in the endfire direction at 6.5 GHz.

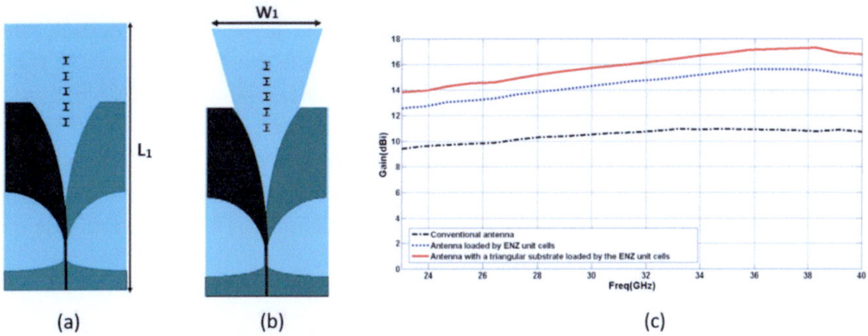


**Fig. 6.9** **a** Geometry of a Vivaldi antenna with **a** single-layer ENZ metamaterial, **b** multilayer ENZ metamaterial, **c** antenna gain enhancement [12]

**Fig. 6.10** Configuration of rounded bow–tie antenna along with MNZ media

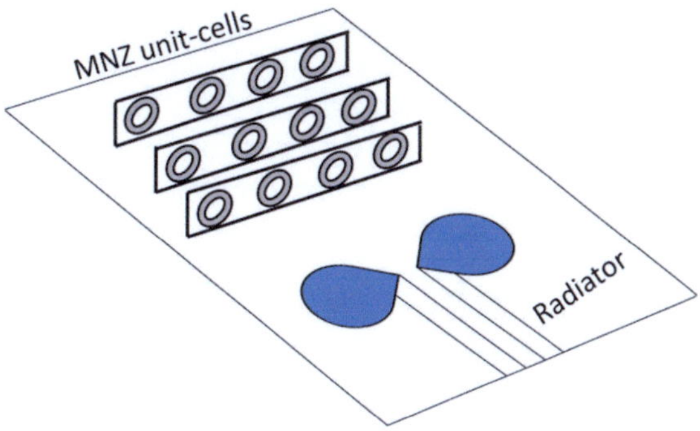

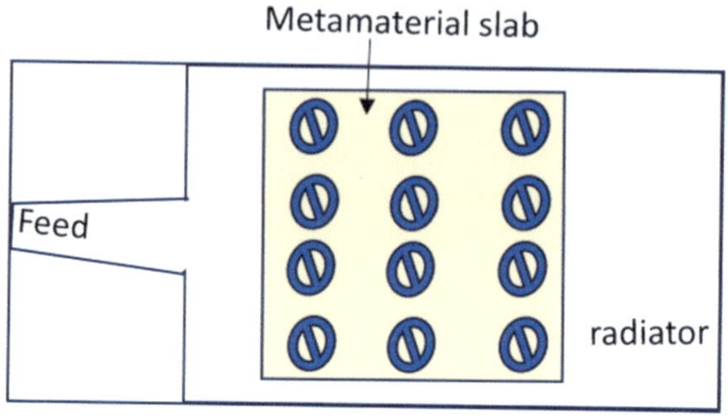


**Fig. 6.11** SIW-SA antenna [14]

#### 6.2.2.3 Low-Impedance Metamaterial

MTM structures with the low-impedance property can enhance the gain of planar antennas. MTMs are periodic metallic structures responsible for inducing electric and/or magnetic dipoles by an incident electromagnetic wave. These dipoles alter the effective characteristics of the material. The intrinsic impedance ($\eta$) of an MTM is given by $\sqrt{\mu/\varepsilon}$. Thus, Very Large Epsilon (VLE) or MNZ exhibits a low-impedance property. This allows for reducing the overall size of the antenna system by diminishing the air gap between the antenna and the MTM layer. In these designs, the real part of $\varepsilon$ should be sufficiently large to satisfy the condition of $\varepsilon >> 1$ for a VLE material or the $\mu$ should be extremely small for realizing the MNZ material, which finally yields a low-impedance material with $\eta << 1$. The cross-circular loop resonator (CCLR) MTM structure was proposed by Pandit et al. in [14] to achieve the low-impedance property. An array of low-impedance CCLR unit cells is embedded above the substrate-integrated waveguide fed-slot antenna (SIW-SA) to increase the gain by 5.8 dB at 9.73 GHz, as illustrated in Fig. 6.11.

#### 6.2.2.4 Lensing Property of Metamaterials

Lens antennas have gained considerable attention due to their advantages in providing wideband and high-gain performance through focusing the radiation into a narrow beamwidth in the required direction. The traditional dielectric lenses suffer from the high-profile structures as well as complex and expensive fabrication process. As an alternative, artificial materials like MTMs are used as lenses, which are integrated with antennas to convert the spherical waves into the planar ones. Various MTM structures act as metalens organized in an array form of single or multilayer, in which they can be etched on the substrate or placed in front/above the antenna's radiating part. The MTM lens is designed using a planar structure, which leads to a considerable enhancement of the antenna gain based on the different characteristics of the refractive index. The unique properties of the MTMs can manipulate the propagation of electromagnetic waves resulting in the freedom to design different

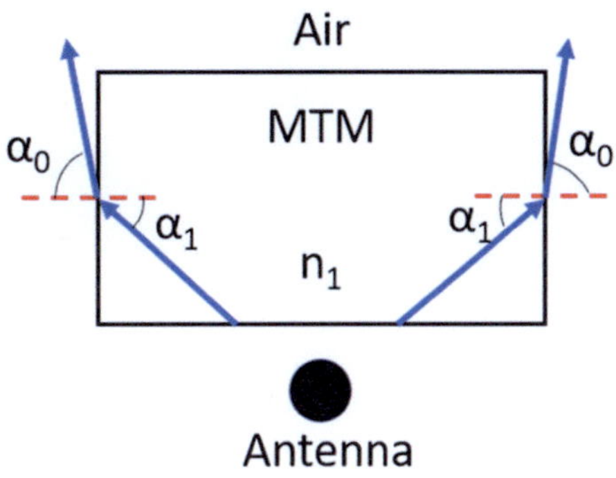


**Fig. 6.12** Mechanism of gain enhancement based on HRI approach

types of lenses such as the high refractive index (HRI) lens ($n_{MTM} > n_{air}$), GRIN, and NRI.

***MTM Lens Based on HRI***

MTMs with a refractive index higher than that of the air ($n_{MTM} > n_{air}$) act as metalens to focus the radiation on the direction of emission where the beam can aggregate in the endfire direction. The principle of refractive index law can explain the mechanism of gain enhancement. Figure 6.12 shows the wave propagation in both mediums: MTM and air. From the Snell's law, when the incident angle $\alpha_1$ is fixed, the high value of the refractive index leads to a larger refractive angle $\alpha_0$. To achieve such a property, the refractive index of the proposed MTM layer should be larger than 1 ($n > 1$) at the desired frequency. Thus, the energy passing through the MTM structure is converged and focused in the direction of emission. The *E*-field is utilized to show the lensing property where MTM slabs convert the spherical waves into the planar waves and reduce the half-power beamwidth (HPBW).

Researchers in [15] introduced parallel-line unit cells embedded into the aperture of the antipodal tapered slot antenna (ATSA), as illustrated in Fig. 6.13. These unit cells act as a conventional lens for beam focusing where the gain was improved by 1.3–3.6 dB over the range of 6–19 GHz. In Fig. 6.13 the refractive index of the considered unit cell is larger than 1 ($n > 1$) below 19 GHz. Thus, the energy transferred through the MTM structure is converged, thereby improving the antenna gain.

GRIN lenses have attracted considerable attention owing to their ability to focus the radiation. These lenses have been proposed as alternatives to the established dielectric ones, where the refractive index is different throughout the lens rather than being dependent on the interfaces of the dielectric material to manipulate and shape the radiation beam. Among the most known lenses are Luneburg and Maxwell fisheye [16]. The gradient refractive index distribution of the former is expressed by $n_r = \sqrt{2 - \left(\frac{r}{R}\right)^2}$, whereas the distribution of the latter is given by $n_r = \frac{n_0}{1+\left(\frac{r}{R}\right)^2}$. In both formulas, *R* and *r* are the lens radius and the distance from any point to the lens center, respectively. Both lenses have the capability to focus a point on their surfaces

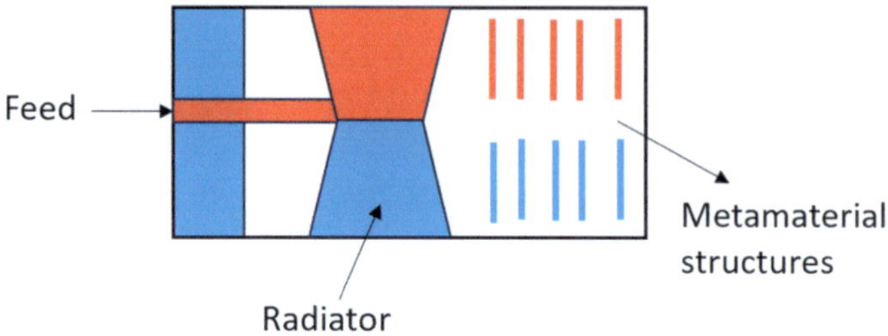


**Fig. 6.13** ATSA loaded with metamaterial gradient index (GRIN) lens based on MTMs

to the diametrically opposite side. Achieving GRIN with the proper refractive index distribution is the main concern as they are artificial materials and need different ways to be synthesized. Alternatively, the distribution of the refractive index can be realized based on MTMs, where it is achieved by geometric tailoring of the neighboring unit cells [17]. The homogeneous MTMs can be achieved by identical unit cells arranged in an array form, which are used in a wide range of applications. On the other hand, if the unit cells are not identical, the inhomogeneous medium, here the GRIN, can be realized. The inherently non-resonant, wide bandwidth, and low-loss behavior of GRIN-based MTMs enable many practical applications, such as the realization of the various lenses to enhance the performance of antennas in terms of the high gain and directivity, efficiency, beam deflection, low sidelobe, and broad bandwidth. The idea of utilizing GRIN to manipulate the propagation of electromagnetic waves was introduced by Smith et al. in [18], where the SRR was used to tilt the microwave beam. This medium acts as metalens to transform the spherical waves to the planar ones, thereby enhancing the antenna gain and directivity.

In [19], a gradient refractive index (GRIN) metamaterial lens is designed and integrated with monopole antenna shown in Fig. 6.14a. The lens is designed by using non-resonant parallel-line metamaterial. This non-resonant structure covers UWB frequency range, which has the capability to enhance the gain of the proposed antenna. The peak gain of the basic monopole is increased approximately by five times (5.3 dB) at 8.8 GHz for GRIN lens. The effective refractive index value becomes low by decreasing the length of the parallel metallic lines. The effective impedance values are near unity when the values of metallic lines are becoming low. The schematic design of the GRIN lens by using parallel-line metamaterial is shown in Fig. 6.14b. The refractive index distribution in GRIN lens is very important to create a minimum reflection. According to the refractive index values, the arrangement of the metamaterial unit cell in the GRIN lens along the $x$- and $y$-directions is shown in Fig. 6.14c, d, respectively. The GRIN metamaterial lens is placed above the basic monopole to amplify the EM wave in the direction of the $x$-axis.

***MTM Lens Based on Negative Refractive Index (NRI)***

The NRI is another type of beam focusing lens, in which these materials act as a convergent planar lens used for focusing electromagnetic waves, thereby enhancing

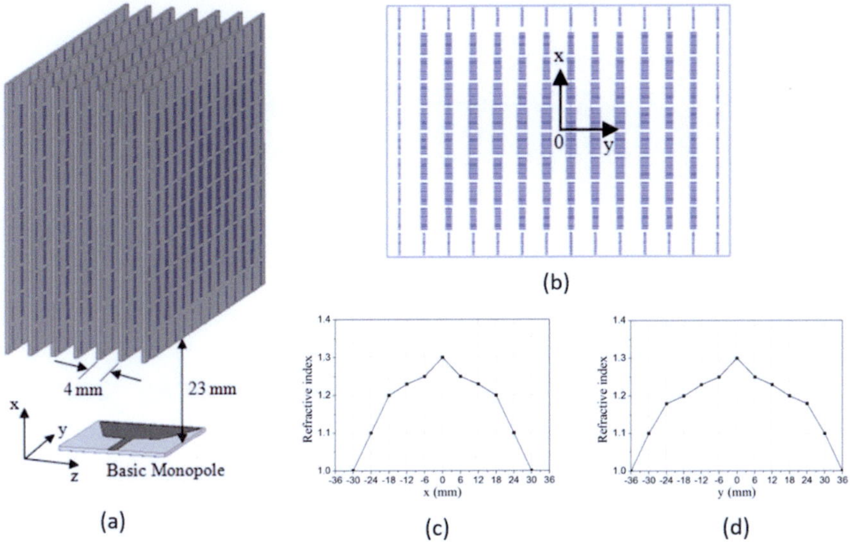


**Fig. 6.14** **a** Arrangement of the GRIN with basic monopole, **b** view of GRIN lens arrangement, **c** refractive index distribution along $x$-direction, **d** refractive index distribution along $y$-direction [19]

the gain performance of the antennas. In [20] a Vivaldi antenna along with negative refractive index metamaterial is used to enhance the antenna gain. From Fig. 6.15a, the real part of the refractive index is seen to be negative. Single-layer and multiple layers' NRI MTMs were assembled into the Vivaldi antenna as shown in Fig. 6.15c. The gain of the single-layered MTM antenna increases approximately by 2.5 dB, while the multiple layered antenna has the gain increase by 4 dB.

## 6.3 Gain Enhancement Using Artificial Surfaces

In recent years, enhancing antenna features using artificial materials/surfaces like various resonators, artificial magnetic conductors (AMCs), electromagnetic bandgap (EBG) structures, and frequency-selective surfaces (FSSs) has gained significant attention. Antenna gain enhancement using such artificial surfaces is discussed below.

### *6.3.1 Resonator Structures*

The electric resonators, such as electric (ELC) resonators and complementary split-ring resonators (CSRRs); magnetic resonators, such as split-ring resonators (SRRs); or composite right/left-handed transmission lines (CRLH TLs) have found immense

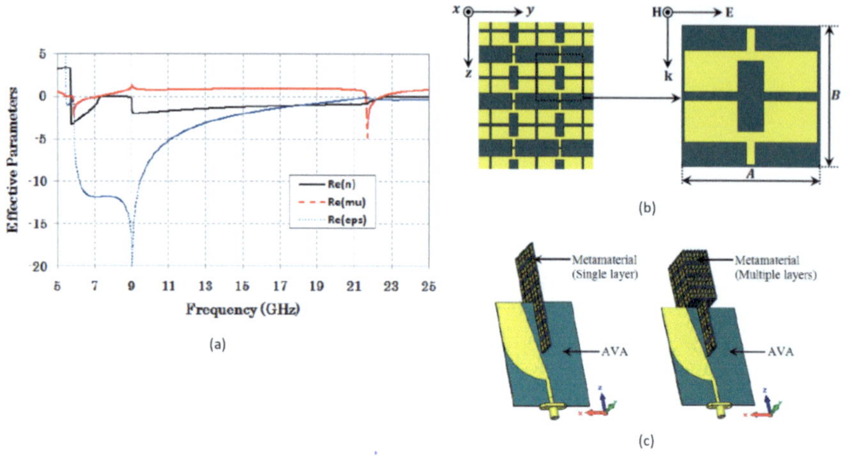


**Fig. 6.15** **a** Retrieved effective parameters, **b** NIM configuration and its unit cell, **c** AVA incorporated with NIM [20]

potential in enhancing antenna gain. The proposed antenna in [21] is composed of a bow–tie-shaped electric dipole, a loop antenna that works as a magnetic dipole, a microstrip-to-coplanar stripline transition balun, and the H-shaped resonator (HSR) structures that are used for gain enhancement, all of which are printed in the same plane perpendicular to the ground which is shown in Fig. 6.16.

The unloaded antenna has a scattered beam, which results in bad radiation patterns and low gains at high frequencies. So, a method of focusing the beam is needed to improve the performance of the antenna. Based on Snell's law $n_s \sin\theta_s = n_0 \sin\theta_0$ which is shown in Fig. 6.17, when the refractive index ($n_s$) of the substrate gets larger, with the same incident angle ($\theta_s$), the refracting angle $\theta_0$ gets larger, so the beam can aggregate in the endfire direction and a high gain in the endfire direction can be obtained. In order to obtain a substrate with a larger refractive index, a metamaterial that consists of the substrate given above and the HSR structure is employed here. Compared with the unloaded structure, a gain increment of about 0.5–6.0 dB over the whole band is obtained. Gain enhancement in [22] is achieved by loading with

**Fig. 6.16** Configuration of the proposed antenna

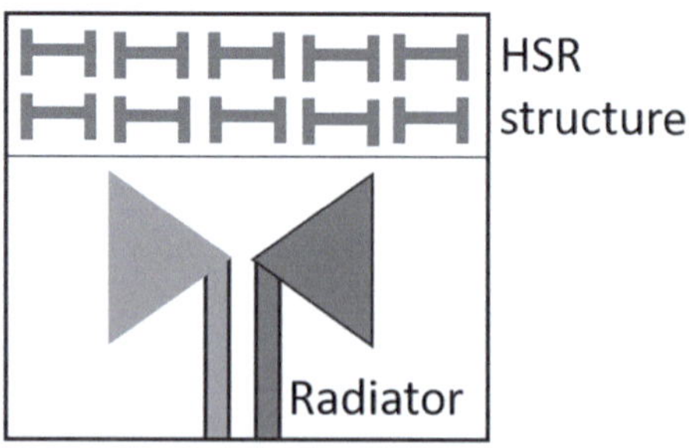

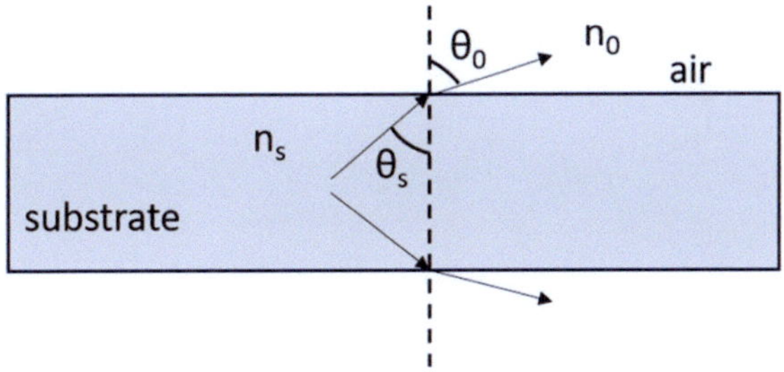


**Fig. 6.17** Electromagnetic wave transmitting sketch map

two rows of ISRs symmetrically in the endfire direction while maintaining the wide-band performance of the periodic endfire antenna. The ISR can create the strongest resonance when the *E*-field is aligned parallel with the *xy*-plane and the *H*-field is aligned with the center axis ($z$) of the ISR. The ISRs can be considered as directors for the antennas. The ISR-loaded antenna owns gain of about 6.5–9.5 dB in the whole working band (5.1–8.3 GHz), which is about 1.5–3.5 dB more than the unloaded case.

### 6.3.2 *Artificial Magnetic Conductors*

AMCs and HISs have also been used as the ground plane of the planar microstrip antennas to significantly enhance the antenna gain. The conventional ground plane made of a good electric conducting material is impenetrable and provides a fairly good shielding, and it is normally approximated by a perfect electric conductor (PEC). However, it introduces an image with a negative current to that of a source, which is placed horizontally above the ground plane. On the other hand, in order for the planar antennas to be low profile, they have to be placed horizontally and very close to the ground plane beneath them. Therefore, current cancelation of the source antenna with its image consequently results in a poor input impedance matching, which in turn prevents the antenna to radiate efficiently. On the other hand, AMCs and HISs not only maintain appropriate shielding for the electronics beneath the antennas and redirecting the wave radiation to the upper half-space, but they also can provide good input impedance matching. As it is known, the image current introduced by a perfect magnetic conductor (PMC) is in parallel and in phase with its source, which is located horizontally above it. This is used to improve the antenna input impedance characteristics, when AMCs or HISs are employed as the antenna ground plane. In [23] gain enhancement of a multiband AMC-based antenna is presented as shown in Fig. 6.18. The designed AMC unit cell consists of four metallic nested-rings with four lumped capacitors. It exhibits the characteristic of four zero phases of the reflection coefficient. The proposed AMC-backed antenna works well in the three frequency bands of 2.13–2.87, 3.22–4.75, and 5.54–5.86 GHz with high radiation gains. In comparison to the multiband antenna without AMC, the AMC-backed

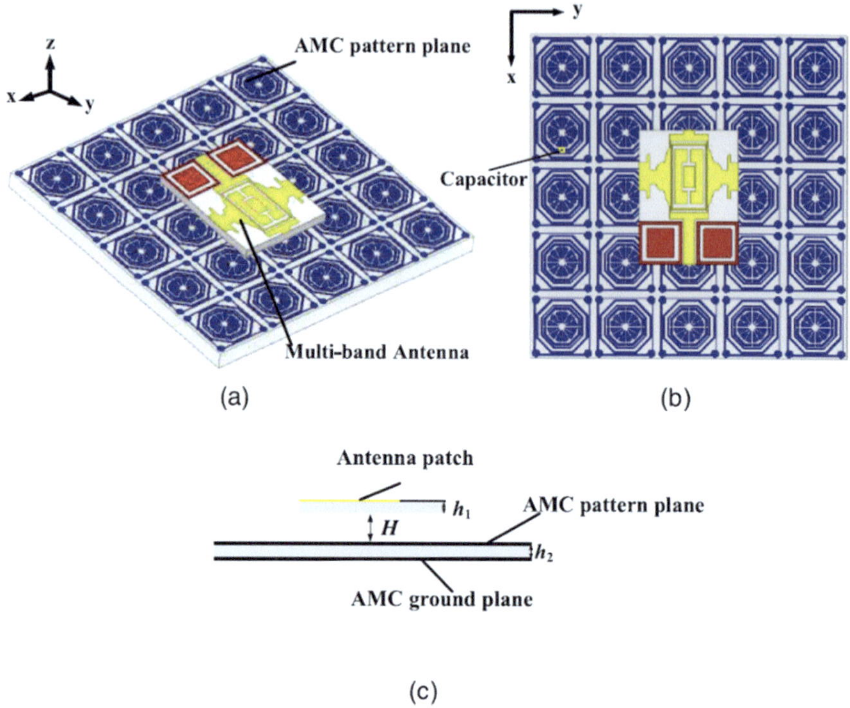


**Fig. 6.18** Topology of the antenna system: **a** 3D view, **b** top view, **c** side view [23]

antenna achieves a gain enhancement by the amounts of 4.93 dB, 5.92 dB, 5.54 dB, and 4.95 dB at the frequencies of 2.45 GHz, 3.5 GHz, 4.6 GHz, and 5.8 GHz, respectively. In addition to this, the proposed AMC-backed antenna system is low profile and suitable for the applications of WLAN, WiMAX, as well as 5G mobile communication system.

### *6.3.3 Electromagnetic Bandgap Structures*

Electromagnetic bandgap structures are defined as artificial periodic (or sometimes non-periodic) objects that prevent/assist the propagation of electromagnetic waves in a specified band of frequency for all incident angles and all polarization states. EBG structures are usually realized by periodic arrangement of dielectric materials and metallic conductors [24]. Electronic bandgap (EBG) structure techniques have been applied to improve the electromagnetic characteristics of antennas. Circular periodic EBG substrate to enhance the gain of a DRA as shown in Fig. 6.19 is reported in [25]. The EBG substrate is based on the mushroom-like structure that is constructed as a combination of two periodic structures. The first one is composed of metal rings and

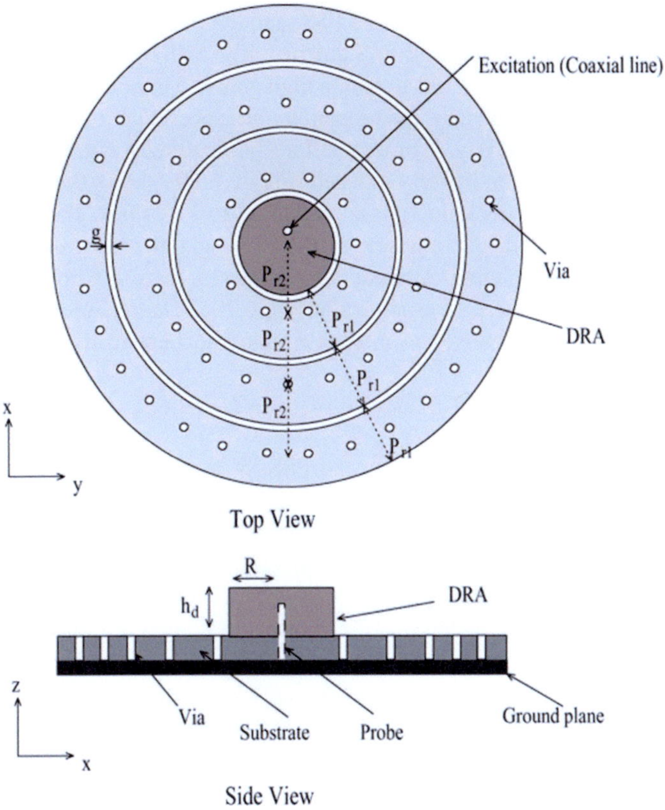


**Fig. 6.19** Dielectric resonator surrounded by a circular mushroom-like substrate [25]. Reprinted with permission from IEEE

the second one of vertical metal vias, which are placed so that they form radial and circular periodic structures. The periods of these two structures have been optimized and a gain enhancement of 3 dB can be achieved. In [26], a modified chessboard structure based on slotted electromagnetic bandgap (EBG) is proposed. The slots integrated on the EBG act as parasitic radiators and are of great effect to change radiation pattern. By combining the parasitic radiators with the chessboard reflector, the forward gain of the designed antenna improves by more than 2.5 dB within the working frequency band from 2.85 to 3.95 GHz.

### *6.3.4 Frequency-Selective Surface*

In this subsection, FSS-based structures to enhance antenna gain are discussed. There are two ways to introduce the FSS within the antenna structure. The FSS added above the antenna is called transmission-based FSS, and the FSS added below the antenna structure is called reflection-based FSS antenna. The FSS array reflector presented in [27] is placed under the antenna to obtain high gain and directional radiations over UWB frequencies. For the FSSs, the reflection phase is crucial for the improvement of antenna gain and directivity. In order to offer a phase reflection to the antennas, the phase of the FSS has to be linearly reduced. Figure 6.20a shows the proposed FSS that achieved a linearly declining phase from 3.05 to 11.9 GHz. The directivity of the proposed UWB antenna is improved by the FSS that reflects waves in phase with radiated waves by the antenna. As frequency increases, there is a linear decrease in the reflection phase of the FSS as shown in Fig. 6.20b. The reflected waves for the constructive interference above the antenna's radiating aperture are controlled by

$$\varphi_{\mathrm{FSS}} = -2\beta S = 2n\pi \quad n = \cdots - 2, -1, 0, 1, \ldots, \tag{6.6}$$

where $\varphi_{\mathrm{FSS}}$ is the reflection phase by the FSS, $S$ is the height between the FSS reflector and the radiator, and $\beta$ is the propagation wave constant in the open space. When $S$ is close to 10 mm ($\lambda/2$), where lambda ($\lambda$) is the wavelength equivalent to the frequency, the reflected plane $\varphi_{\mathrm{FSS}}$ turns to zero. The reflection phase of the FSS equals to zero at 5 GHz. A parametric study was conducted and it was decided that at a height of 10 mm, which is ($\lambda/4$), it provides a good response for the entire bandwidth.

The FSS reflector is printed on an FR4 single layer with dielectricconstant ($\varepsilon_r$) of 4.5, loss tangent (tan $\delta$) of 0.02, a thickness of 1.6 mm, and a dimension of 6 mm × 6 mm and consists of 100 periodic elements (10 × 10). The resonance of FSS structures can be characterized by size, geometry, and spacing between the elements. They can reflect, transmit, or absorb the electromagnetic radiation. The

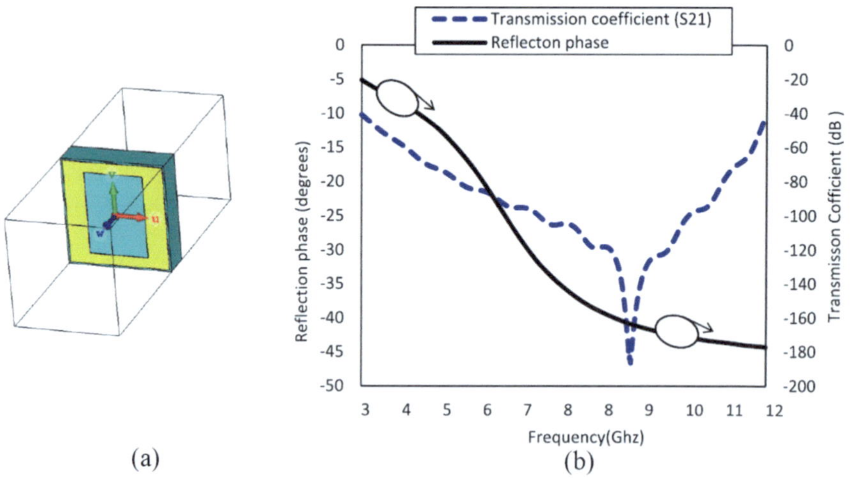


**Fig. 6.20** **a** Suggested unit cell, **b** predicted reflection phase and transmission coefficient [27]

gain enhancement ability of FSS is shown in Fig. 6.21 where the gain is enhanced from 1.65 to 7.87 dB at UWB frequencies. The compact and low profile of the whole antenna with FSS is beneficial for UWB application.

Second technique to enhance the antenna gain by ensuring reduction in the antenna back radiation as shown in Fig. 6.22 is reported in [28]. The FSS cell is composed of two metallic rings connected to a rectangular strip with a thickness of 0.035 mm. The FSS unit cell has a size of 5 mm × 5 mm. The FSS-based reflectors are designed at the same fundamental mode of the antenna and added under the antenna at a specific distance to achieve the required gain enhancement. The antenna gain is increased by around 5.3 dBi as shown in Fig. 6.23.

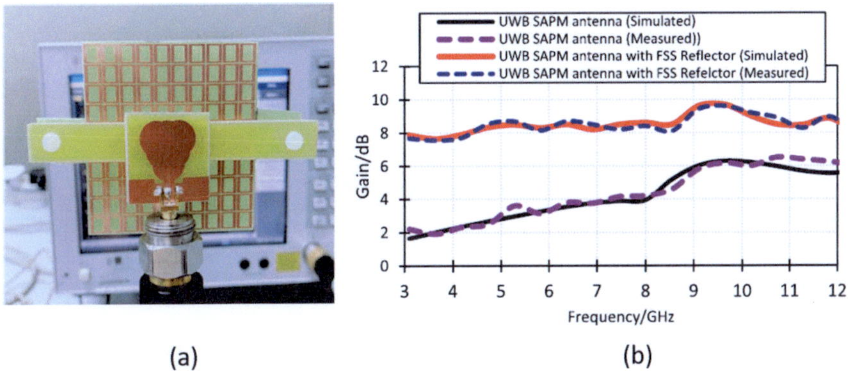


**Fig. 6.21** **a** UWB SAPM antenna with metallic single-layer FSS reflector, **b** gain versus frequency characteristics with and without FSS [27]

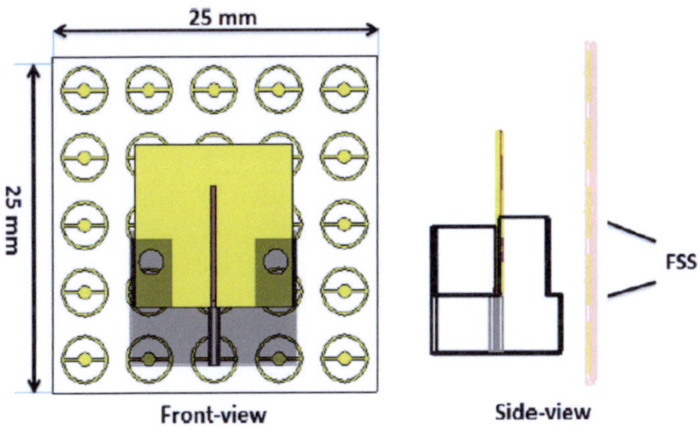


**Fig. 6.22** Perspective view of the suggested antenna attached with the FSS [28]

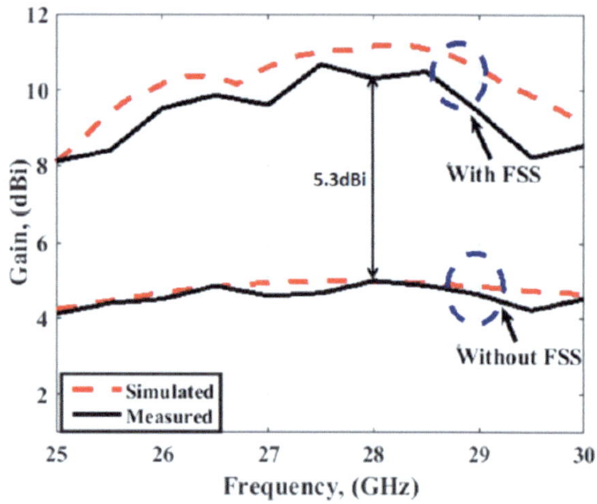


**Fig. 6.23** Suggested slot antenna attached with FSS realized gain [28]

## 6.4 Gain Enhancement Using Dielectrics

### *6.4.1 Dielectric Lens*

In recent years, lens antennas have brought significant improvements to antenna performance, thus attracting much attention. Conventional lens antennas are made up of antennas loaded with homogeneous dielectrics. These dielectrics were molded into desired shapes and coupled in front of the antenna aperture to correct the radiated electromagnetic (EM) waves, thereby improving the antenna gain. The dielectric lenses provide an effective way to increase the directivity of a source antenna by reshaping the spherical wave front into a planar one. All-dielectric metasurfaces are particularly attractive due to the absence of ohmic losses that normally occur in their metallic counterparts. Microwave lenses are usually bulky components, whose dielectric material is tailored to achieve a desired phase profile [29]. Fortunately, the advent of additive manufacturing technologies, or three-dimensional (3D) printing, has increased the degree of freedom in the design of electromagnetic structures significantly. In [30] an extended length hemispherical lens was chosen, due to its simple design. It consists of a low-cost, low-loss, and easy-to-fabricate 3D-printed integrated lens antennas (ILAs) for 5G broadband wireless communications in the 28-GHz frequency range. The ILA design consists of an extended hemispherical lens, fed by two different types of source antennas, a substrate-integrated waveguide (SIW) slot antenna array and a microstrip patch antenna (MPA) array. The fabrication process features commercially available desktop 3D printer, CEL Robox® RBX02-DM, using standard FDM process. The ILAs were fabricated with a reduced infill percentage of 50%, which speeds up prototyping time, and simultaneously decreases dielectric loss, manufacturing cost, and weight. The lens antennas provide considerable gain enhancement of 6 dB, for two different types of source antennas—a $2 \times 1$

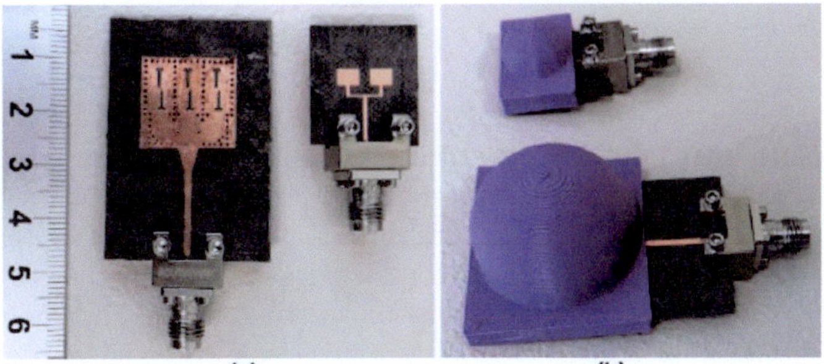


**Fig. 6.24** Fabricated prototypes of ILAs **a** source antennas, **b** 3D-printed dielectric lens on top of source antennas [30]

MPA array and a SIW slot antenna array. Figure 6.24 demonstrates that the achieved broadband gain is relatively flat over the frequency band of interest, 15.5 ± 0.5 dBi for the SIW ILA and 12 ± 0.5 dBi for the MPA ILA [30].

Another 3D-printed metalens is proposed in [31] for gain enhancement in the Ka band. The metalens is designed with ABS plastic cylindrical posts and this demonstrated the first all-dielectric, 3D printed, microwave metalens with diffraction efficiency in excess of 70%. In addition to this, the metalens increases the directivity of the receiving antenna by reducing its HPBW to approximately 3° related to both azimuthal and elevation planes. Furthermore, the metalens is still able to focus the beam into the receiving antenna even when rotated, with a 50° and 40° span at azimuthal and elevation planes, respectively. This design also demonstrated that it is possible to design functional metasurfaces with a phase control of $\pi$ $[-\pi/2–\pi/2]$, allowing low-permittivity materials to be used while still maintaining electrically small thicknesses. The proposed low-loss and low-cost all-dielectric metalens, with a measured gain as high as 7.5 dB at 32.5 GHz, outperforms conventional all-dielectric lenses.

In [32], a compact microwave lens designed using 13 layers of pure dielectric sheets in a parallel and periodic way is proposed for high-gain enhancement of linearly polarized antennas. The proposed microwave lens has a very simple structure and a compact volume of only $1.7\lambda_0 \times 1.6\lambda_0 \times 1.7\lambda_0$ and is easy to fabricate. To demonstrate the gain enhancement capability of the proposed microwave lens, a linearly polarized 3D-patch antenna consisting of a radiating patch with dual feed is designed and used as the source antenna. The antenna generates a polarized signal with high purity. The microwave lens can significantly increase the boresight gain of the patch antenna by 11.9 dB from 5.1 dBi to 17 dBi at 8.5 GHz and significantly reduce the beamwidth of the radiation patterns in both the *E*- and *H*-planes. The microwave lens antenna has a much higher gain enhancement with a much smaller volume.

### 6.4.2 Dielectric Superstrates

Superstrate techniques with the ability to achieve high gain and directivity have attracted much attention. Dielectric slab superstrate (DSS) is very attractive for superstrate-loaded antennas because of simple structure and ease of fabrication. The DSS antenna with the side length larger than $2\lambda_0$, where $\lambda_0$ is the free-space wavelength at the designed frequency, can be considered as resonant cavity antenna (RCA). Its design is based on the multireflections between interfaces with infinite sizes. In [33], the investigation of the DSS antennas with the side length around $1.5\lambda_0$ shows that the leaky-wave effects and the edge diffractions of the superstrate result in the gain enhancement. The primary feed is a WR-62 waveguide aperture, and parallel to the antenna, a small dielectric puck is placed having a thickness of 1.25 mm and permittivity of 10.51 at a distance of 10.7 mm. The puck (the PRS) allows the fields to be radiated through the puck itself (the leaky-wave effect) or be diffracted at the periphery of the puck, resulting in a wideband high-gain performance. The antenna performance varies significantly when the superstrate is a small dielectric puck that allows the fields to reach the edges before leaking out through the puck itself. This effect was found to be principal in antennas with pucks of smaller sizes and circular shapes, resulting in a considerable enhancement of broadside gain and bandwidth. A superstrate to achieve a gain enhancement for a dual-polarized large antenna array as shown in Fig. 6.25 is proposed in [34]. The DSS is placed in front of the dual-polarized aperture-coupled patch antenna and has proved to be available for gain enhancement for both polarizations. The DSS can realize a gain enhancement for both polarizations of the single feed antenna without much influence on the impedance bandwidth. The DSS can increase the gain of $2 \times 2$ array to 18.3 dBi and 18.4 dBi for two polarizations. The piecewise homogeneous DSS and source antenna array hold the advantages of high gain, large aperture efficiency, and easy processability. The DSS can be effective for $4 \times 4$ and $8 \times 8$ arrays as well.

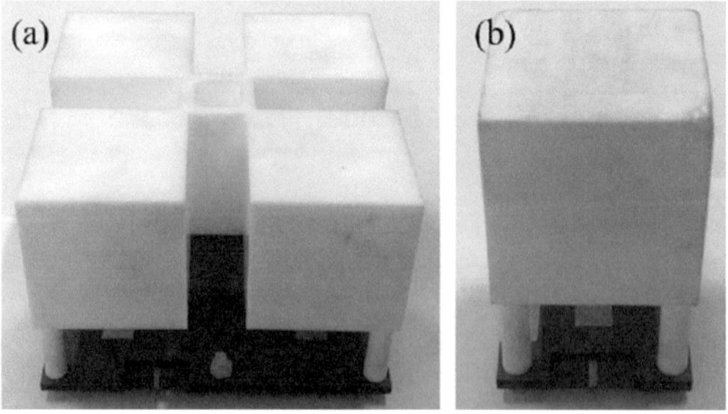


**Fig. 6.25** **a** Fabricated DSS with the $2 \times 2$ patch antenna array. **b** The fabricated DSS with the single patch antenna [34]

Antenna gain enhancement using a holey dielectric superstrate on a microstrip patch antenna is presented in [35]. The superstrate consists of a dielectric layer with vertical holes drilled over it. Antenna gain is increased by creating an in-phase electric field area on the upper surface of the superstrate. The in-phase field is created by changing the radius of the holes in the superstrate to control the effective permittivity of the superstrate. It is observed that the value of the effective permittivity is inversely proportional to the radius of the holes. With respect to holes with a radius of 1, 2, 3, and 4 mm, the values of effective permittivity are calculated as 9.5, 8.3, 6.5, and 4.5, respectively. It achieves a 2.3-dB gain enhancement when the superstrate is added over the microstrip antenna at the design frequency of 5.8 GHz. Another holey superstrate that is used for gain enhancement is proposed in [36]. A 2 × 2 square high dielectric dense patch antenna is excited through aperture coupling feed method. The superstrate dielectric layer used has a relative permittivity of 10.2 and a thickness of 0.64 mm. The dielectric substrate layer is perforated by a set of identical periodical circular holes of diameter D (with period P) placed along $x$- and $y$-axes. Drilling holes on the dielectric layer can change the effective permittivity of the material. An effective relative permittivity of the holey superstrate layer is calculated as

$$\varepsilon_{\text{eff}} = \varepsilon_{\text{r}}\left(1 - \frac{\pi}{2}\left(\frac{D}{D+g}\right)^2 + \frac{\pi}{2}\left(\frac{D}{D+g}\right)^2\right), \tag{6.7}$$

where $D$ is the hole diameter and $g$ is the distance between the holes.

A high gain of 16 dBi and radiation efficiency of about 92% are achieved. The antenna gain and radiation efficiency are almost flat over the antenna bandwidth. A patch antenna gain enhancement based on phase compensation concept is presented in [37]. Putting a superstrate as flat artificial lens over normal patch antenna, phases of distributed electric field arrows will be same, leading to patch antenna gain enhancement. By introducing a novel criterion named phase pattern consideration for phase compensation method, a new shaped superstrate for this method is designed. Since the design strategy for the superstrate is extracting from phase pattern of distributed electric field, antenna gain is increased from 6.9 to 10.2 dB.

## 6.5 Gain Enhancement Using Fabry–Pérot Cavity

Fabry–Pérot antennas are also called partially reflective surface (PRS) antennas, 2D leaky-wave antennas, electromagnetic bandgap (EBG) resonator antennas, resonant cavity antennas, and FP resonator antennas. Typically, FPRAs are formed by placing a primary radiator, such as a printed patch or a waveguide aperture, in a cavity between a partially reflective surface (PRS) and a metal ground plane, as shown in Fig. 6.26. The conducting ground plane behind the patch antenna acts as a shield against backward radiation. Ground plane and PRS act as parallel plate, where multiple reflections take place in between [38]. Radiated waves from the patch antenna strike the PRS

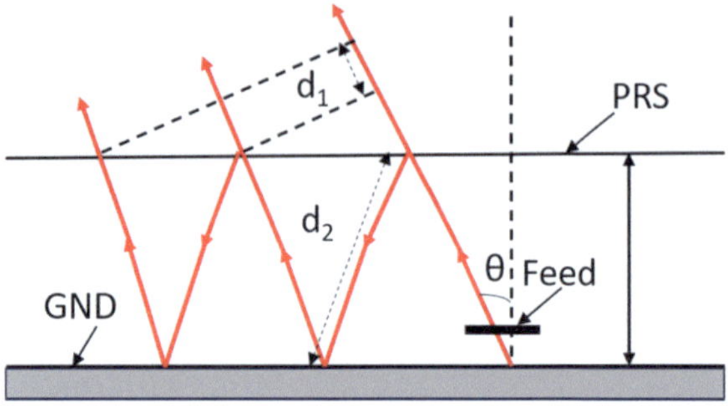


**Fig. 6.26** Ray-tracing model of the antenna

at different angles. Total internal reflection of waves occurs when incidence angle is greater than the critical angle. As the angle of incidence increases, reflection by PRS inside FPC decreases. Due to this, strong radiation from the center of PRS and weaker toward margin sides are observed. Multiple reflections inside the cavity cause increase in antenna gain in the direction of boresight ($\theta = 0°$). The cavity height is selected typically close to half of resonating wavelength to achieve constructive interference for waves bouncing between PRS and ground plane. The resonance of the cavity results in a cophasal and tapered field distribution along the lateral direction. The waves that leak out in the normal direction from the PRS cause an increase in radiation property of the antenna.

According to the ray-tracing analysis in [39] and shown in Fig. 6.26, the resonance condition of the cavity can be formulated as

$$f_0 = \frac{c}{4\pi h}(\varphi_{\text{PRS}} + \varphi_{\text{GND}} - 2N\pi), \quad N = 0, 1, 2, \ldots, \tag{6.8}$$

where $f_0$ is the resonance frequency; $h$ is the depth of the cavity; $\varphi_{\text{PRS}}$ and $\varphi_{\text{GND}}$ are the reflection phases of the PRS and the ground plane, respectively. The number of modes excited in the FPC depends on location of MPA and cavity dimensions. As $N$ increases, the antenna is operating at higher modes, and consequently, the sidelobes become higher. If the reflection phase of the PRS $\varphi = \pi$ and the antenna operates at the first resonant frequency, the distance h should be half a wavelength. With $\varphi_{\text{GND}} = \pi$, the formulation demonstrates that if reflection phase of the PRS increases with resonance frequency $f_0$ (namely, with a positive slope versus frequency), FPRAs will produce a wideband resonance accompanied by an enhanced gain bandwidth.

FPC antennas with different reflecting surfaces to enhance antenna gain are mentioned in the literature. Some methods to enhance the gain of the FPC antenna involve tuning either the phase of the near field of the aperture of the antenna or the transmission magnitude of the PRS. The non-uniform phase distribution on the antenna aperture plane is mainly caused by the non-uniform phase distribution in the internal cavity of the antenna. A method to enhance gain based on tuning the reflection phase of the PRS to correct the non-uniform phase in the resonant cavity of the antenna is reported in [40]. The metasurface with a controllable reflection phase is designed to correct the phase distribution to a nearly uniform distribution, thereby enhancing the antenna gain. The unit cell of the metasurface is a simple substrate with two patches as shown in Fig. 6.27. The top layer patch is sufficiently large to provide high reflectivity, whereas the bottom layer patch has a variable size to tune the reflection phase of the unit cell. The electromagnetic waves emanating from the radiator experience multiple reflections and transmissions in the cavity. The maximum directivity at the broadside is achieved when the waves exiting the cavity are in phase.

$$\varphi_{\mathrm{s}} + \varphi_{\mathrm{d}} - \frac{4\pi h}{\lambda_0} = 2n\pi, \quad N = 0, \pm 1, \pm 2 \ldots . \tag{6.9}$$

In the equation above, $\lambda_0$ denotes the free-space wavelength at the operating frequency, and $\varphi_{\mathrm{s}}$ and $\varphi_{\mathrm{d}}$ are the reflection phases of the PRS and the ground plane, respectively. The first resonance ($N = 0$) is adopted to maintain a low profile for the antenna.

The patch antenna along with the PRS is shown in Fig. 6.28. A 2-dB gain enhancement is obtained compared to the reference antenna. The maximum gain of the antenna reaches 19.7 dBi at 10 GHz.

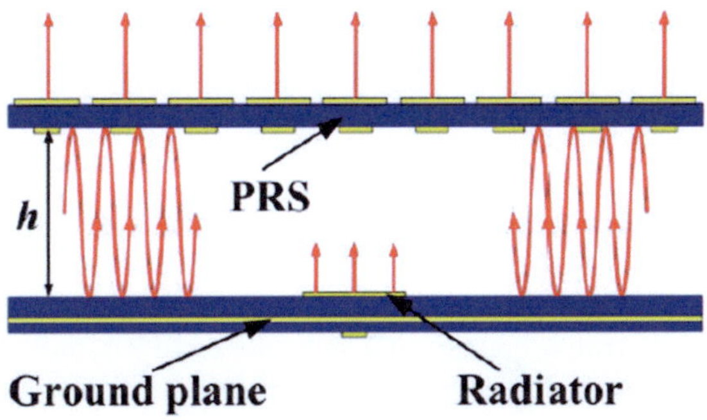


**Fig. 6.27** Schematic of FPC antenna [40]

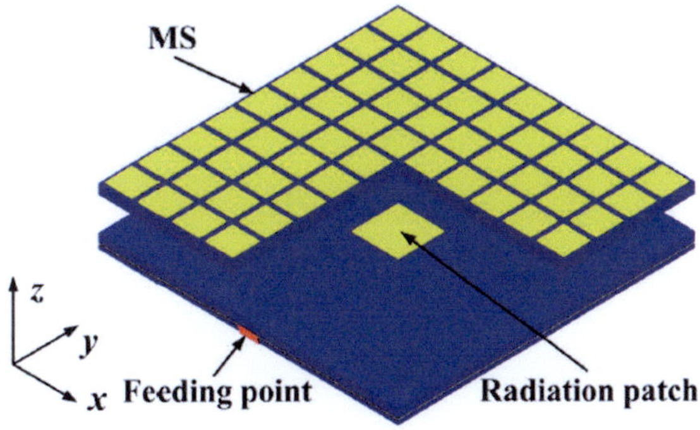


**Fig. 6.28** Geometry of proposed FPC antenna [40]

In early stages, the FP resonant cavity had high $Q$-value characteristics, which enabled FP antennas to exhibit high gain and narrow-band properties. With the advancement of FP antenna research, they can achieve properties such as wide bandwidth, low profile, multiband operation, which provide them with a broad application prospect in the field of wireless communication.

For wideband FP antennas, various methods for enhancing the 3-dB gain bandwidth have been proposed. The PRS structures of wideband FP antennas can be divided into non-uniformly printed PRSs [41] and uniformly printed PRSs [42] and all-dielectric PRS structures [43]. Wideband FP antennas, which employ a uniform printed PRS with positive reflection phase gradient, exhibit a sharp gain drop in the high-frequency operating band. This is due to the fact that the reflection phase of PRSs drops significantly toward high frequencies. In [44], a hybrid reflection method is proposed to realize wideband gain enhancement for an FP antenna. This method is realized by the cooperation of two FSS units on a hybrid partially reflecting surface (HPRS). The two FSS units are etched on both sides of a single-layer dielectric substrate with chessboard arrangement to form the reflective surface (RS) and hybrid surface (HS) of the HPRS. The gain enhancement properties are achieved by using two kinds of FSSs that have different positive reflection phase gradient bands and reflection amplitudes on the RS. A slot-fed patch antenna is used as the primary radiator. The hybrid reflection method is effective in improving gain enhancement with a peak gain of 17.08 dBi at 9.5 GHz. In [45], a double-layered superstrate design is presented as shown in Fig. 6.29, which is composed of two electrically thin substrate layers with the same or different permittivities. The reflection phase of the superstrate structure $\varphi_{\text{PRS}}$ increases with frequency. A wideband FPRA as shown in Fig. 6.30 working in the X-band is designed with the proposed double-layered superstrate structure. The thickness of each superstrate layer is less than $\lambda_d/4$ making corresponding FPRA designs cost-effective and flexible. The gain of the feed antenna is increased by 6 dB by the double PRS structures.

To realize high-gain CP radiation by FPRA, there are three popular methods. The first method is adopting a CP antenna as the feed of the FPRA [46]. The second

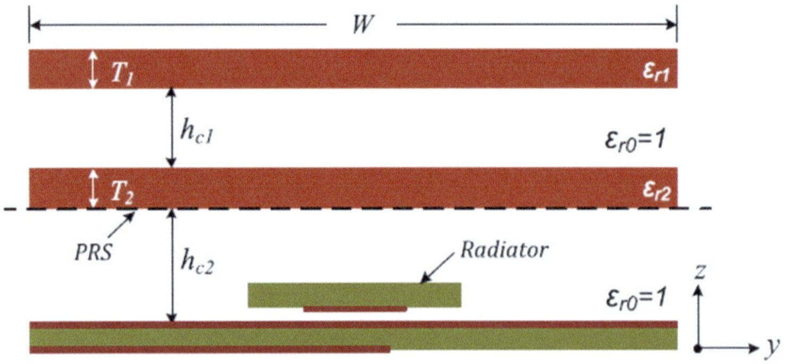


**Fig. 6.29** General configuration of the FPRA [45]

**Fig. 6.30** Prototype of the wideband FPRA structure [45]

method involves exciting two orthogonal LP waves in the resonant cavity, usually fed by a 45° tilted LP patch or a patch with truncated corners, for two orthogonal polarizations [47]. The last way is using the LP feed and realizing polarization conversion depending on other parts of the antenna [48]. However, most CP FPRAs suffer from the disadvantage of a narrow working band, which limits the application of the antenna to a great extent. The narrow band of the FPRA is determined by the resonant condition of the antenna. A very effective way to further enhance the gain is by adding a phase correction structure above the FPRA [49]. In [50], a wideband CP FPRA with excellent gain and AR performance is presented. First, a PRS with high reflectivity and a positive reflection phase gradient are proposed. By combining this PRS with a well-designed wideband CP patch antenna, a wideband CP FPRA is constructed. Then, a receiver–transmitter MS with high transmissivity in a wide frequency band is designed. The transmission phase of the meta-atom adopted by the MS can be tuned by changing the width of the receiver and transmitter patches. Then, the MS is employed to transmit the phase distribution on the aperture field of the CP FPRA to nearly uniform, which can enhance the gain of the antenna. Most importantly, the MS can tune the transmission phase of two orthogonally polarized waves independently. Thus, the phases of $x$- and $y$-polarized waves can also be corrected to maintain a 90° phase difference at every position, which will make the antenna obtain good AR performance. After the phase correction of the MS, the gain and AR performances of the CP FPRA are both improved significantly with a peak gain of 18.3 dBic achieved at 10.2 GHz.

Researchers in [51] proposed a dual-wideband high-gain FPCA based on a novel PRMS. The designed PRMS consists of isotropic MS unit cells, which are optimized by loading magnetic coupling enhancement layers and specific etching. The PRMS unit cell exhibits the positive gradient reflection phases with quasi-identical slope

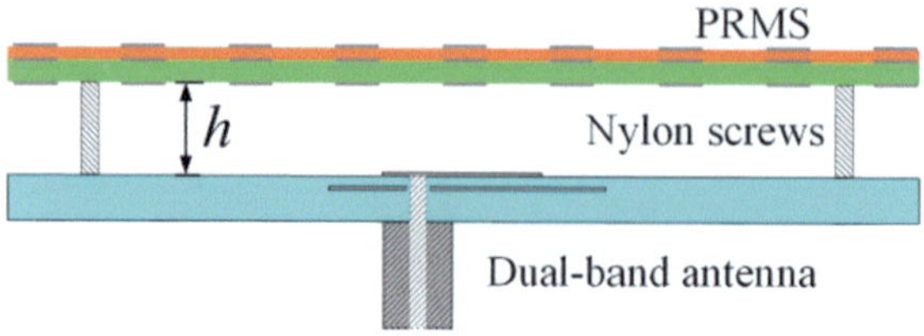


**Fig. 6.31** Side section view of the dual-wideband high-gain FPCA [51]

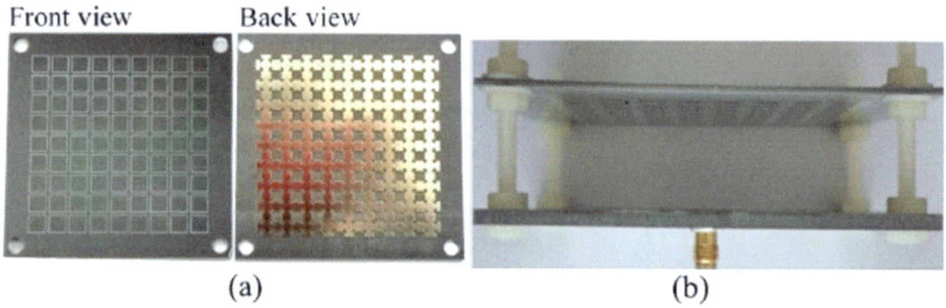


**Fig. 6.32** Fabricated **a** PRMS, **b** FPCA [51]

over two widebands of 5.5–6.5 and 11.5–12.5 GHz. The dual-wideband FPCA is constructed by loading the PRMS of 9 × 9 unit cells above the dual-band microstrip antenna, as shown in Fig. 6.31, and the fabricated prototype is shown in Fig. 6.32. To achieve a dual-band FPCA, the reflection coefficient of the desired PRMS should be as close as possible to 1 in both the widebands. To satisfy this, the PRMS unit cell is designed according to the following steps:

(a) PRMS unit cell with positive gradient reflection phase over a wideband.
(b) PRMS unit cell with positive gradient reflection phase over two widebands.
(c) PRMS unit cell with positive gradient reflection phase over two widebands with almost the same slopes, which maintains the FP resonance condition on both widebands.

## 6.6 Other Gain Enhancement Techniques

### *6.6.1 Reflectors*

The gain or directivity of omnidirectional antennas (monopole and dipole) can be enhanced in a specific direction by reducing its radiation in other directions simply with the help of metal reflectors without increasing their size, unlike in antenna arrays. Although metallic reflectors improve the front-to-back ratio and gain, among other parameters, their functioning is optimal over a relatively smaller frequency band.

Such problems can be overcome with frequency-dependent reflectors, such as a patch-type frequency-selective surface (FSS), which have the additional advantages of multiple frequency bands for better impedance matching and reduction of the front-to-back ratio, leading to decreased back radiation [52]. The performance of a pair of closely spaced monopole antennas with resonating frequencies at 3 and 5.5 GHz is enhanced by using an FSS-based corner reflector as reported in [53]. The reflector consists of two patch-type FSS screens kept at an angle of $90^0$. The reflector operating at 5.5 GHz enhances the antenna bandwidth ($-10$ dB) by 24% and the gain by 8–10 dBi over 4.5–6.5 GHz while maintaining its omnidirectional radiation at around 3 GHz. The electric field is enhanced by the FSS at the desired frequencies.

### 6.6.2 Directors

To improve antenna gain and radiation pattern, a director can be placed on top of the dipole. The common method is to use a metal strip director [54], which can be further expanded to multiple directors with periodic spacing. A 60-GHz antenna with novel ladder-like multidirectors is presented in [55]. The three-stage multidirectors are introduced and investigated for gain improvement, sidelobes' reduction, and bandwidth extension. By using the proposed directors, the fabricated antenna prototype features a bandwidth extension of more than 30%, good polarization characteristics, and a gain improvement of 2.2–3.4 dB across the band. To improve the azimuth gain, instead of the traditional way of increasing the profile, a low-profile omnidirectional director is exploited in [56], which is basically a passive magnetic dipole consisting of an annular side-coupling open cavity. Several directors co-operating with the driven element form the proposed antenna, whose working principle is like the Yagi–Uda antenna. Compared with that of the driven element only, the omnidirectional azimuth gain is significantly improved within the bandwidth, with the maximum gain being enhanced from $-$ 4.2 to 3.72 dBi at 5.34 GHz.

To further improve the antenna gain, with a broad bandwidth, several parasitic elements served as director cell are added to the shorter end of the PLPDA antenna according to the principle of the quasi-Yagi antenna [57]. So this novel antenna can be called printed Log-Yagi dipole array antenna (PLYDA), which consists of several active elements as the drivers, several shorter passive elements as the directors, and the metallic plate as the reflector. In [58], several parasitic dipole elements acting as director cell are proposed, which exhibits good performance to enhance the gain of the PLYDA antenna over the operating frequency range of 21–40 GHz. The proposed PLYDA antenna is fed by HMSIW structure for wideband and high-gain radiation. The proposed PLYDA antenna exhibits gain enhancement of 0.2–2 dB in comparison with the equal size PLPDA antenna over the operating frequency range 21–40 GHz. PLPDA antennas with parasitic cell to overcome the disadvantages above and further enhance the gain of PLPDA antenna are reported in [59]. The parasitic cell consisting of several bow–tie directors and parasitic patches is added to the shorter end of the log-periodic dipole on both top and bottom layers of the dielectric substrate. First, double-sided bow–tie directors with the same size and uniform spacing can help to enhance the antenna gain according to the principle of the quasi-Yagi antenna and the bow–tie antenna. Furthermore, based on the principle of the array antenna, the parasitic patches are added to increase the PLPDA antenna gain by expanding the transverse radiation aperture of the antenna. It can be observed that the PLPDA antenna gain with the parasitic cell is increased about 1 dBi compared with the antenna with the single-sided rectangle directors. The gain of the presented antenna is 7.3–12.5 dBi over the whole operating frequency range of 40–50 GHz.

### *6.6.3 Shorting Pins*

Another method to enhance the gain is to concentrate the radiation of an antenna array within a fixed radiation aperture, by using some vertical structures. In [60], multi-layered PCB was used to construct dielectric cubes or vertically stacked Yagi-like directors, in order to concentrate more radiation toward the boresight of the antenna array. An efficient approach to improve the gain performance is presented at 60 GHz for narrow-band applications in [61]. This approach is also based on the radiation concentration, by applying vertical metallic walls along the $H$-plane of a waveguide slot antenna array. The vertical metallic walls can be realized and co-fabricated in a standard low-temperature cofired ceramic (LTCC) process, to ensure high reliability in high-gravity environments for space applications. The proposed approach can efficiently enhance the gain performance within their operational bandwidth.

When compared with the conventional antenna arrays without metallic walls, the proposed approach can achieve gain enhancement of about 4.4 dB for the $(1 \times 8)$ 1D case and $(4 \times 8)$ 2.5–3.1 dB for the 2D case, across the impedance bandwidth. Both antenna arrays can efficiently transmit data within a 1000 MHz (59.5–60.5 GHz) bandwidth for some narrow-band applications. A two-layer substrate-integrated magnetoelectric dipole antenna is proposed in [62]. The weakening of gain is caused by the conflict of neighboring resonances, which come from two different sources of the magnetic dipole.

Next, the shorting pins are proposed to tune the conflicting sources for gain improvement. Meanwhile, the shorting pins increase the impedance bandwidth, reduce the operating frequency, and enhance the gain on the low operating band. Then, the high-pass property of the vertical split-ring resonator (VSRR) on the wide region is presented to improve the front-to-back ratio (FBR) on the wideband. Moreover, the gains at some frequencies are also enhanced. The proposed antenna achieves wide bandwidth with stable gain. The shorting pins enhance the gain up to about 6.5 dB by tuning the conflicting resonances at the high operating band for a stable gain ranging from 7.8 to 9.8 dBi. An implantable circularly polarized (CP) antenna with an enhanced gain is proposed and investigated in [63]. The proposed antenna consists of a pin-loaded patch and via walls around the patch. The pin-loaded patch has a pair of orthogonal normal modes instead of degenerated modes, and two open-ended slots etched on the patch make resonant frequencies of the two modes close and introduce a 90° phase difference. The via walls protect the antenna against gain loss due to lossy tissues. The proposed antenna has a gain improvement of 1.5 dB due to the via walls.

### 6.6.4 Substrate Engineering

The highly desirable features of microstrip antennas, such as performance, flexibility, simplicity, and low fabrication cost, make them very popular for many applications. Their conventional design consists of a metallic patch (usually square or circular) on top of a dielectric substrate. When the antenna is energized, waves can propagate in all directions. Some of these waves, referred to as surface waves, travel in the substrate and do not contribute to the main radiation. Therefore, some energy is lost in these directions, limiting the maximum achievable gain between 5 and 9 dB for these radiating elements [64]. Moreover, when these elements are used in an array, surface waves are primarily responsible for scan blindness [65]. These drawbacks limit the range of applications in which these elements can be used. Over the years, several alternatives have been proposed to overcome the gain limitation, usually at the expense of decreasing the impedance bandwidth.

The suppression of electromagnetic waves leads to a significant enhancement of the maximum gain when a microstrip patch antenna is placed above one of these surfaces [66]. However, the drawback is complexity of construction because many of these designs include a large number of vias. When a wave encounters an interface between two materials, part of the energy is transmitted, and some is reflected [67]. The reflections, combined with the incident field, create constructive or destructive interferences depending on the characteristics of the materials. If this basic concept is applied to the design of microstrip antennas, the characteristics of the substrate can be modified, by using two or more materials, to reflect the surfaces waves. The reflected field can then enhance the maximum antenna gain by reducing the surface-wave propagation in the substrate.

A substrate design that uses a ferrite ring to enhance the gain of a microstrip patch antenna without compromising the bandwidth is proposed in [68]. This design forces constructive interference between the incident and reflected fields inside the substrate and the ferrite. The interference between the incident and reflected waves is created by the permittivities of both materials, the permeability of the ferrite, and the distance from the patch periphery to the ferrite ring. An unbiased ferrite, with relative permeability of 14, relative permittivity of 10, dielectric loss tangent 0.0017, and magnetic loss tangent 0.0391, is used. The results of the patch antenna above a conventional substrate (i.e., no ferrite ring) and above the hybrid substrate (which includes the ferrite ring) are compared. Enhancements of about 4 dB in gain and nearly 1% in impedance bandwidth are obtained by using the hybrid substrate instead of the conventional one.

### 6.6.5 Orthogonal Radiating Choke

A design method of complementary gains is presented to overcome the gain reduction caused by the cancelation of reverse currents in the middle of the band of interest, where design equation of the excited element of the antenna is presented in [69]. This objective is achieved by accurately analyzing the resonant modes of the two parallelly connected antennas of the excited element in order to achieve an overall wideband radiator. An orthogonal radiating choke (ORC) is presented as shown in Fig. 6.33 to remove the gain depression due to the leakage of electromagnetic waves toward the upper portion of the band of interest through simultaneously enhancing the broadside gain and choking the endfire traveling wave. Each one patch of the ORC acting as both a $TM_{10}$ mode microstrip antenna and a choke of endfire traveling wave is used in this design. With the help of the ORC, the gain in $+/-z$-directions is enhanced by 13.1 dB at 6 GHz compared to that without the ORC.

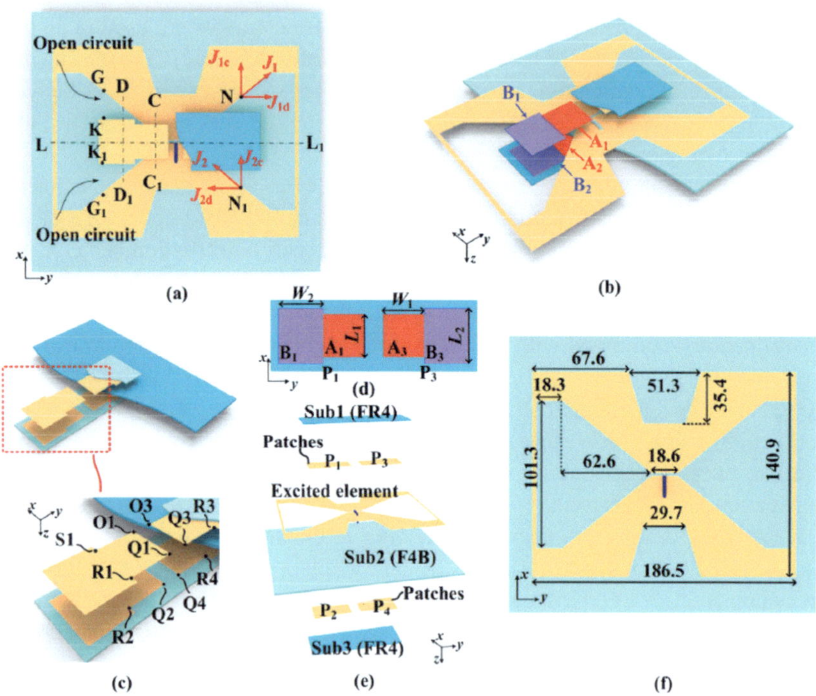


**Fig. 6.33** Configuration of the proposed antenna **a** front view, **b** global perspective view, **c** partial perspective view, **d** radiating patches of P1 and P3, **e** explosion view, and **f** excited element [69]

## 6.7 Conclusion

Different antennae gain enhancement techniques are described in this chapter. First, the gain enhancement using metamaterial structure is described. In this section, a detailed description of metamaterials is provided followed by the gain enhancement theory. Gain enhancement using different types of metamaterials like ZIM, ENZ/MNZ, and low-impedance metamaterials was described. This is followed by discussion on metamaterial lenses and how different dielectric values of the lenses affect the gain enhancement feature. Gain enhancement using artificial surfaces like resonator, AMC, EBG, and FSS structures is also discussed. Gain enhancement using various structures that are fully dielectric is also covered in this chapter. High-gain antennas are also designed using Fabry–Perot cavity that makes use of repeated reflections within a cavity structure. Finally, other gain enhancement techniques like using reflectors, directors, shorting pins, and substrate engineering are also briefly described.

## References


1. Balanis, C.: A Antenna Theory: Analysis and Design. Wiley, Germany (2012)
2. Esmail, B.A.F., Majid, H.A., Abidin, Z.Z., Dahlan, S.H., Himdi, M., Dewan, R., Rahim, M.K., Al-Fadhali, N.: Reconfigurable radiation pattern of planar antenna using metamaterial for 5G applications. Materials **13**(3), 115 (2020)
3. Jiang, H., Si, L.-M., Hu, W., Lv, X.: A symmetrical dual-beam bowtie antenna with gain enhancement using metamaterial for 5G MIMO applications. IEEE Photon. J. **11**(1), 19 (2019)
4. Pendry, J.B., Holden, A.J., Robbins, D.J., Stewart, W.J.: Magnetism from conductors and enhanced nonlinear phenomena. IEEE Trans. Microw. Theory Techn. **47**(11), 20752084 (1999)
5. Smith, D.R., Padilla, W.J., Vier, D.C., Nemat-Nasser, S.C., Schultz, S.: Composite medium with simultaneously negative permeability and permittivity. Phys. Rev. Lett. **84**(18), 41844187 (2000)
6. Liu, Y., Zhang, X.: Metamaterials: a new frontier of science and technology. Chem. Soc. Rev. **40**(5), 2494–2507 (2011)
7. Esmail, B.A.F., Koziel, S., Szczepanski, S.: Overview of planar antenna loading metamaterials for gain performance enhancement: the two decades of progress. IEEE Access **10**, 27381–27403 (2022)
8. Milias, C., et al.: Metamaterial-inspired antennas: a review of the state of the art and future design challenges. IEEE Access **9**, 89846–89865 (2021)
9. Pendry, J.B., Holden, A.J., Robbins, D.J., Stewart, W.J.: Low frequency plasmons in thin-wire structures. J. Phys. Condens. Matter **10**(22), 47854809 (1998)
10. Niu, X., Hu, X., Chu, S., Gong, Q.: Epsilon-near-zero photonics: a new platform for integrated devices. Adv. Opt. Mater. **6**(10), 1–36 (2018)
11. Zhou, B., Li, H., Zou, X.Y., Cui, T.-J.: Broadband and high-gain planar Vivaldi antennas based on inhomogeneous anisotropic zero-index metamaterials. Prog. Electromagn. Res. **120**, 235–247 (2011)
12. El-Nady, S., Mahmoud Zamel, H., Hindy, M., Zekry, A.A., Attiya, A.L.: Gain enhancement of a millimeter wave antipodal vivaldi antenna by epsilon-near-zero metamaterial. Progress Electromagnet. Res. C **85**, 105–116 (2018)
13. Dadgarpour, A., Kishk, A.A., Denidni, T.A.: Gain enhancement of planar antenna enabled by array of split-ring resonators. IEEE Trans. Antennas Propag. **64**(8), 3682–3687 (2016)

14. Pandit, S., Mohan, A., Ray, P.: A low-profile high-gain substrate-integrated waveguide-slot antenna with suppressed cross polarization using metamaterial. IEEE Antennas Wirel. Propag. Lett. **16**, 1614–1617 (2017)
15. Chen, L., Lei, Z., Yang, R., Fan, J., Shi, X.: A broadband artificial material for gain enhancement of antipodal tapered slot antenna. IEEE Trans. Antennas Propag. **63**(1), 395–400 (2015)
16. Ma, H.F., Cai, B.G., Zhang, T.X., Yang, Y., Jiang, W.X., Cui, T.J.: Three-dimensional gradient-index materials and their applications in microwave lens antennas. IEEE Trans. Antennas Propag. **61**(5), 2561–2569 (2013)
17. Zhang, S., Arya, R.K., Whittow, W.G., Cadman, D., Mittra, R., Vardaxoglou, J.C.: Ultra-wideband at metamaterial GRIN lenses assisted with additive manufacturing technique. IEEE Trans. Antennas Propag. **69**(7), 3788–3799 (2021)
18. Smith, D.R., Mock, J.J., Starr, A.F., Schurig, D.: 'Gradient index metamaterials. Phys. Rev. E, Stat. Phys. Plasmas Fluids Relat. Interdiscip. Top. **71**(3), 1–6 (2005)
19. Singha, R., Vakula, D.: Directive beam of the monopole antenna using broadband gradient refractive index metamaterial for ultra-wideband application. IEEE Access **5**, 9757–9763 (2017)
20. Alhawari, A.R.H., Ismail, A., Mahdi, M.A., Abdullah, R.S.A.R.: Antipodal vivaldi antenna performance booster exploiting snug-in negative index metamaterial. Progress Electromagnet. Res. C **27**, 265–279 (2012)
21. Wang, H., Liu, S.-F., Chen, L., Li, W.-T., Shi, X.-W.: Gain enhancement for broadband vertical planar printed antenna with H-shaped resonator structures. IEEE Trans. Antennas Propag. **62**(8), 4411–4415 (2014)
22. Cao, W., Zhang, B., Liu, A., Yu, T., Guo, D., Wei, Y.: Broadband high-gain periodic endfire antenna by using I-shaped resonator (ISR) structures. IEEE Antennas Wirel. Propag. Lett. **11**, 1470–1473 (2012)
23. Gong, Y., Yang, S., Li, B., Chen, Y., Tong, F., Yu, C.: Multi-band and high gain antenna using AMC ground characterized with four zero-phases of reflection coefficient. IEEE Access **8**, 171457–171468 (2020)
24. Fan, Y., Yahya, R.-S.: Electromagnetic Band Gap Structures in Antenna Engineering. Cambridge University Press (2009)
25. Denidni, T.A., Coulibaly, Y., Boutayeb, H.: Hybrid dielectric resonator antenna with circular mushroom-like structure for gain improvement. IEEE Trans. Antennas Propag. **57**(4), 1043–1049 (2009)
26. Han, Z.-J., Song, W., Sheng, X.-Q.: Gain enhancement and RCS reduction for patch antenna by using polarization-dependent EBG surface. IEEE Antennas Wirel. Propag. Lett. **16**, 1631–1634 (2017)
27. Al-Gburi, A.J.A., Ibrahim, I.B.M., Zeain, M.Y., Zakaria, Z.: Compact size and high gain of CPW-fed UWB strawberry artistic shaped printed monopole antennas using FSS single layer reflector. IEEE Access **8**, 92697–92707 (2020)
28. Mohamed, H.A., Edries, M., Abdelghany, M.A., Ibrahim, A.A.: Millimeter-wave antenna with gain improvement utilizing reflection FSS for 5G networks. IEEE Access **10**, 73601–73609 (2022)
29. Laribi, M., Hakem, N.: Gain-bandwidth enhancement of 60 GHz DRA using lens technique. In: Proceedings of IEEE International Symposiyum Antennas Propagation, pp. 385–386 (2016)
30. Malik, B.T., Doychinov, V., Zaidi, S.A.R., Robertson, I.D., Somjit, N.: Antenna gain enhancement by using low-infill 3D-printed dielectric lens antennas. IEEE Access **7**, 102467–102476 (2019)
31. Pepino, V.M., da Mota, A.F., Martins, A., Borges, B.-H.V.: 3-D-printed dielectric metasurfaces for antenna gain improvement in the Ka-band. IEEE Antennas Wireless Propag. Lett. **17**(11), 21332136 (2018)
32. Li, Q.L., Cheung, S.W., Wu, D., Yuk, T.I.: Microwave lens using periodic dielectric sheets for antenna-gain enhancement. IEEE Trans. Antennas Propag. **65**(4), 2068–2073 (2017)
33. Al-Tari, M.A., Anagnostou, D.E., Amert, A.K., Whites, K.W.: The puck antenna: a compact design with wideband, high-gain operation. IEEE Trans. Antennas Propag. **63**(4), 18681873 (2015)

34. Zhang, H., Yu, Y.-H., Wu, W., Fang, D.-G.: Dual-polarized antenna gain enhancement using dielectric slab superstrate and its application in forming an array. IEEE Access **9**, 80632–80638 (2021)
35. Kim, J.H., Ahn, C.-H., Bang, J.-K.: Antenna gain enhancement using a holey superstrate. IEEE Trans. Antennas Propag. **64**(3), 1164–1167 (2016)
36. Asaadi, M., Sebak, A.: Gain and bandwidth enhancement of 2 × 2 square dense dielectric patch antenna array using a holey superstrate. IEEE Antennas Wirel. Propag. Lett. **16**, 1808–1811 (2017)
37. Sharifian Mazraeh Mollaei, M.., Zanganeh, E., Feshki Farahani, M.: Enhancement of patch antenna gain using cylindrical shell-shaped superstrate. IEEE Antennas Wireless Propagat. Lett. **16**, 2570–2573 (2017)
38. Trentini, G.V.: Partially reflecting sheet arrays. IEEE Trans. Antennas Propag. **AP-4**(4), 666–671 (1956)
39. Wang, N., Zhang, C., Zeng, Q., Wang, N., Xu, J.-D.: New dielectric 1-D EBG structure for the design of wideband resonator antennas. Progress Electromagnet. Res. **141**, 233–248 (2013)
40. Xie, P., Wang, G., Li, H., Gao, X.: A novel methodology for gain enhancement of the Fabry-Pérot antenna. IEEE Access **7**, 176170–176176 (2019)
41. Liu, Z., Zhang, W., Fu, D., Gu, Y., Ge, Z.: Broadband Fabry-Pérot resonator printed antennas using FSS superstrate with dissimilar size. Microw. Opt. Technol. Lett. **50**(6), 1623–1627 (2008)
42. Ge, Y., Esselle, K.P., Bird, T.S.: The use of simple thin partially reflective surfaces with positive reflection phase gradients to design wideband, low-profile EBG resonator antennas. IEEE Trans. Antennas Propag. **60**(2), 743–750 (2012)
43. Hashmi, R.M., Esselle, K.P.: A class of extremely wideband resonant cavity antennas with large directivity-bandwidth products. IEEE Trans. Antennas Propag. **64**(2), 830–835 (2016)
44. Liu, Z., Liu, S., Zhao, X., Kong, X., Huang, Z., Bian, B.: Wideband gain enhancement and RCS reduction of Fabry-Perot antenna using hybrid reflection method. IEEE Trans. Antennas Propag. **68**(9), 6497–6505 (2020)
45. Wang, N., Talbi, L., Zeng, Q., Xu, J.: Wideband Fabry-Perot resonator antenna with electrically thin dielectric superstrates. IEEE Access **6**, 14966–14973 (2018)
46. Vaidya, A.R., Gupta, R.K., Mishra, S.K., Mukherjee, J.: Righthand/left-hand circularly polarized high-gain antennas using partially reflective surfaces. IEEE Antennas Wireless Propag. Lett. **13**, 431–434 (2014)
47. Orr, R., Goussetis, G., Fusco, V.: Design method for circularly polarized Fabry-Pérot cavity antennas. IEEE Trans. Antennas Propag. **62**(1), 19–26 (2014)
48. Calleau, A., García-Vigueras, M., Legay, H., Sauleau, R., Ettorre, M.: Circularly polarized Fabry-Pérot antenna using a hybrid leaky-wave mode. IEEE Trans. Antennas Propag. **67**(9), 5867–5876 (2019)
49. Meriche, M.A., Attia, H., Messai, A., Mitu, S.S.I., Denidni, T.A.: Directive wideband cavity antenna with single-layer meta-superstrate. IEEE Antennas Wireless Propag. Lett. **18**(9), 1771–1774 (2019)
50. Xie, P., Wang, G., Zou, X., Zong, B.: Gain and AR improvements of the wideband circularly polarized Fabry-Perot resonator antenna. IEEE Trans. Antennas Propag. **69**(10), 6965–6970 (2021)
51. Lv, Y.-H., Ding, X., Wang, B.-Z.: Dual-wideband high-gain Fabry-Perot cavity antenna. IEEE Access **8**, 4754–4760 (2020)
52. Pasian, M., Monni, S., Neto, A., Ettorre, M., Gerini, G.: Frequency selective surfaces for extended bandwidth backing reflector functions. IEEE Trans. Antennas Propag. **58**(1), 43–50 (2010)
53. Chatterjee, A., Parui, S.K.: Performance enhancement of a dual-band monopole antenna by using a frequency-selective surface-based corner reflector. IEEE Trans. Antennas Propag. **64**(6), 2165–2171 (2016)
54. Kaneda, N., Deal, W.R., Qian, Y., Waterhouse, R., Itoh, T.: A broadband planar quasi-yagi antenna. IEEE Trans. Antennas Propag. **50**(8), 1158–1160 (2002)

55. Lu, L., Ma, K., Meng, F., Yeo, K.S.: Design of a 60-GHz quasi-yagi antenna with novel ladder-like directors for gain and bandwidth enhancements. IEEE Antennas Wirel. Propag. Lett. **15**, 682–685 (2016)
56. Lin, S., Liao, S., Yang, Y., Che, W., Xue, Q.: Gain enhancement of low-profile omnidirectional antenna using annular magnetic dipole directors. IEEE Antennas Wirel. Propag. Lett. **20**(1), 8–12 (2021)
57. Sor, J., Qian, Y., Itoh, T.: Coplanar waveguide fed quasi-Yagi antenna. Electron. Lett. **36**, 1–2 (2000)
58. Zhai, G., Cheng, Y., Yin, Q., Zhu, S., Gao, J.: Gain enhancement of printed log-periodic dipole array antenna using director cell. IEEE Trans. Antennas Propag. **62**(11), 5915–5919 (2014)
59. Chu, Q.-X., Li, X.-R., Ye, M.: High-gain printed log-periodic dipole array antenna with parasitic cell for 5G communication. IEEE Trans. Antennas Propag. **65**(12), 6338–6344 (2017)
60. Ghassemi, N., Wu, K., Claude, S., Zhang, X., Bornemann, J.: Low-cost and high-efficient W-band substrate integrated waveguide antenna array made of printed circuit board process. IEEE Trans. Antennas Propag. **60**(3), 1648–1653 (2012)
61. Chu, H., Chen, J.-X., Guo, Y.-X.: An efficient gain enhancement approach for 60-GHz antenna using fully integrated vertical metallic walls in LTCC. IEEE Trans. Antennas Propag. **64**(10), 4513–4518 (2016)
62. Du, Z., Feng, Q.: Gain improvement and front-to-back ratio enhancement for substrate-integrated magnetoelectric dipole antenna. IEEE Antennas Wirel. Propag. Lett. **21**(1), 59–63 (2022)
63. Xu, L.-J., Jin, X., Hua, D., Lu, W.-J., Duan, Z.: Realization of circular polarization and gain enhancement for implantable antenna. IEEE Access **8**, 16857–16864 (2020)
64. Balanis, C.A.: Antenna Theory: Analysis and Design, 3rd edn. Wiley, New York, NY, USA (2005)
65. Pozar, D., Schaubert, D.: Scan blindness in infinite phased arrays of printed dipoles. IEEE Trans. Antennas Propag. **AP-32**(6), 602–610 (1984)
66. Boutayeb, H., Denidni, T.A., Mahdjoubi, K., Tarot, A.-C., Sebak, A.-R., Talbi, L.: Analysis and design of a cylindrical EBG-based directive antenna. IEEE Trans. Antennas Propag. **54**(1), 211–219 (2006)
67. Balanis, C.A.: Advanced Engineering Electromagnetics, 2nd edn. Wiley, New York, NY, USA (2012)
68. Rivera-Albino, A., Balanis, C.A.: Gain Enhancement in microstrip patch antennas using hybrid substrates. IEEE Antennas Wirel. Propag. Lett. **12**, 476–479 (2013)
69. Chi, L., Qi, Y., de Paulis, F., Zhang, Y.: A Compact wideband antenna with an orthogonal radiating choke for broadside gain enhancement. IEEE Open Journal of Antennas and Propagation **3**, 902–910 (2022)

# Chapter 7
# Gain Enhancement and Equalization of WLAN Antennas

## 7.1 Introduction

In this chapter, the design requirements and guidelines for pattern diversity antennas with multiple ports for WLAN applications are discussed. The discussion is limited to one-, two-, and three-port antenna systems, but the designs could be extended to antennas with higher number of ports. The antenna systems presented in this chapter are designed for a carrier frequency of 5.2 GHz with adequate bandwidth. The design tweaks presented in this chapter also enhance the pattern integrity of the individual antennas thereby ensuring high-gain bandwidth. The antenna systems presented in this chapter also have a wide coverage, making them suitable for practical deployment. The first design is a single-port wideband cavity antenna whose gain is enhanced using a metasurface design. The second design is a two-element antenna with a pattern diversity at 0° and + 90° followed by a three-element antenna with a pattern diversity of 0°, + 45°, and − 45°. The fourth design is a three-port antenna consisting of a dipole and patch that provides endfire and broadside coverage, respectively. In the fifth design discussed in this chapter, the radiation pattern of the antenna switches between 0°, + 25°, and − 25°.

Different types of metamaterial structures are used for gain enhancement in each design. Each metamaterial structure has different inherent properties like transmissive type, reflective type, and partially reflective type. This chapter illustrates how metamaterial structures are designed and integrated with antenna designs for enhancement of certain performance metrics. Metamaterial structures placed as superstrate structures or in the same plane as the antennas are also explained. An illustration of metamaterials as superstrates is given in Fig. 7.1. The expected radiation patterns when the antenna is turned on are also illustrated for each design. As the radiation patterns are oriented in different directions, the metamaterial structures have the ability to enhance the gain in oblique incidences also along with broadside directions. Metamaterial designs have a simple structure and provide inexpensive



S. K. Koul et al., *Antenna Systems for Modern Wireless Devices*, Signals and Communication Technology, https://doi.org/10.1007/978-981-97-3369-9_7

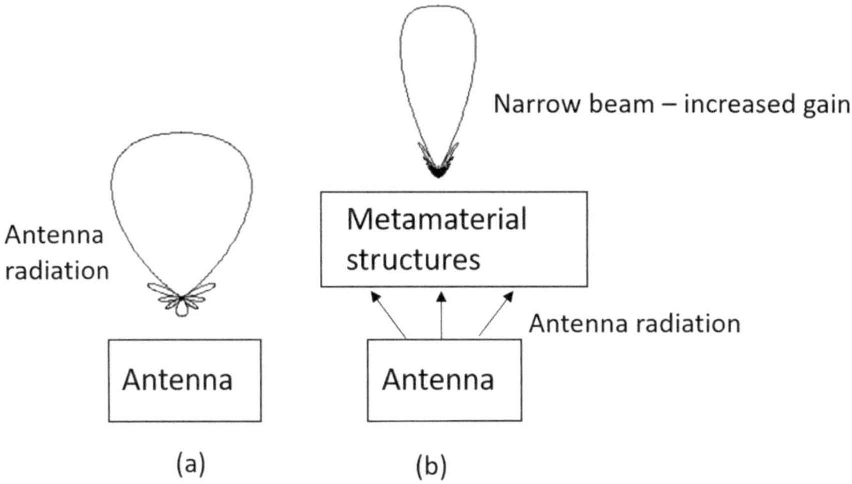


**Fig. 7.1** Illustration of gain enhancement using metamaterial structures

fabrication with desirable characteristics. The integrated module provides inexpensive fabrication with desirable characteristics of pattern diversity and minimal electrical footprint for WLAN. By the end of this chapter, the reader will have an idea about different ways to design and conduct a detailed analysis of a pattern diversity antenna design.

## 7.2 Gain Enhancement of 3D-Printed Wideband Cavity Antenna

Various gain enhancement methods are proposed in the literature. In contrast to the previous works, where researchers have used metasurfaces as superstrates or planar surfaces, the proposed design consists of a metasurface placed inside the cavity antenna for enhancing the gain. Metasurfaces placed around the antenna or as superstrates increase the antenna size, whereas in the proposed design, there is no change in the antenna size.

### *7.2.1 Wideband Cavity Antenna Design*

The schematic of the proposed wideband cavity antenna is shown in Fig. 7.2. It is designed on polylactic acid (PLA) substrate with $\varepsilon_r = 2.58$ and $\tan\delta = 0.015$. The size of the cavity structure is $0.62\lambda \times 0.62\lambda \times 0.1\lambda$. It is made using 0.5 mm thick

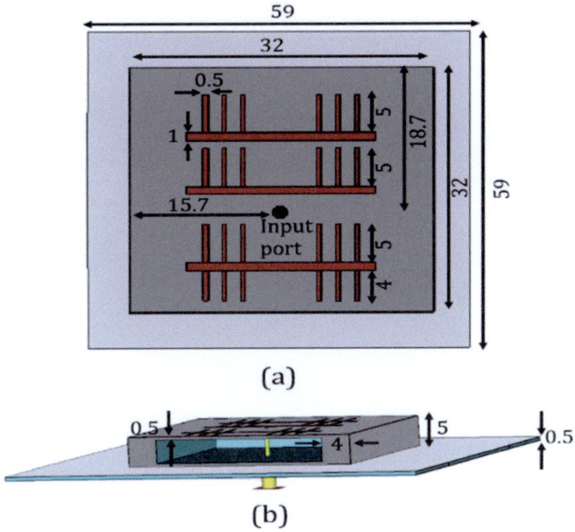


**Fig. 7.2** Schematic of the wideband cavity antenna **a** top view, **b** isometric view (all dimensions are in mm) [1]. Reprinted with permission from IEEE

PLA substrate enclosed all around with an opening at one side. The whole structure is covered using copper tape of 80$\mu$ thickness.

The ground plane size is 1.15λ × 1.15λ. PLA substrate of thickness 0.5 mm with copper on both sides is used as the ground plane. A coaxial feed is used to connect the input port to the top of the cavity structure. Figure 7.2a shows the various slots created at the metal layer to improve the bandwidth of the antenna.

### *7.2.2 Metasurface Design*

The objective here is to enhance the gain of the cavity antenna. To the proposed antenna in Sect. 7.2.1, a metasurface is placed inside the cavity structure as shown in Fig. 7.3. The size of the metasurface is 0.2λ × 0.35λ. It is placed in the left side gap between the input coaxial feed and the cavity wall, touching the left side cavity wall. It is placed at a height of 3.5 mm from the bottom cavity wall. The metasurface is placed very close to the top wall of the cavity with only a gap of 0.5 mm between them. Placing the antenna closer to the top wall ensures more gain enhancement. A gap of 0.8 mm is kept between the back wall of the cavity and the metasurface. This gap ensures that there is no detuning of the antenna from its resonance frequency.

The metasurface is also designed using a PLA substrate of 0.5 mm thickness. The design consists of 10 mm × 0.4 mm (0.2λ × 0.008λ) narrow metal strips repeated at 0.4 mm intervals along the substrate. The metasurface is placed strategically inside the cavity structure such that the placement can provide maximum gain enhancement to the antenna. The return loss and gain characteristics of the antenna before and after the placement of the metasurface inside the cavity are shown in Fig. 7.4. The bandwidth of the cavity antenna is from 5.2 to 6.2 GHz. After the metasurface is

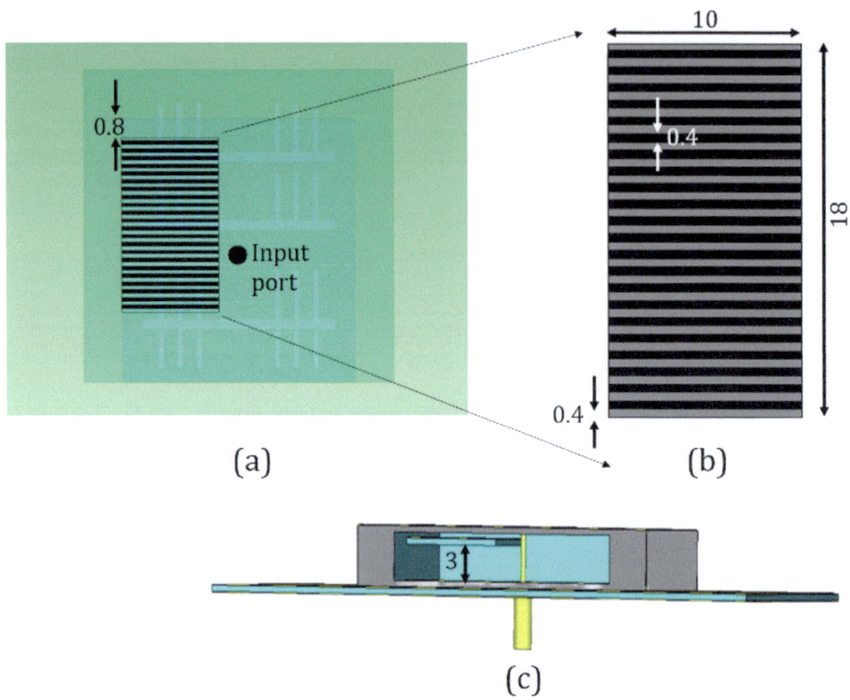


**Fig. 7.3** Schematic of the wideband cavity antenna with metasurface **a** top view, **b** metasurface, **c** isometric view (all dimensions are in mm) [1]. Reprinted with permission from IEEE

placed inside the cavity, the antenna bandwidth gets shifted to 5.07–6.08 GHz. There is a shift in the resonant frequencies of the antenna, but the total antenna bandwidth remains the same.

The gain of the proposed antenna is 4.2 dBi at 5.8 GHz which gets enhanced to 6.6 dBi. There is also significant gain enhancement in the 5.6–5.8-GHz frequency range with a 4.2 dB enhancement at 5.6 GHz. The radiation pattern of the gain-enhanced antenna at 5.8 GHz is shown in Fig. 7.5. The 3-dB beamwidth of the antenna is $80^0$ and the front-to-back ratio is 12.6 dB. A 3D-printed wideband cavity antenna is presented. A metasurface layer is placed inside the cavity to enhance the antenna gain. Without affecting the total antenna bandwidth, the metasurface enhances the gain by 2.4 dB at 5.8 GHz. The proposed design can be used for WLAN applications.

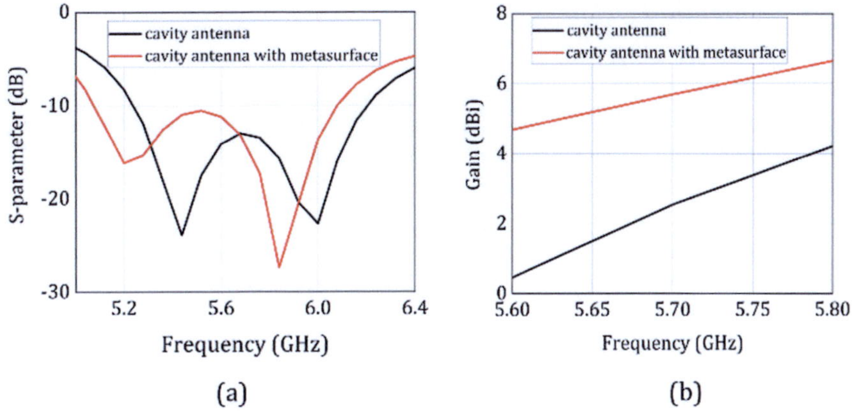


**Fig. 7.4** **a** Return loss characteristics of the cavity antenna, **b** gain of the cavity antenna [1]. Reprinted with permission from IEEE

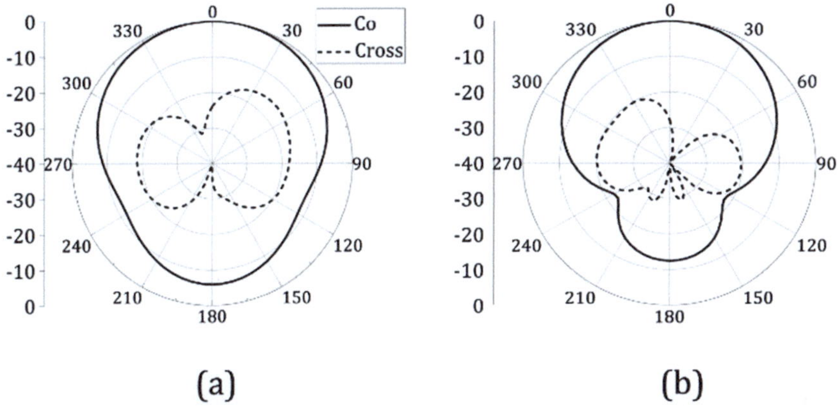


**Fig. 7.5** Radiation pattern of the cavity antenna with metasurface at **a** 5.4 GHz, **b** 5.8 GHz [1]. Reprinted with permission from IEEE

## 7.3 Common Aperture Two-Element Dipole Antenna Design

The proposed antenna is a two-element dipole antenna resonating at 5.2 GHz with a pattern diversity at 0°, 90° with an individual gain of 10 dBi. The gain enhancement for individual ports is equalized by strategically loading complementary transformed H-shaped resonator structures in the endfire due to localized change in the refractive index.

### 7.3.1 Antenna Design

The antenna design that establishes the pattern diversity requirement is presented first. The geometry of the proposed antenna design consists of two-dipole antennas placed perpendicular to each other as illustrated in Fig. 7.6. The substrate used for the antenna design is Rogers RO4350B. The substrate dielectric constant $\varepsilon_r$ is 3.66 and thickness is 0.508 mm or 20 mil. For designing an antenna, the dielectric constant of the substrate must be low [2]. Low dielectric constant helps to maintain minimal surface-wave modes. The thickness of 20 mil ensures proper mechanical stability for the antenna, with minimal dielectric loss.

The two-dipole arms are printed on opposite sides of the substrate. One arm and a microstrip feedline are printed on one side of the substrate, and the other arm and ground plane are printed on the other side. The width of the dipole arm is 1.14 mm, which is matched to 50 Ω at the operating frequency and length of the dipole arm has a length of 11.8 mm ($\lambda/4$) in each direction, where $\lambda$ is the wavelength in free space for the resonant frequency of 5.2 GHz. The ground plane dimension of the dipole is 8 mm × 20 mm ($0.14\lambda \times 0.35\lambda$). The dipole ground plane acts as a reflector. The distance between the dipole arms and the truncated ground plane is $0.22\lambda$, which is based on the Yagi antenna's director spacing. The complete two-port antenna module

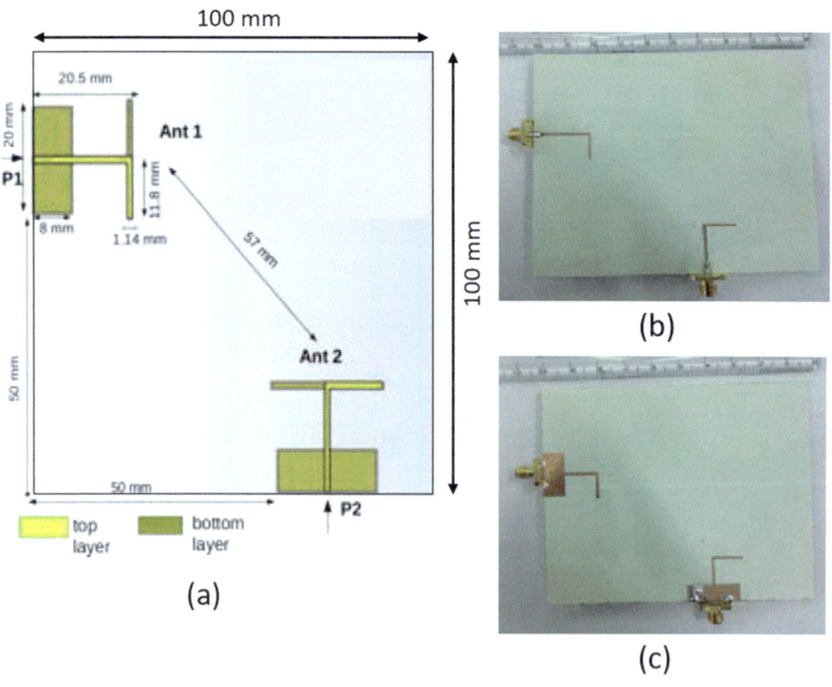


**Fig. 7.6** **a** Configuration of the proposed two-dipole antenna design, **b** fabricated design top layer, **c** fabricated design bottom layer [3]. Reprinted with permission from IEEE

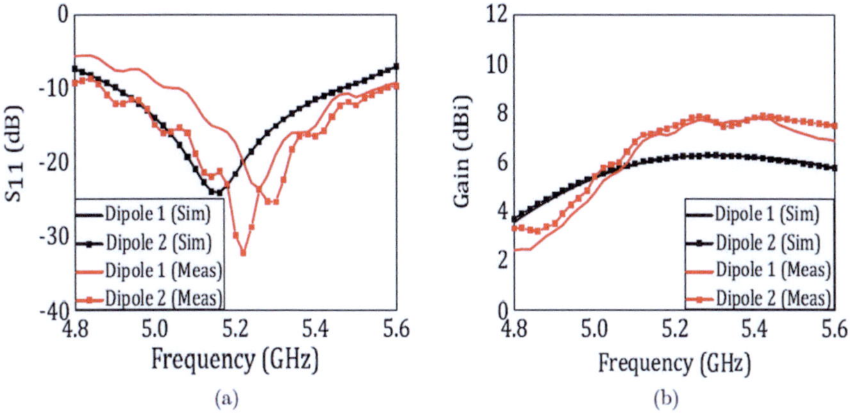


**Fig. 7.7** **a** Return loss characteristics of the proposed design, **b** gain of the proposed design [3]. Reprinted with permission from IEEE

has an area of 100 × 100 mm$^2$ (1.75λ × 1.75λ). The two-dipole antennas are spaced at a distance of 57 mm (1λ). As the two antennas are placed perpendicular to each other, it provides good isolation of $|S_{21}| < -20$ dB. This is also the ideal distance for the antenna to be placed as the antennas placed closer than 0.4λ cause the antenna gain to reduce due to increased coupling between them.

For each port, the simulated and measured input reflection coefficients are shown in Fig. 7.7a. The formula for the resonant frequency of a double-sided printed dipole antenna is explained in [4]. The simulated impedance bandwidth for the two antennas is 560 MHz (4.9–5.46 GHz) which is 10%, centered at 5.2 GHz, the antenna resonant frequency. A slight frequency shift for the measured resonance can be attributed to fabrication tolerances, that would introduce a small lead inductance resulting in an additional stepped impedance to the radiating element. The gain of the two-dipole antennas is 6.20 dBi as presented in Fig. 7.7b. The radiation pattern along the E-plane (*xy*-plane) of the two dipoles at the frequency of operation is shown in Fig. 7.8. The simulated half-power beamwidth (HPBW) of antenna 1 is 71° with a front-to-back ratio of 3 dB, and for antenna 2, it is 60° and 3 dB, respectively.

## *7.3.2 Two-Port Antenna Designs*

MIMO antennas should have a high gain, compact structure, and good isolation between the ports in the operating frequency band. The MIMO system integrates multiple antennas at the transmitter and receiver sides. Multiple antennas ensure an improved data rate and system capacity without a need for any extra spectrum [5]. Several designs have been reported in the literature. The antenna design in [6] operates in UWB with a notch in the 5.1–5.8 GHz. The two antennas are placed

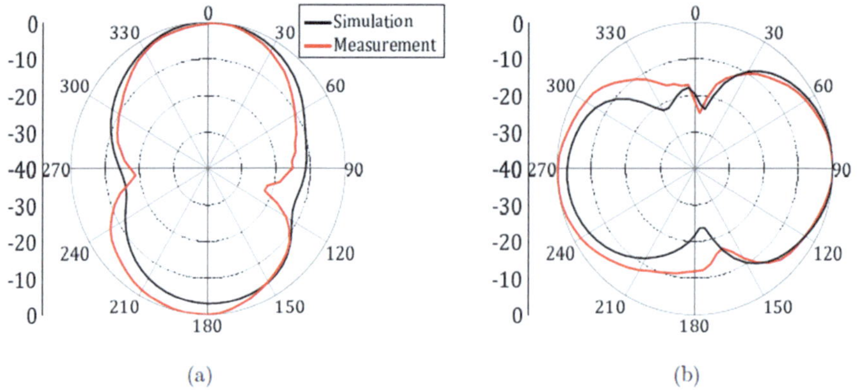


**Fig. 7.8** Radiation pattern at the operating frequency of 5.2 GHz **a** in *xy*-plane when port 1 is excited, **b** in *xy*-plane when port 2 is excited [3]. Reprinted with permission from IEEE

parallel to each other and use a T-shaped ground stub with a slot to reduce the mutual coupling between the two antennas as illustrated in Fig. 7.9. The antenna in [7, 8] uses orthogonally placed monopoles for ultra-wideband applications. Even though the perpendicular placement ensures good isolation, both designs use a narrow slot in the ground plane to reduce the mutual coupling further. But all these antennas have a gain of less than 5 dBi, and the dielectric resonator in [8] has a non-planar structure, which makes the integration with commercial communication systems difficult compared to planar structures.

For the presented antenna, the objective is to improve the gain and front-to-back ratio of both the perpendicularly mounted antennas in the endfire direction. For that, a unit-cell structure is designed that resonates at the operating frequency of the antenna. To make the design compact, the metamaterial aperture is shared by both antennas for gain improvements in their respective endfire directions.

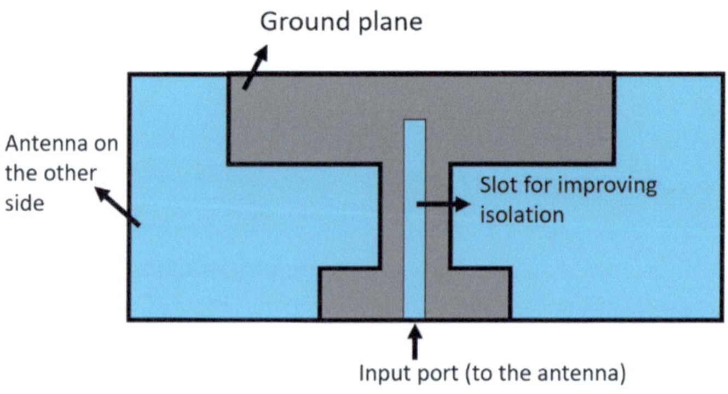


**Fig. 7.9** Isolation improvement design

### 7.3.3 Metamaterial Design

Figure 7.10a depicts the layout of the modified complementary H-shaped unit-cell structure. The initial H-shaped unit-cell structure is designed to resonate at 4.5 GHz. The parallel slots cut across the design created a frequency offset of 700 MHz, and it resonates at 5.2 GHz.

To extract the *S*-parameters as well as refractive index, perfect electric conductor (PEC) and perfect magnetic conductor (PMC) boundary conditions were assigned in the *yz*- and *xy*-planes of the proposed unit cell. The two ports are assigned in the y-direction. The parameters of the unit cell are extracted using the full-wave EM simulation software Computer Simulation Technology (CST) Microwave Studio. The transmission and reflection coefficients of the unit-cell structure, whose orientation is relative to the incident E-field, at 0° and 90° are shown in Fig. 7.11. It shows that the unit cell acts as a transmission type in the operating frequency of 5.2 GHz at 0° unit-cell alignment or when the input field in the *y*-direction. Here, the simulated $|S_{11}|$ for the unit cell is − 41 dB and $|S_{21}|$ is − 0.09 dB at 5.2 GHz. When the unit cell is aligned at 90° or when the input field in the *x*-direction, there is a shift in the resonant frequency, the transmission coefficient is − 0.24 dB, and the reflection coefficient is − 14 dB which is higher than the 0° aligned unit cell. This phenomenon is due to the inherent asymmetry in the proposed structure. The polarization of the EM wave used for the unit-cell simulation is matched to the polarization of the incident wave propagating from the dipole antenna in the same direction.

From Fig. 7.12, both permittivity and permeability are negative for the unit-cell structure in both 0° and 90° alignments. A medium with both permittivity and permeability less than zero ($\varepsilon < 0$, $\mu < 0$) is designated as double-negative (DNG) medium. Therefore, the proposed unit cell is a DNG unit cell. This class of materials is not found in nature, but they are physically realizable [9]. The DNG medium was first proposed by Pendry whose analysis shows that near-field waves experience exponential growth in the medium [10]. Due to the negative phase delay experienced by the

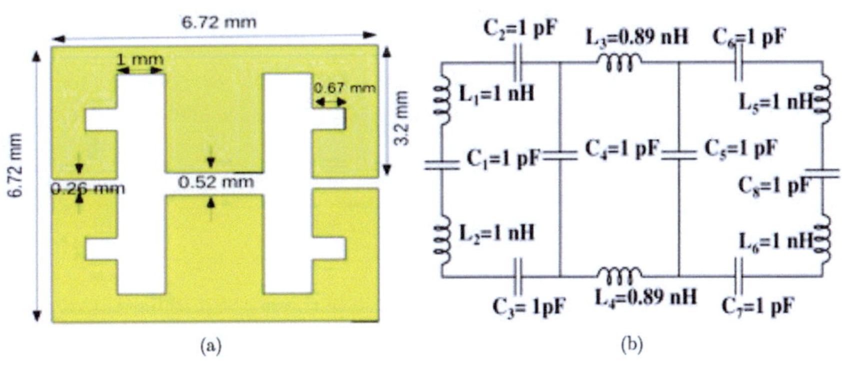


**Fig. 7.10** **a** Configuration of the proposed unit cell, **b** equivalent circuit structures. Reprinted with permission from IEEE [3]

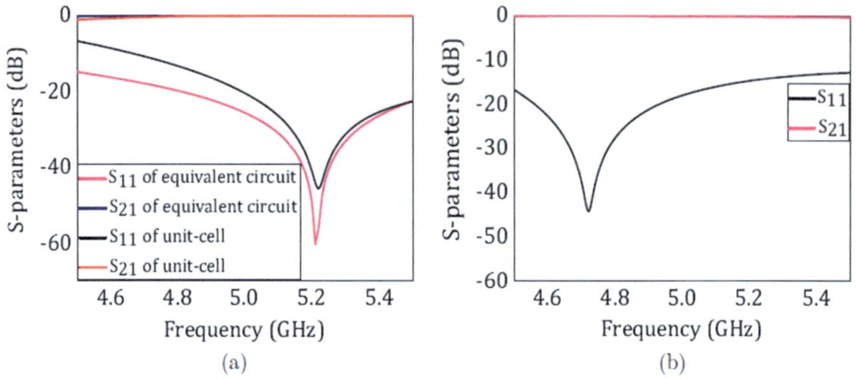


**Fig. 7.11** *S*-parameters of the proposed unit-cell structures in **a** 0° and **b** 90° alignment

**Fig. 7.12** Extracted permittivity and permeability of the proposed unit-cell structures in 0° and 90° alignment

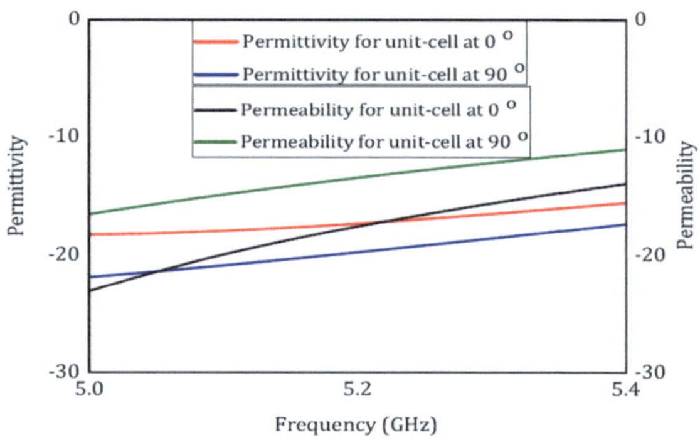


propagating waves, Pendry's lens is able to focus the entire beam [11]. Ziolkowski and Heyman [12] demonstrated that waves in general DNG media coalesce into localized beam fields that channel the power flow. Even though a perfect lens is not practically realizable, it makes the waves focus along the axis of propagation which enhances the gain.

Next, the unit cells are arranged with the two-port antenna design discussed in the previous section and its influence in enhancing the gain of the antennas is analyzed.

### *7.3.4 Antenna Design with Metamaterials*

As the objective is to enhance the gain of the antenna along with equalizing the gain at the two ports, each unit cell needs to be arranged carefully. Initially, all the unit cells are arranged with antenna 1 facing all the unit cell at 0°, while antenna 2 facing the unit cells at 90° and vice versa. In these cases, the antenna facing the 0° unit cell had a gain improvement of 4.35 dB, while the antenna facing the 90° unit cell had a

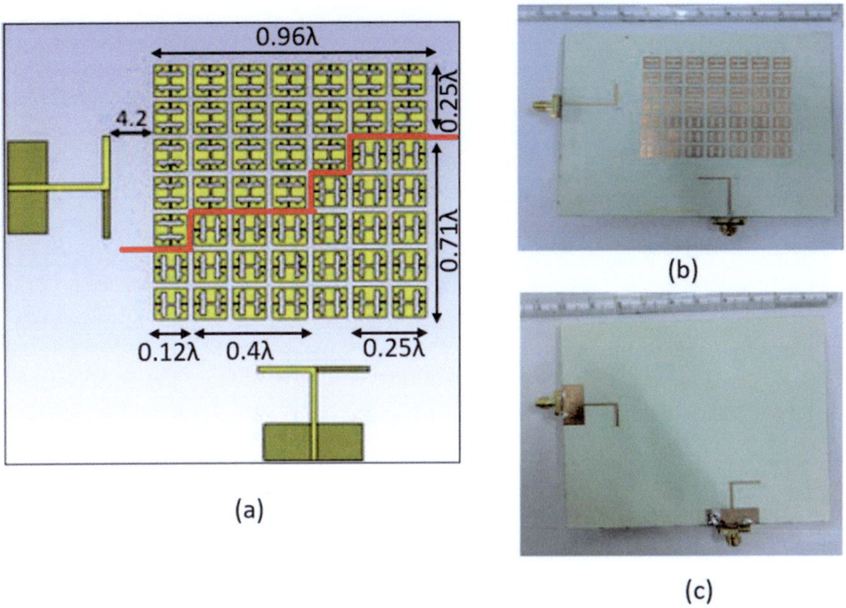


**Fig. 7.13** **a** Configuration of the proposed two-dipole design with unit-cell structures, **b** fabricated design top layer, **c** fabricated design bottom layer [3]. Reprinted with permission from IEEE

gain improvement of 2 dB. The difference in the effective aperture for its respective polarization in the two cases results in two different gain improvements for both antennas. Combining the two instances, the structure proposed in Fig. 7.13 ends in equivalent gain improvement for both the antennas.

In the proposed unit-cell arrangement, the unit cells are arranged such that 24 unit cells in a partial pyramidal arrangement are placed facing antenna 1 at 0° alignment, while the other 25 unit cells in another partial pyramidal pattern are placed facing antenna 2 at 0° alignment. The distance between two adjacent unit cells is 1.4 mm (0.025λ). From the simulations for a range of distances, it was observed that when the unit cells are close to each other, some energy reflects backward, causing the gain to reduce in the forward direction. This arrangement is unique such that there is a gain enhancement of 3.6 dB at the operating frequency for both the antennas resulting in gain flattening with a common aperture.

The distance between the dipole antenna and unit-cell structure in Fig. 7.13a is fixed at 4.2 mm. It was observed that when the unit cells are brought closer to the dipole, the dipole antenna detunes from the designed frequency of 5.2 GHz. This is due to the interference caused from the coupling of EM waves from the unit-cell structure. When the unit-cell structure is moved further away from the dipole, and when the distance between them becomes more than 12 mm (0.2λ), the gain of the antenna starts to reduce. The distance between the dipole and unit cells was chosen as 4.2 mm for a compact antenna design with the required performance characteristics.

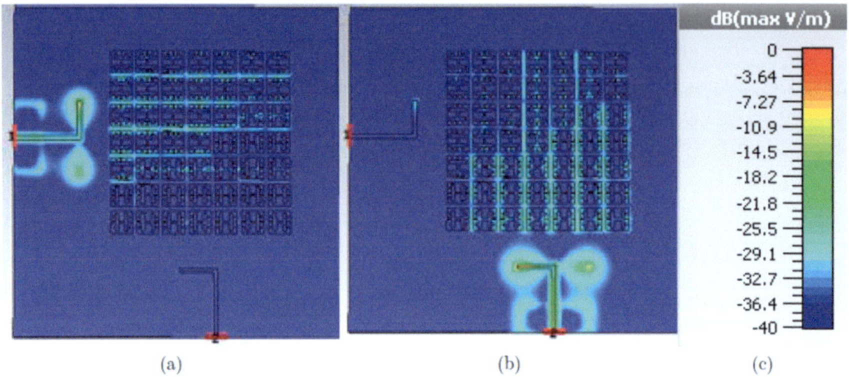


**Fig. 7.14** E-field of the antenna when **a** port 1, **b** port 2 are excited [3]. Reprinted with permission from IEEE

The radiated electric field from the antenna illuminates the unit cells as shown in Fig. 7.14. It makes the unit cells behave like a parasitic element to the antenna beam, thus enhancing its gain.

Most metamaterial loading designs in the literature focus only on improving the gain, like the loading technique proposed in [13]. Here, the metamaterial structure is placed in the common endfire direction of both antennas rather than around the antennas, which might result in different antenna gain enhancements and take up more area. The proposed arrangement in Fig. 7.13 is optimized based on how the electric field illuminates the individual unit cell at the 0° and 90° orientations. The electric field interactions of the individual unit cells were observed and the direction of the unit cell was changed to the 0° orientation when the gain had to be increased and changed to 90° orientation when the gain had to be reduced so that the final objective of equal gain enhancement is achieved. In this study, the antenna was simulated with different arrays of unit cells of similar arrangement and their performances are summarized in Table 7.1.

The entire metamaterial structure used has an area of 55 $\times$ 55 mm$^2$ (0.95$\lambda$ $\times$ 0.95$\lambda$). For the same unit-cell area, a simulation was performed using Yagi-Uda directors to compare the gain enhancement obtained using the unit-cell structure. In case 1, the Yagi-Uda directors are spaced at 0.2$\lambda$. Directors with decreasing length of 5% from the previous director were put along the endfire direction of a single dipole. The gain obtained at the operating frequency was 8.69 dBi resulting in a gain improvement of 2.39 dB. In case 2, the directors were placed at the endfire direction

**Table 7.1** Performance of different array sizes

| Array size | 2 × 2 | 3 × 3 | 4 × 4 | 5 × 5 | 6 × 6 | 7 × 7 |
|---|---|---|---|---|---|---|
| Gain enhancement in Ant. 1 | 0.32 | 0.7 | 1.41 | 2.13 | 3.06 | 3.76 |
| Gain enhancement in Ant. 2 | 0.29 | 0.77 | 1.27 | 2.21 | 3.03 | 3.73 |

of both the dipoles resulting in a reduction of gain to 3.95 and 3.72 dBi at both dipole antennas. This gain reduction is due to destructive interference of the waves when the directors are placed orthogonal to each other. The proposed metamaterial structure yields a much higher gain than Yagi-Uda case for the given aperture size. Additionally, simultaneous gain enhancement of the orthogonal antennas would not be possible with a Yagi-like topology.

The simulated and measured return losses of the two antennas after metamaterial loading are shown in Fig. 7.15a. The simulated 10-dB impedance bandwidth for the two antennas is 540 MHz (4.98–5.52 GHz). The simulated resonance of the antennas is at 5.22 GHz, while the measured resonance of antenna 1 is at 5.26 GHz and antenna 2 is at 5.26 GHz resulting in a frequency shift of 40 MHz. It also confirms that the metamaterial loading does not influence the return loss of the antennas. After using metamaterials, the simulated gain improved to 10 dBi for both the antennas, as shown in Fig. 7.15b. It concludes that the metamaterial loading has improved the front-to-back ratio by 4.5 dB.

Figure 7.16a shows the mutual coupling of the two-dipole antennas used in the proposed design. A strong coupling between the two antennas will affect the system performance. By improving the mutual coupling, the gain of the antenna is improved. From the figure, it is clear that when the unit-cell structures are introduced, there is an improvement in the mutual coupling between the two antennas. The introduction of the unit-cell structure resulting in gain improvement ensures less electromagnetic interaction between the antennas, thereby improving their isolation. Table 7.2 details the mutual coupling between the antennas and its improvement. The radiation efficiency of the antenna without and with the unit cells is depicted in Fig. 7.16b.

Figure 7.17 shows the radiation pattern along the $E$-plane ($xy$-plane) and $H$-plane ($xz$-plane) of the two dipoles at the frequency of operation. The simulated half-power beamwidth (HPBW) of antenna 1 is 46° with a front-to-back ratio of 7.4 dB, and for antenna 2, it is 49° and 7.9 dB, respectively.

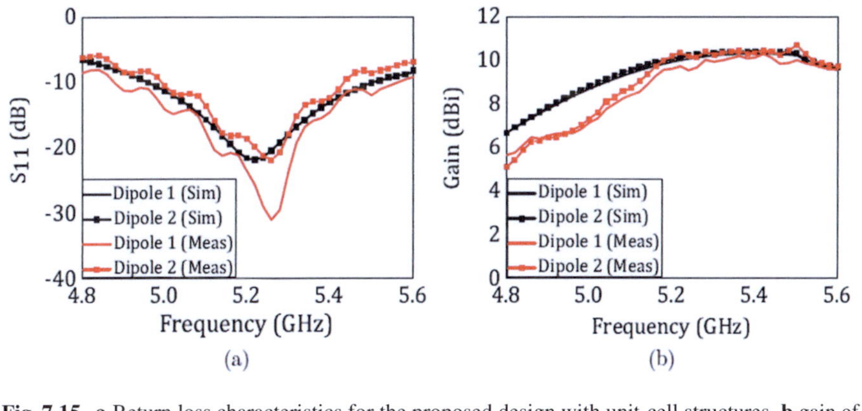


**Fig. 7.15** **a** Return loss characteristics for the proposed design with unit-cell structures, **b** gain of two-dipole antennas of the proposed design with unit-cell structures [3]. Reprinted with permission from IEEE

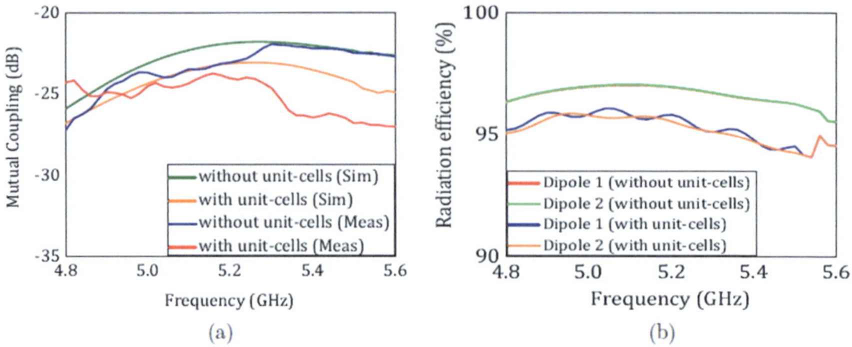


**Fig. 7.16** **a** Mutual coupling between the two-dipole antennas, **b** radiation efficiency of the two-dipole antennas [3]. Reprinted with permission from IEEE

**Table 7.2** Mutual coupling between the two antennas

| Mutual coupling results—$S_{21}$ | Without unit cells (dB) | With unit cells (dB) | Improvement (dB) |
|---|---|---|---|
| Simulation | – 21.87 | – 23.11 | 1.24 |
| Measurement | – 23.05 | – 24.01 | 1 |

The envelope correlation coefficient (ECC) graph for the operating bandwidth is shown in Fig. 7.18. The ECC is the square of correlation coefficient, that describes how much the communication channels are isolated or correlated with each other. The computation of envelope correlation for a two-antenna system is described in [14]. Ideally the value of ECC should be zero. For the proposed design, as all the values are below 0.002, it indicates a good diversity performance of the antenna.

### 7.3.5 Antenna Measurement

The reflection coefficient of the antenna was measured using Anritsu MS2028C Vector Network Analyzer. The far-field measurements are conducted inside an anechoic chamber. When port 1 is excited, port 2 is terminated with a 50 Ω load and vice versa.

The gain transfer method is used for measuring the gain of the antenna. The ETS-Lindgren® 3115 Model double-ridged waveguide horn antenna is used as the standard gain antenna for the measurement. The horn antenna, which works as the transmitting antenna, is connected to one port of the VNA and the antenna for measurement, which works as the receiving antenna, is connected to the second port of VNA. The VNA is used to measure $S_{21}$, ratio of received power, Pr, and transmitted power, Pt.

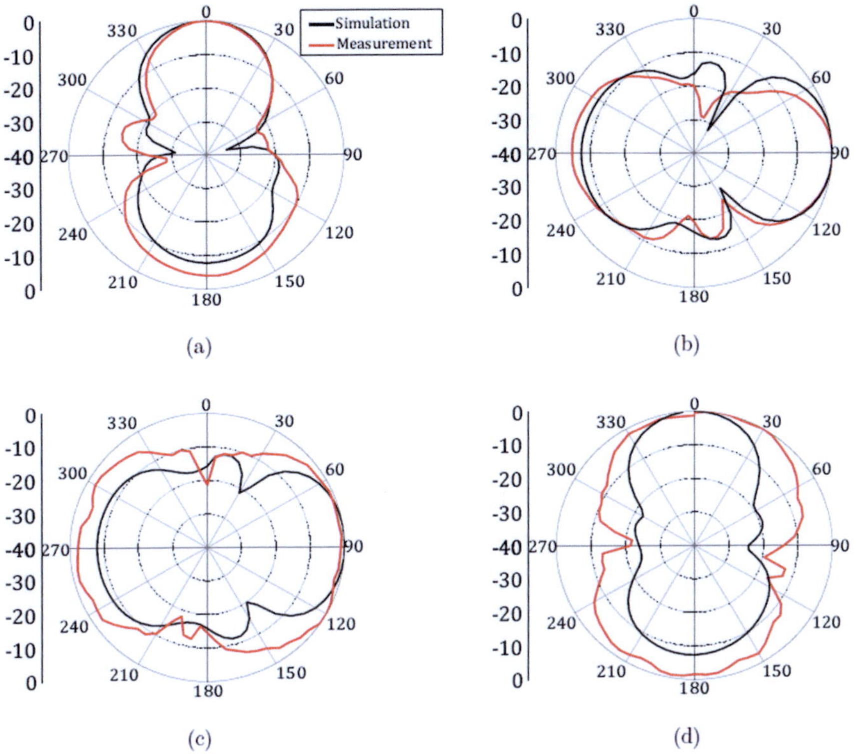


**Fig. 7.17** Radiation pattern at the operating frequency of 5.2 GHz **a** in *xy*-plane when port 1 is excited, **b** in *xy*-plane when port 2 is excited, **c** in *yz*-plane when port 1 is excited, **d** in *yz*-plane when port 2 is excited [3]. Reprinted with permission from IEEE

**Fig. 7.18** Simulated ECC for antenna ports 1 and 2

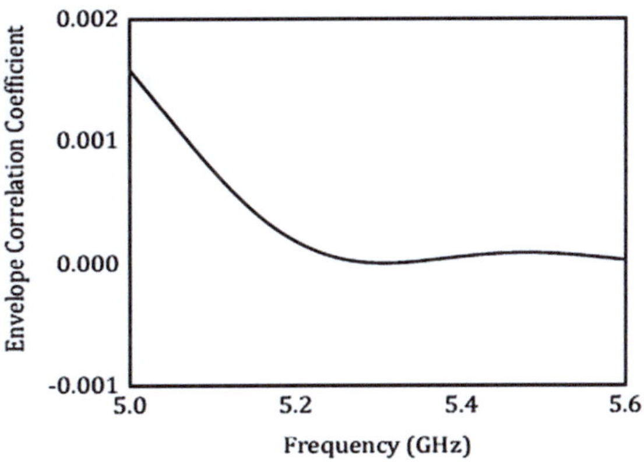

$$|S_{21}|^2 = \left(\frac{\lambda}{4\pi R}\right)^2 G_{\mathrm{HORN}} G_{\mathrm{AUT}}. \tag{7.1}$$

There is a reasonable agreement between the simulation and measurement results; however, slight discrepancies are due to the connector soldering and fabrication tolerances which were not taken into account during simulation. A solderless end launch connector would give a better result.

## 7.4 Common Aperture Three-Element Dipole Antenna Design

The proposed antenna is a two-element dipole antenna resonating at 5.2 GHz with a pattern diversity at 0°, + 45°, and − 45° with an individual gain of 10 dBi.

### *7.4.1 Antenna Design*

The geometry of the three-port antenna consists of three-dipole antennas placed at an angular offset of $45^0$ with each other. The three-element antenna system consists of three-dipole antennas arranged as shown in Fig. 7.19. The substrate used for the antenna design is Rogers RO4350B with a dielectric constant of 3.66 and thickness of 0.508 mm. The spacing between three antennas is 28 mm (0.5λ). Keeping the antennas closer results in higher mutual coupling, which reduces the gain of the individual antennas. The antenna placed at port 1 radiates a beam at − 45°, the antenna at port 2 radiates a beam at + 45°, and the antenna at port 3 radiates a beam at 0°. This design provides a wider beam coverage at the endfire direction.

For each port, the simulated and measured return loss characteristics are shown in Fig. 7.20a. The simulated impedance bandwidth for the three antennas is 530 MHz (4.94–5.47 GHz) which is 10% of 5.2 GHz for $|S_{11}| < -10$ dB. The simulated resonance of the three antennas is at 5.18 GHz, while the measured resonance of antenna 1 is at 5.32 GHz, antenna 2 is at 5.32 GHz, and antenna 3 is at 5.34 GHz. The dipole antenna at port 1 radiates along the endfire direction ($\varphi = 45°$) with a simulated gain of 6.54 dBi, while the dipole antenna at port 2 radiates along the endfire direction ($\varphi = -45°$) with a simulated gain of 6.37 dBi and the dipole antenna along port 3 radiates along the endfire direction ($\theta = 90°$, $\varphi = 0°$) with a simulated gain of 6.36 dBi, respectively, at 5.2 GHz as depicted in Fig. 7.20b. Figure 7.21 shows the radiation pattern along the $E$-plane ($xy$-plane) of the three dipoles at the frequency of operation. The simulated half-power beamwidth (HPBW) of antenna 1 is 71° with a front-to-back ratio of 3 dB, and for antenna 2, it is 6° and 3 dB, respectively. The simulated half-power beamwidth (HPBW) of antennas 1, 2, and 3 is 86°, 86°, and 94° with a front-to-back ratio of 4.2 dB, 4 dB, and 5.5 dB, respectively.

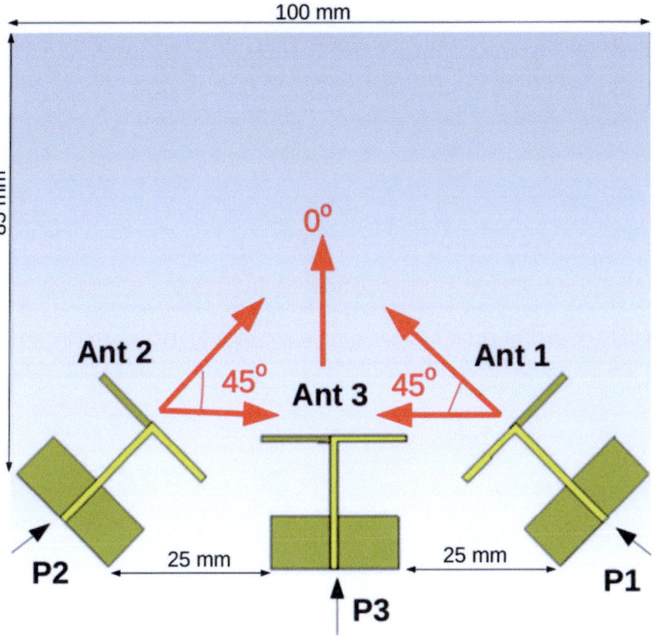


**Fig. 7.19** Configuration of the proposed three-dipole antenna design

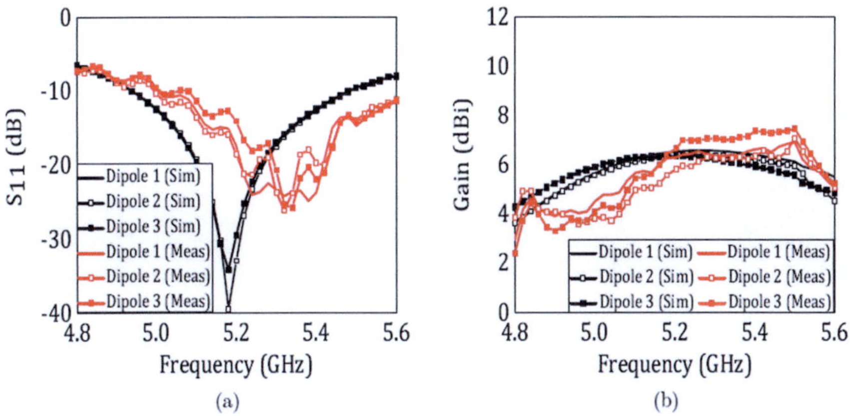


**Fig. 7.20** **a** Return loss characteristics for the three-dipole antenna design, **b** gain of three-dipole antennas of the proposed design

## *7.4.2 Antenna Design Integrated with Metamaterials*

For obtaining equal gain between the antennas, the unit cells must be arranged as shown in Fig. 7.22. A 6 $\times$ 7 array of the proposed unit cells was printed onto the

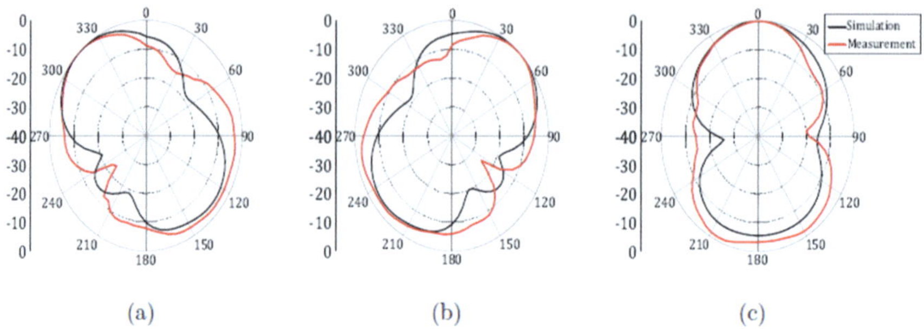


**Fig. 7.21** Radiation pattern at the operating frequency of 5.2 GHz in *xy*-plane when **a** port 1 is excited, **b** port 2 is excited, and **c** port 3 is excited

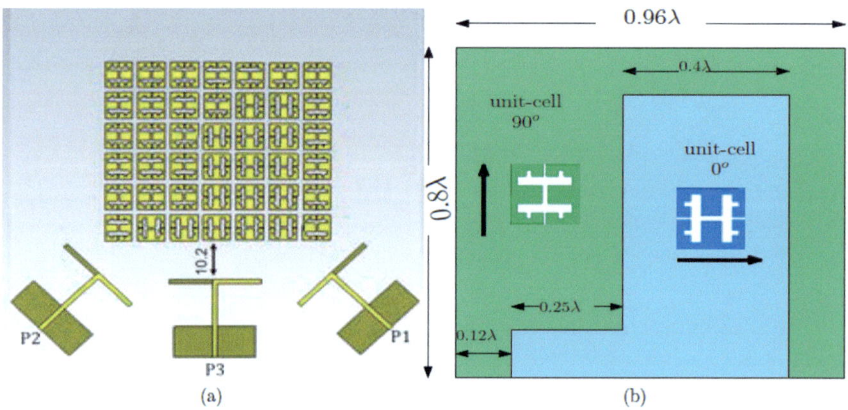


**Fig. 7.22** **a** Configuration of the proposed three-dipole design with unit-cell structures, **b** arrangement of the unit-cell structures (all dimensions are in mm)

substrate and placed in a common position of all the three antennas. Here, 26 unit cells are arranged in 90° orientation and 16 unit cells are arranged in 0° orientation. The entire metamaterial structure has an area of $55 \times 47\text{mm}^2$ ($0.95\lambda \times 0.82\lambda$). For the same unit-cell area, Yagi-Uda directors were placed to compare the gain enhancement obtained using the unit-cell structure. The directors were placed at the endfire direction of the three dipoles resulting in a gain of 2.5 dBi for antennas 1 and 2 and 6.5 dBi for antenna 3. For the given aperture, the proposed metamaterial structure yields a much higher gain than Yagi directors, proving the utility of metamaterial structures.

The spacing between the metamaterial structures and the antennas is 10.2 mm. It was observed that a lesser spacing between them caused detuning of the antennas due to coupling between the unit cells and the dipole antenna. Placing the unit-cell structures far away would cause the gain of the antennas to reduce. After optimization, the spacing was finalized to be 10.2 mm.

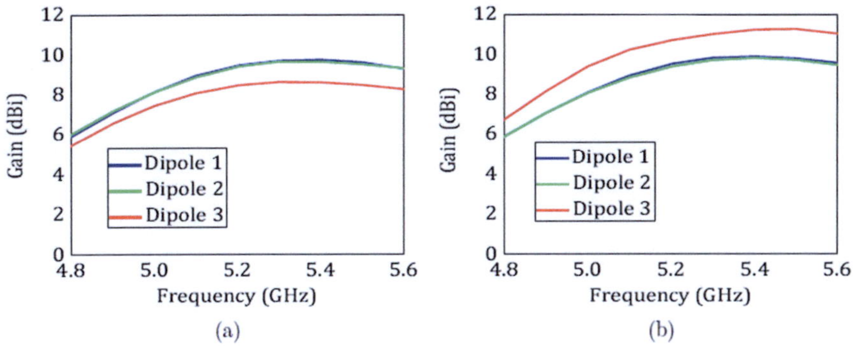


**Fig. 7.23** **a** Only 90° oriented unit cells are used, **b** only 0° oriented unit cells are used

To achieve the objective of gain equalization across the three antenna ports, the unit cells must be arranged in the way as shown in Fig. 7.22. To obtain this configuration, initially, all the unit cells are arranged in 90° orientations. In this case, the gain of dipole 3 is 8.7 dBi and the gain obtained at other dipole antennas is 9.7 dBi, as shown in Fig. 7.23a, resulting in 1-dB gain variation between the ports. Then, the unit cells are arranged in 0° orientations. In this case, the gain of dipole 3 is 11.2 dBi and the gain obtained at other dipole antennas is 9.8 dBi, as shown in Fig. 7.23b, resulting in a 1.4-dB gain variation between the ports.

The simulated and measured return loss characteristics of the two antennas after metamaterial loading are shown in Fig. 7.24a. The simulated impedance bandwidth for antennas 1 and 2 is 530 MHz (4.93–5.46 GHz) and for antenna 3 is 520 MHz (5.04 MHz–5.56) for $|S_{11}| < -10$ dB. The simulated resonance of antennas 1 and 2 is at 5.18 GHz and antenna 3 is at 5.25 GHz, while the measured resonance of antenna 1 is at 5.26 GHz, antenna 2 is at 5.28 GHz, and antenna 3 is at 5.26 GHz. It was noted that the $0^0$ unit-cell orientation resulted in a very high gain for dipole 3 and 90° orientation resulted in a lower gain for dipole 3. The difference in the effective aperture for its respective polarization in the two cases results in two different gain improvements for port 3. After arranging the unit cells at 0° orientation, some unit cells were rotated to 90° orientations to reduce the port 3 antenna gain. After some iterations of the unit-cell arrangements, the proposed unit-cell arrangement was obtained that resulted in the gain of dipole 3 at 10 dBi and the gain of dipole 2 and 3 to be 9.8 dBi as shown in Fig. 7.24b resulting in only 0.2–0.4-dB variation in gain between the antennas. Here 26 unit cells are arranged in 90° orientation and 16 unit cells are arranged in 0° orientation to obtain equal gain enhancement.

Figure 7.25a shows the mutual coupling between antenna 2 and antenna 3 from the proposed design. From the figures, we see that when the unit-cell structures are introduced, there is an improvement in the mutual coupling between the two antennas.

Table 7.3 details the mutual coupling between antenna 2 and antenna 3 and its improvement after the unit-cell loading. The mutual coupling between antenna 1

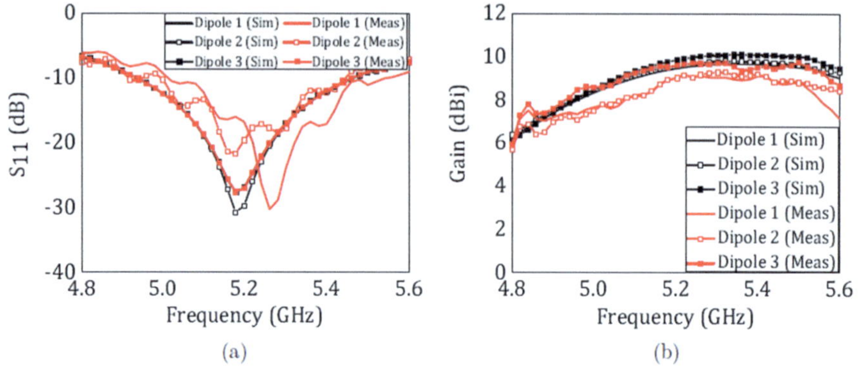


**Fig. 7.24** **a** Return loss characteristics for the proposed design with unit-cell structures, **b** gain of three-dipole antennas of the proposed design with unit-cell structures

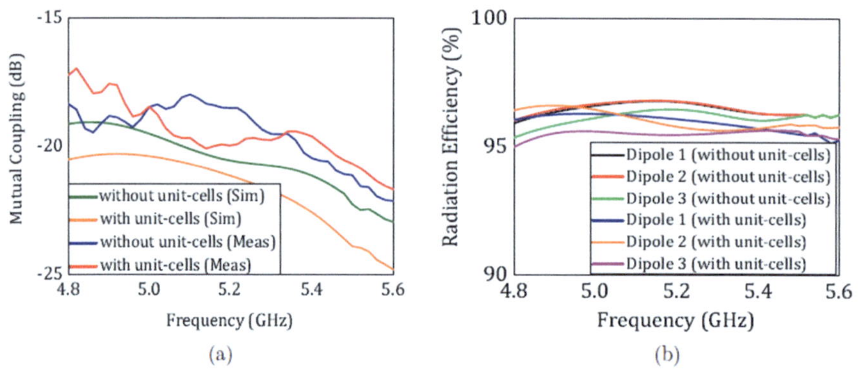


**Fig. 7.25** **a** Mutual coupling between the two-dipole antennas, **b** radiation efficiency of the three-dipole antennas

**Table 7.3** Mutual coupling between antennas 2 and 3

| Mutual coupling results—$S_{32}$ | Without unit cells (dB) | With unit cells (dB) | Improvement (dB) |
|---|---|---|---|
| Simulation | – 20.54 | – 21.13 | 0.6 |
| Measurement | – 18.51 | – 19.96 | 1.45 |

and antenna 2 is also like the coupling between antennas 2 and 3. The antenna has good isolation without the use of any isolation enhancement techniques. This makes the design much simpler as the inclusion of various isolation techniques can make the system very complex. Good isolation is an important parameter as it reduces the correlation between the antenna ports which is vital to any WLAN antenna designs.

Figure 7.25b depicts the radiation efficiency of the antenna without and with the unit-cell arrangement. The efficiency of the dipole antennas in all the cases is more than 95%. Figure 7.26 shows the radiation pattern along the *E*-plane (*xy*-plane) and *H*-plane (*xz*-plane) of the two dipoles at the frequency of operation. The simulated half-power beamwidth (HPBW) of antennas 1, 2, and 3 is 54°, 54°, and 60° with a front-to-back ratio of 7.3 dB, 7.3 dB, and 8.6 dB, respectively. This shows that the metamaterial loading has improved the front-to-back ratio by 3 dB.

The ECC graph for the operating bandwidth is shown in Fig. 7.27. All the values across the three antenna ports are below 0.01 which indicates a good diversity performance of the antenna. Figures 7.28 and 7.29 depict the fabricated prototype of the antenna design.

The performance of the proposed antenna is compared with the existing designs in the literature and is tabulated in Table 7.4. The proposed antenna provides a very high gain for the coverage of 90° in the broadside direction which is higher than the existing designs. In existing antenna designs, the gain of the antennas is less if there is any gain equalization throughout the different ports as seen in [15] or there is a high gain variation across the ports as seen in [16, 17]. The proposed antenna is able to achieve high gain along with gain equalization across the ports.

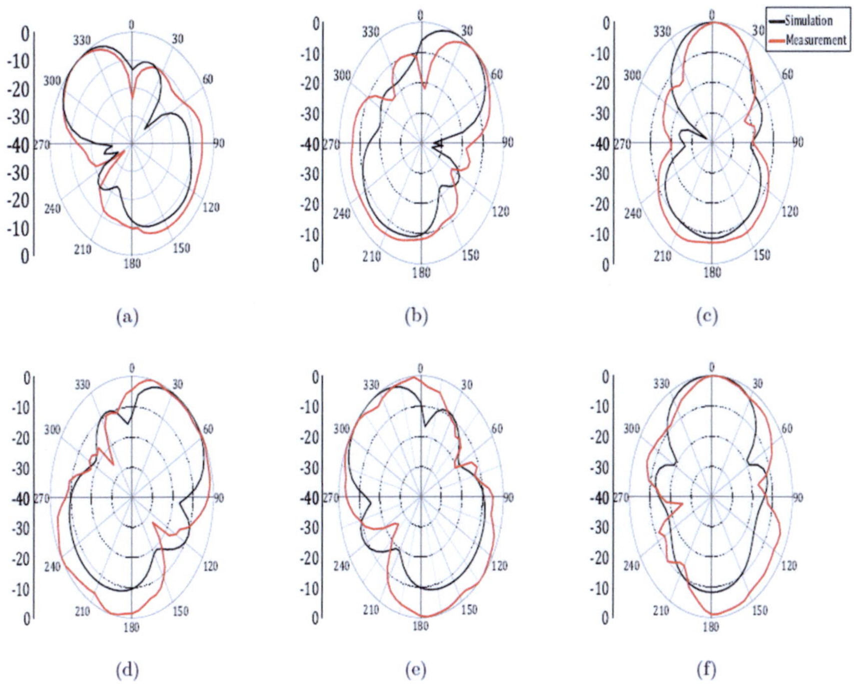


**Fig. 7.26** Radiation pattern at the operating frequency of 5.2 GHz in *xy*-plane when **a** port 1 is excited, **b** port 2 is excited, **c** port 3 is excited and in *xz*-plane when **d** port 1 is excited, **e** port 2 is excited, **f** port 3 is excited

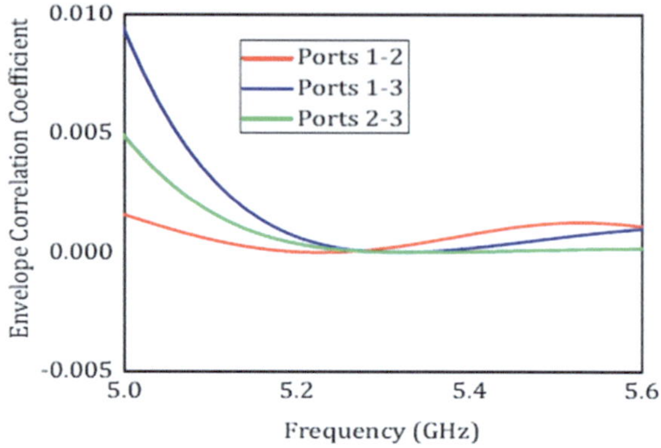


**Fig. 7.27** Simulated ECC for the three-dipole antenna design

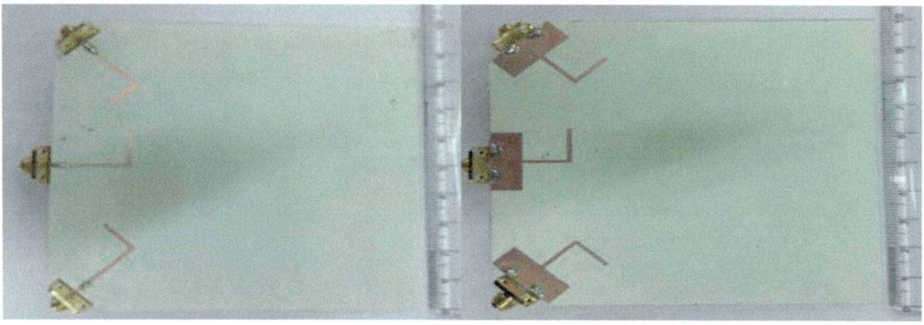

(a) top view (b) bottom view

**Fig. 7.28** Fabricated three-dipole antenna design without unit-cell structures

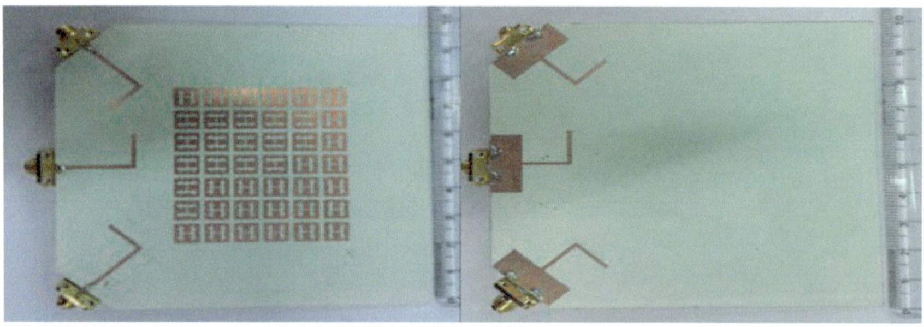

(a) top view (b) bottom view

**Fig. 7.29** Fabricated three-dipole antenna designs with unit-cell structures

**Table 7.4** Comparison of the proposed three-dipole structure with other antenna designs

| Ref. | Freq. (GHz) | Freq. range (GHz) | Size | Element spacing | Isolation (dB) | Gain (dBi) | Patterns at (°) |
|---|---|---|---|---|---|---|---|
| [15] | 2.4 | 2.4–2.48 | 0.23λ × 0.23λ | 0.2λ | 24 | 2–2.3 | 0°, + 120°, – 120° |
| [16] | 5.8 | 5.5–5.85 | 1λ × 0.56λ | 0.14λ | 18 | 0.8–3.5 | 0°, + 90°, – 90° |
| [17] | 6.0 | 5.45–6.4 | 1.25λ × 0.3λ | 0.28λ | 15 | 4.3–5.3 | 0°, + 90°, – 90° |
| Proposed Antenna | 5.2 | 4.94–5.47 | 1.74λ × 1.74λ | 0.43λ | 20.5 | 9.8–10 | 0°, + 45°, – 45° |

## 7.5 Gain-Equalized Three-Port Antenna with Superstrate Loading

The previous section described two/three-port antennas with gain enhancement and gain equalization across the ports at 0°, + 90° and 0°, + 45°, – 45°. The antenna beams of these designs are in the same azimuth plane. This section describes a three-port antenna that radiates at 0°, + 90°, – 90° in the azimuth and elevation plane and consists of two dipoles and a patch. Along with gain equalization across the ports, the proposed antenna has an improved beam coverage and polarization diversity which makes it a good candidate for WLAN applications.

### *7.5.1 Motivation*

Modern wireless communication technology is being built on multiple-input-multiple-output antenna systems. Such technology needs high data rates along with low power level operation. Being a multiport antenna system, isolation between the antenna ports is also of great concern. Various isolation enhancement techniques have been proposed in the literature like using parasitic elements [18], orthogonal feeding [19], and decoupling structures [20]. However, use of such techniques in a MIMO antenna can make the system more complex. To overcome these issues, [16] proposed a pattern and polarization diversity antenna for WLAN frequency having good isolation without the use of any isolation enhancement techniques. The design consists of two dipoles and a patch antenna at the center as shown in Fig. 7.30 and radiates beams at 0°, + 90°, – 90°. But the gain of the three antennas is in the range of 1.8–4 dBi which is low for WLAN applications. A high-gain antenna will ensure the reliability of the communication link.

To overcome the situation of low gain, the proposed design uses some metamaterial structures to enhance and equalize the gain across the ports. A reflective-type

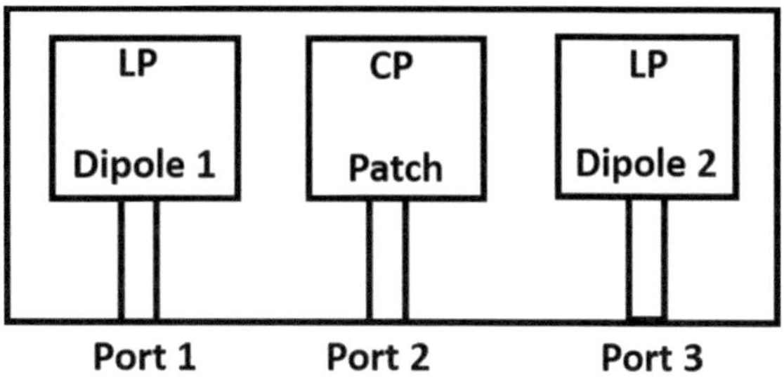


**Fig. 7.30** Three-port antenna illustration

metamaterial is used with dipole antenna to improve the gain in the endfire direction and a transmission-type unit cell is used above the patch as superstrate to improve the gain in the broadside. The superstrate material is designed and placed such that the gain throughout the frequency band and across the ports is equal. This ensures that the antenna provides maximum coverage across the hemispherical antenna plane. As a result, the information coming from all sides of the antenna is captured and no information is lost. This helps to improve the quality of transmission between a transmitter and receiver making it a good candidate for WLAN applications.

## *7.5.2 Antenna Design*

The three-element antenna design consists of two dipoles at the two edges of the substrate along with a patch at the center as shown in Fig. 7.31. The substrate used is Rogers RO4350B with dielectric constant $\varepsilon_r = 3.66$ and 20 mil thickness. The low dielectric constant helps to maintain minimal surface-wave modes. The thickness of 20 mil ensures proper mechanical stability to the antenna.

Initially, a conventional circular patch with coaxial feed was designed to model an antenna structure that is compact and optimized to operate at 5.2 GHz. A circular slot along with two rectangular slots was introduced in the patch to further improve the impedance bandwidth of the antenna. The patch antenna is fed using a coaxial feed which aids in an electrically compact design with high pattern integrity and helps in easier impedance matching [22]. Thereafter, the two dipoles are designed, and their two arms are printed on the opposite sides of the substrate along with a coaxial fed input. The input feed point for the three antennas is optimized for 50Ω impedance matching at the operating frequency. The entire three-port antenna system has an area of $32 \times 50\ \text{mm}^2$ ($0.56\lambda \times 0.9\lambda$), where $\lambda$ is the free-space wavelength with respect to 5.2 GHz. An adequate finite ground plane of dimension 30 mm × 32 mm ($0.52\lambda \times 0.56\lambda$) is provided at the bottom side of the substrate. The ground plane acts as a reflector and stabilizes the radiation pattern in the required frequency band. In this design, we considered two dipoles and a patch for the three ports. On the contrary, if we consider three dipoles in each ports, the dipole at port 3 with inadequate ground plane will result in poor front-to-back ratio during antenna radiation. If we consider

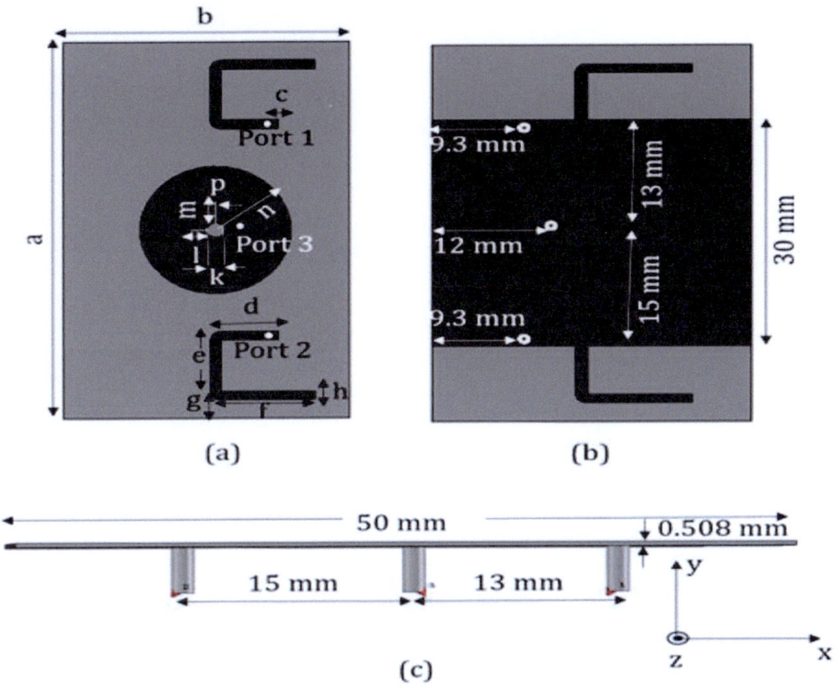


**Fig. 7.31** Schematic of three-element antenna system **a** top view, **b** bottom view, **c** side view [$a = 50$ mm, $b = 32$ mm, $c = 1.37$ mm, $d = 6.5$ mm, $e = 7$ mm, $f = 10.5$ mm, $g = 2.5$ mm, $h = 1.14$ mm, $k = 1.94$ mm, $l = 1.96$ mm, $m = 2.93$ mm, $n = 8.45$ mm, $p = 0.39$ mm] [21]

three-patch antennas, the patches at port 1 and port 2 will not provide sufficient tilt for the patterns to cover the required area, unless the antennas are tilted physically which will make it a non-planar design. For a planar antenna to cover the required area, the solution is to use two dipoles for endfire coverage and a patch for broadside coverage.

### 7.5.3 Antenna Design Theory

The parametric study of the antenna to obtain the desired dimensions is shown in Fig. 7.32.

Here, Fig. 7.32a shows the variation of the resonant frequency of the patch as the radius of the patch is increased ($n1$) and decreased ($n3$) from the optimum radius. Figure 7.32b depicts the variation in the bandwidth of the patch when the radius of the inner slotted circle is changed. The patch shows two resonant modes when the dimension of $k$ is increased ($k1$) and a lower bandwidth when $k$ is reduced ($k3$). The optimum dimension ($k2$) gives the maximum bandwidth for the patch. Figure 7.32c,

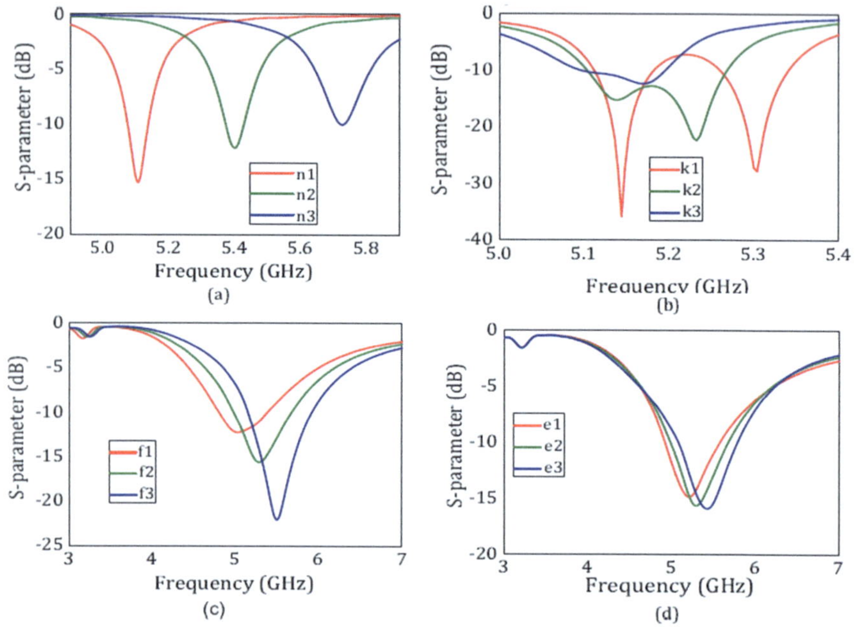


**Fig. 7.32** Parametric study of different dimensions of the antenna structure [21]

d shows how the resonant frequency is affected when the dimension of the dipole is varied. The length of the dipole must be approximately λ/2 at the resonant frequency ($f2$, $e2$). As inferred from the figure, when the length of the dipole is increased ($f1$, $e1$), the resonant frequency reduces, and the resonant frequency increases when the length of the dipole is reduced ($f3$, $e3$). The antennas are initially optimized separately and then moved close to each other at the same time keeping a good mutual coupling between the antennas.

The equivalent circuit for the proposed antenna is represented in Fig. 7.33. The equivalent circuit model for the dipole [23] and the patch [24] is provided along with the impedance matching circuit [25] which is matched to 50Ω input ports. The $C_{op}$ at the patch specifies the feed point impedance and $R_{op}$ is the conduction loss of the substrate. The tank circuits of the patch and the dipole antenna are frequency dependent which provides the resonance for the antenna. The mutual coupling between the antennas is modeled as the coupling capacitance between the dipole and the patch antenna. The basic mathematical formulation for the design of the antennas was obtained from [26]. Further, full-wave method of moments (MoM) analysis to understand the circular patch antenna is described in [27]. The modeling of a coaxial fed patch antenna is also described in [28]. The 50Ω impedance matching for the coaxial inputs for the dipole and patch was determined through parametric study of the antennas.

The simulated and measured input reflection coefficients are shown in Fig. 7.34. The simulated impedance bandwidth for the antenna system is determined by the

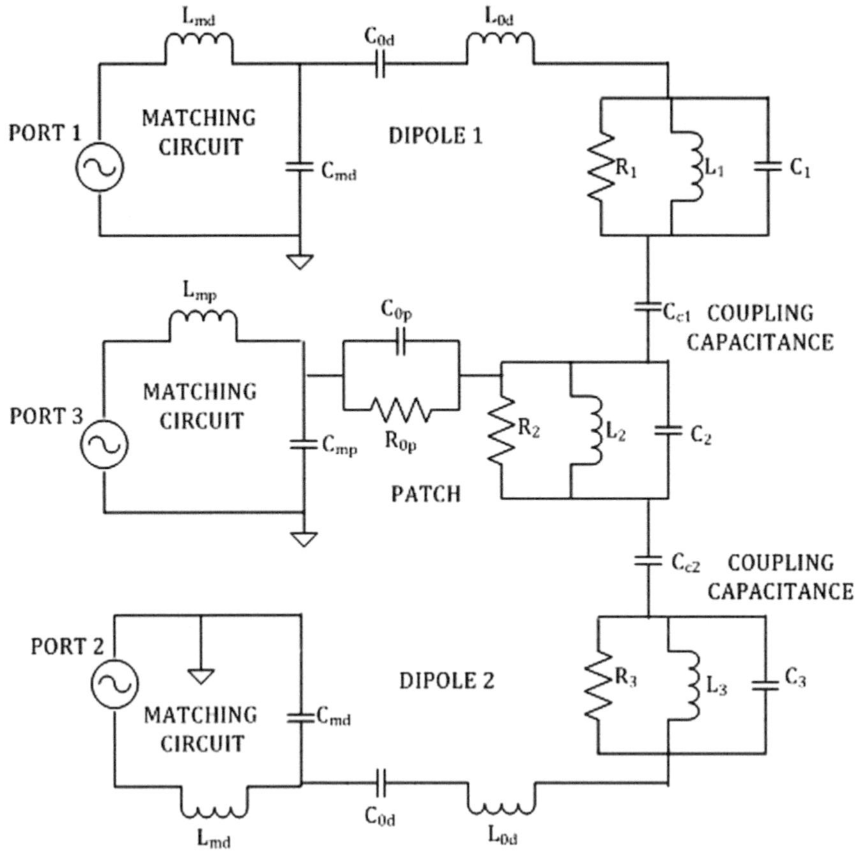


**Fig. 7.33** Equivalent circuit of the antenna [21]

patch whose bandwidth is 150 MHz (5.10–5.26 GHz). The variation in bandwidth is because the bandwidth of $S_{33}$ is determined by patch antenna which has a narrow bandwidth due to its inherent resonating nature compared to a dipole which determines the bandwidth for $S_{11}$ and $S_{22}$. A frequency shift of 200 MHz resonant frequency between measurement and simulation can be attributed to the fabrication tolerances that introduce a small lead inductance resulting in an additional stepped impedance to the radiating element and the tolerance of substrate permittivity.

Many communication systems such as WLAN do not require large instantaneous bandwidth and therefore prefer narrow-band antennas [29]. The half-power beamwidth (HPBW) of the dipole is around 180° with a front-to-back ratio of 11 dB, and for the patch, it is around 100° and 15 dB, respectively. The dipole antennas on port 1 and port 2 radiate along endfire orientation and the patch radiates along the broadside with a gain of 4.5 dBi.

To enhance the gain of the antenna described in the previous section, two different unit-cell structures are designed.

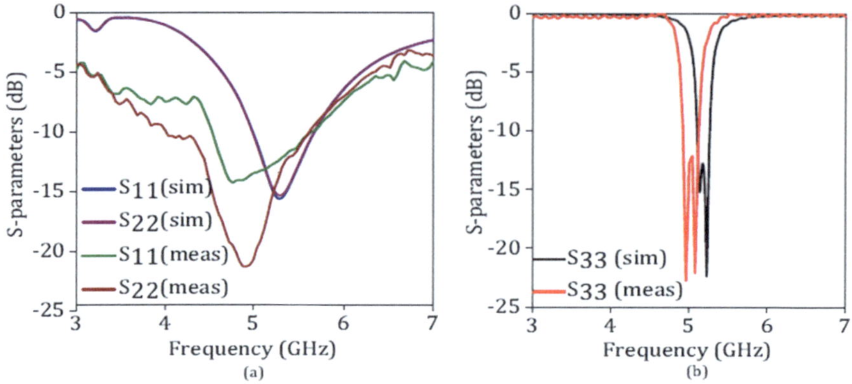


**Fig. 7.34** Input reflection coefficients of dipole and patch without unit-cell structures [21]

### 7.5.4 Metamaterial Design 1

The unit-cell structure as shown in Fig. 7.35a is designed to improve the characteristics of the patch antenna. The S-parameters from Fig. 7.35b conveys that it is a transmission-type unit-cell structure. To extract the *S*-parameters as well as refractive index, PEC and PMC boundary conditions were assigned in the *xz* and *yz*-planes of the proposed unit cell. The two ports are assigned in the *z*-direction. The parameters of the unit cell are extracted using the full-wave EM simulation software CST Microwave Studio®. The magnitude $S_{21}$ of the unit cell at the operating frequency of 5.2 GHz is − 0.1 dB. The high transmission characteristic of the unit cell ensures a higher gain for the antenna by squeezing the radiation from the radiator below the metasurface into a narrow beam. The $S_{11}$ magnitude of − 16 dB makes sure that little power gets reflected back to the antenna by the unit-cell layers.

### 7.5.5 Metamaterial Design 2

The unit-cell structure as shown in Fig. 7.36a is designed to improve the characteristics of the dipoles. The *S*-parameters from Fig. 7.36b conveys that it is a reflective-type unit cell with a return loss of − 1.2 dB at the operating frequency. The PEC and PMC boundary conditions were assigned in the *xz* and *yz*-planes of the proposed unit cell. The two ports are assigned in the *y*-direction. The magnitude $S_{11}$ of the unit cell at the operating frequency of 5.2 GHz is − 1 dB where the unit cells act like a reflector. Also, as the $S_{21}$ magnitude is − 6.6 dB, there is very less transmission of EM waves which helps in squeezing the radiation and improving the gain in forward direction.

The unit cells are strategically integrated with the earlier discussed antenna geometry and characterized along with the antenna design.

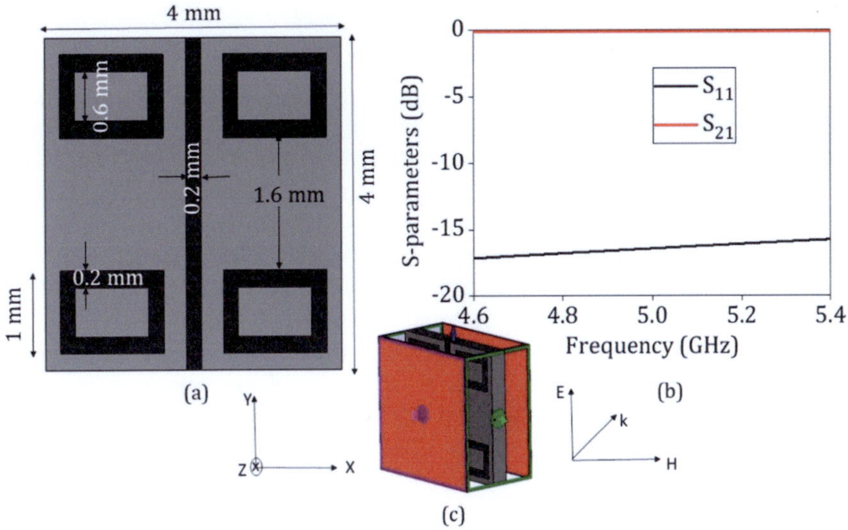


**Fig. 7.35** **a** Configuration of unit cell 1, **b** *S*-parameters of the unit cell, **c** the boundary conditions used in the simulation [21]

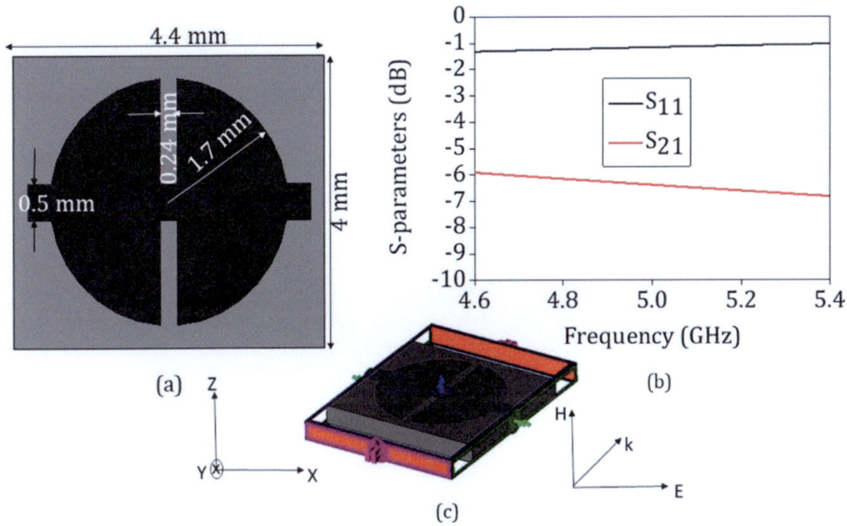


**Fig. 7.36** **a** Configuration of unit cell 2, **b** *S*-parameters of the unit cell, **c** the boundary conditions used in the simulation [21]

### 7.5.6 Antenna Design Using Metamaterials

Figure 7.37 shows the proposed antenna along with the unit-cell structures. Initially, only a single layer of the unit cell 1 as shown in Fig. 7.38a was placed over the patch antenna, where the superstrate layer is maintained at 0.18λ. The unit cell 2 as shown in Fig. 7.38b was arranged beside the dipole antennas. There is gain improvement, but the gain of dipole and patch is equal only at a single frequency of 5.26 GHz. An additional layer of unit cell 1 was placed at a height of 0.04λ from the previous layer over the patch. The extra layer of unit cells makes the gain of three ports to be enhanced to 6 dBi over the entire bandwidth. The gain enhancement using Yagi-Uda directors and reflectors is significantly lesser compared to the gain enhancement obtained using the unit-cell structures. If the height of the superstrate is further increased, the gain of dipoles and the patch begins to reduce. If the superstrate is placed closer to the antennas, it is observed that the gain of the dipole is reduced. When compared to the superstrate design in [30], the proposed design is much more compact in terms of distance from the patch for a minimal gain variation between the antennas.

The electric field distribution before and after the unit-cell loading is shown in Fig. 7.39. For the dipole, the unit cell increases the electric field intensity toward the endfire side which demonstrates the reflective action of the unit cells. When the metasurface is mounted 0.22λ away from the patch, it aids in phase correction consequently leading to gain enhancement. The gap between the patch and the superstrate is significant because it has an influence on the gain of the patch as well as

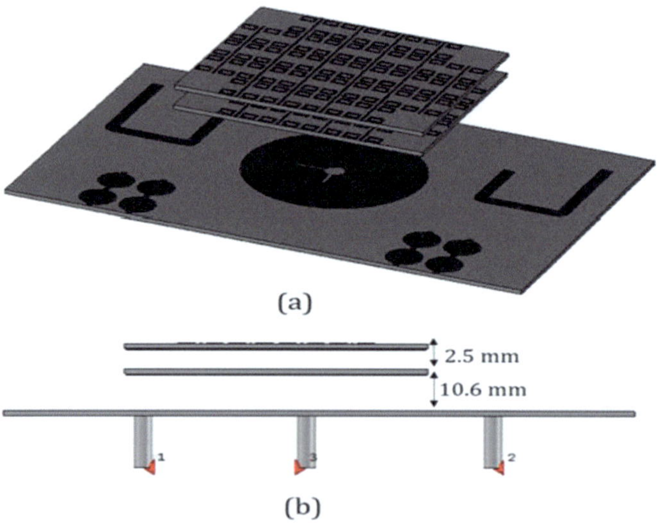


**Fig. 7.37** Configuration of the proposed three-element antenna structure with unit cells **a** top view, **b** side view [21]

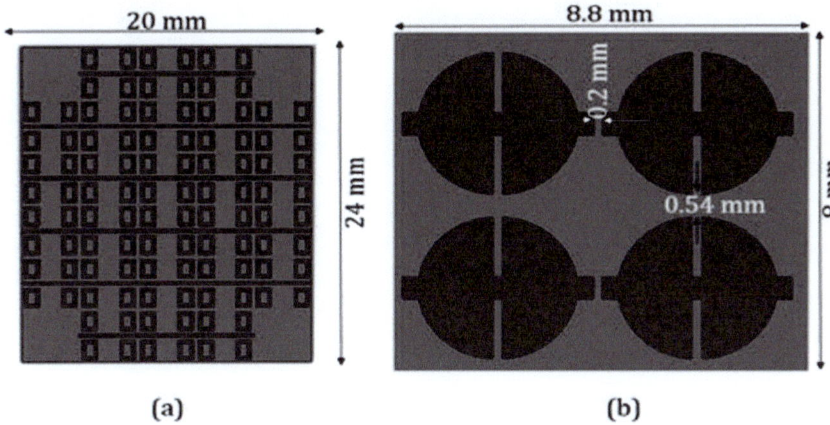


**Fig. 7.38** **a** Arrangement of the unit cell 1 above the patch, **b** arrangement of the unit cell 2 besides the dipole [21]

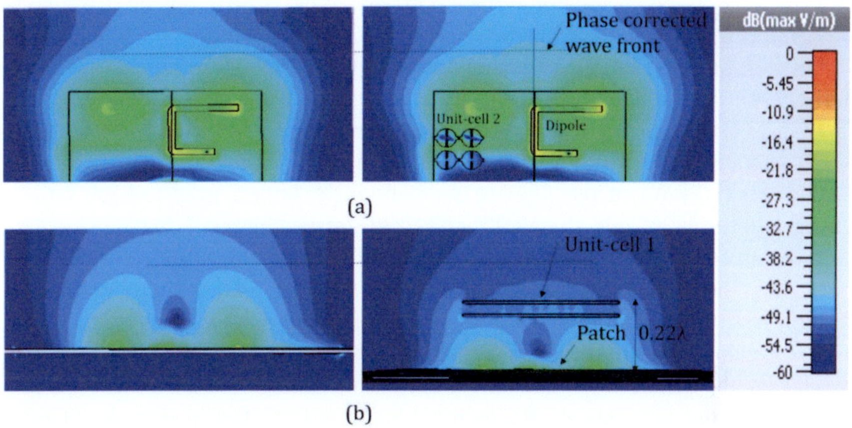


**Fig. 7.39** **a** Electric field distribution for the dipole, **b** electric field distribution for the patch before and after the arrangement of unit cells [21]

the dipole. The appended unit-cell structures are aiding in phase correction of the incoming cylindrical waves. The phase shift contributed by the individual unit-cell structures is same which then adds up for the entire structure. The unit-cell structures will reflect the EM waves in phase at the resonant frequency which then aids in gain enhancement in the corresponding direction.

Here, the metasurface above the patch has periodic, electrically small, unit-cell structures. In a material with positive permittivity and permeability, there is a delay in the phase of the wave as it moves forward [31]. By keeping a metasurface on the other side at a proper distance with an air gap, we can have the phase delay to be

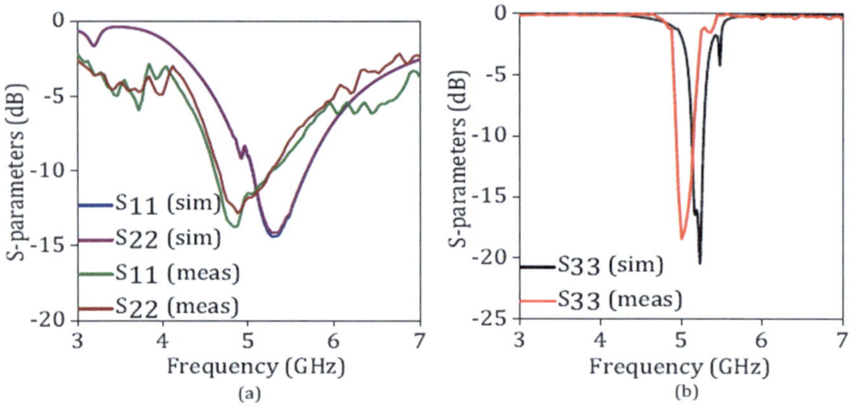


**Fig. 7.40** Input reflection coefficients of dipole and patch after unit-cell loading [21]

reversed such that the phase of the wave after the metasurface is same as the phase of the wave at the patch antenna surface. This makes the wave to radiate like a planar wave rather than a spherical like wave profile [32].

For each port, the simulated and measured input reflection coefficients are shown in Fig. 7.40. The simulated impedance bandwidth for the antenna system is 150 MHz (5.12–5.27 GHz) which presents the fact that there is no significant impedance bandwidth alteration after the integration by the unit cells. The simulated half-power beamwidth (HPBW) of the dipoles is around 120° with a front-to-back ratio of 13 dB, and for patch, it is around 90° and 11 dB, respectively.

Two superstrate layers were chosen to achieve the objective of equal gain throughout the band and across the ports. As seen in Fig. 7.41, the number of superstrate layers influences the gains of both patch and the dipole. If only one layer is used, the gain of the dipole varies across the operating bandwidth. Using more than two layers, the gain is equalized throughout the operating bandwidth. But it is seen that while using three layers, the gain begins to reduce. But in the case of patch antenna, as the number of superstrate layers is increased, the gain also increases. It is only at the instance of two superstrate layers that the required objective of equal gain of 6 dBi throughout the operating bandwidth for both dipole and patch is obtained.

The radiation pattern of the antenna in the *E*-plane before and after the unit-cell loading is shown in Fig. 7.42. The *E*-plane patterns show radiation at 0°, + 90°, − 90°. From the figure, we can infer that the radiation pattern is stable at the antenna resonant frequency. The antenna beamwidth after the unit-cell loading gets compressed compared to the patterns of the antenna before unit-cell loading due to increased gain. The cross-polarization levels are much smaller over the operating frequency band for the dipole antennas. The circular patch is circularly polarized (LHCP) which can be inferred from similar co- and cross-polarization levels of the radiation patterns. There is good agreement between the simulated and measured radiation patterns and slight variations are due to misalignments during measurement.

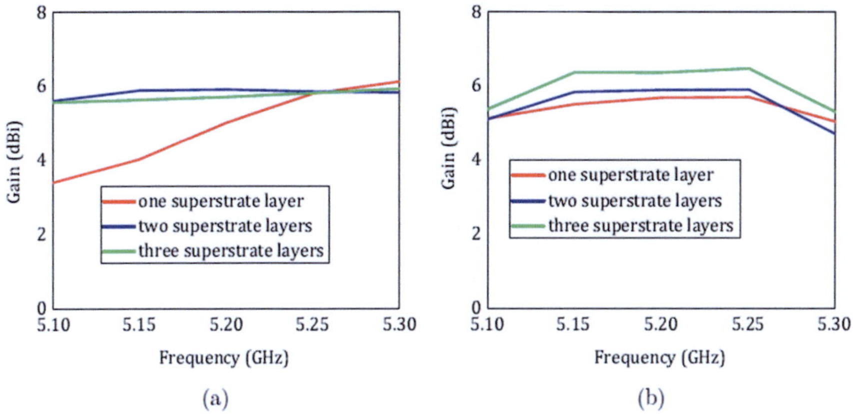


**Fig. 7.41** **a** Gain of the dipole antenna when number of superstrate layers are varied, **b** gain of the patch antenna when number of superstrate layers are varied

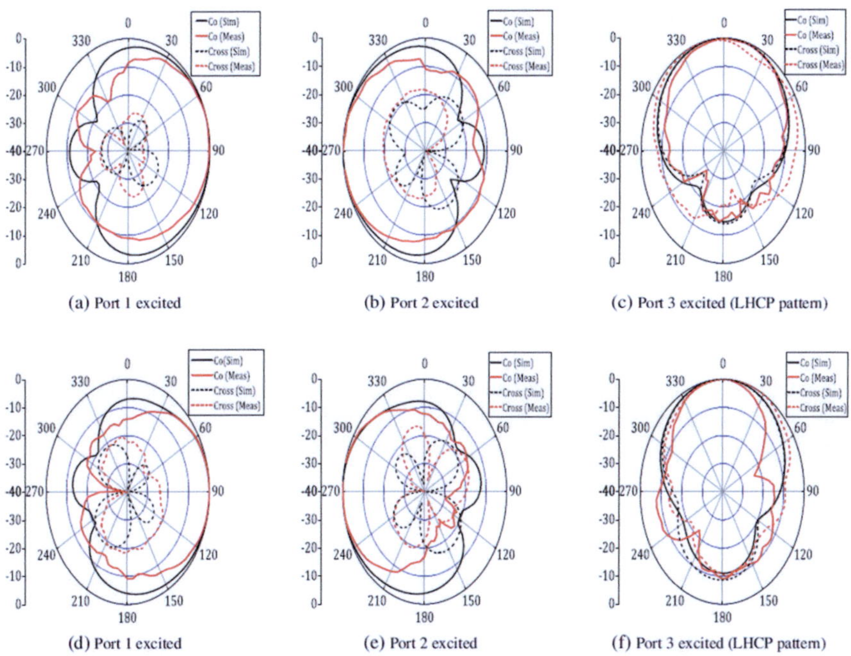


**Fig. 7.42** Radiation pattern at the operating frequency of 5.2 GHz **a**–**c** before unit-cell loading and **d**–**f** after unit-cell loading [21]

The axial ratio of the patch antenna is shown in Fig. 7.43. As the axial ratio values are below 6 dB, the patch antenna could be considered as circularly polarized without any loss of generality.

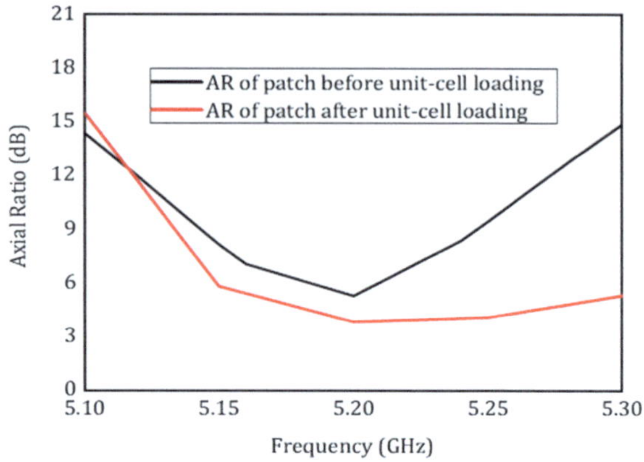


**Fig. 7.43** Axial ratio of the circular patch antenna at port 3 [21]

To understand the resonant mode behavior of the proposed antenna, the simulated surface current distributions at 5.2 GHz are shown in Fig. 7.44. The surface current distribution is seen at dipoles when either port 1 or 2 is switched ON and at the patch when port 3 is switched ON. There is an increased magnitude in the surface current distribution post the integration with unit-cell structures. The surface current distribution of the dipole shows a high gain toward the endfire directions, whereas the surface current distribution of the patch shows a high gain toward the broadside direction. Due to the coupling of energy from the antenna to the unit cells, the surface current distribution of the antenna feed will be changed, and currents will be induced on the metasurface. As a result, the patch antenna acts as the primary microstrip feed, while the metasurface is the real aperture that radiates energy to the free space. Figure 7.45 shows the 3D gain of the proposed antenna at the resonant frequency of 5.2 GHz when different ports are excited after the unit-cell integration. This indicates the direction of maximum radiation for different values of theta and phi. The maximum gain of 6 dBi for the antenna is at 0°, + 90°, and − 90°. We can also infer that the directivity of the antennas is wide enough to provide coverage of 330° across the three ports.

The gain enhancement of the three ports is shown in Fig. 7.46a with a simulated gain of 6 dBi after the introduction of the unit-cell structures. Figure 7.46b shows the simulated radiation pattern of the proposed antenna and how it covers $330^0$ around the antenna structure, making it suitable for WLAN applications. The antennas maintain an isolation of 26 dB at the operating frequency for an electrical distance of 0.25λ between the ports as shown in Fig. 7.47a, and the radiation efficiency of the antennas is shown in Fig. 7.47b. The ECC graph for the operating bandwidth is shown in Fig. 7.48. All the values are below 0.005 which indicates a good diversity performance of the antenna. Figure 7.49 depicts the photograph of the fabricated prototype of the antenna design without and with the unit-cell structures. Three SMA connectors are connected to the input ports of the antenna through coaxial feeding. Anritsu MS2028C Vector Network Analyzer is used for measuring the reflection coefficient of the antenna and the far-field measurements are conducted inside an anechoic chamber. Each time

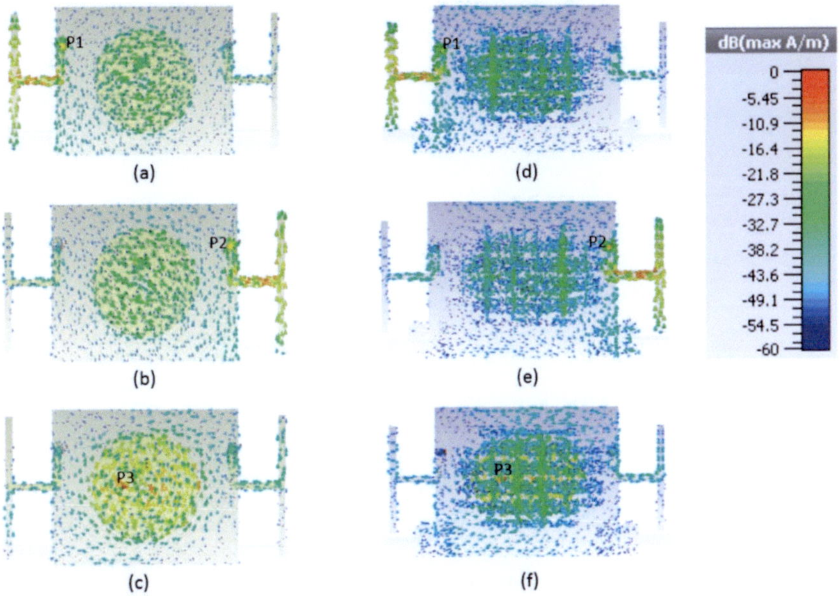


**Fig. 7.44** Surface current distribution of the antenna when **a** port 1, **b** port 2, **c** port 3 are excited before unit-cell loading and **d** port 1, **e** port 2, **f** port 3 are excited after unit-cell loading [21]

a single port is excited and the other ports are terminated with a 50Ω load. The gain transfer method is used for measuring the antenna gain. The ETS-Lindgren 3115 Model double-ridged waveguide horn antenna is used as the standard gain antenna for the measurement. The slight discrepancies seen between the simulation and measurement results are due to the connector soldering and fabrication tolerances which were not taken into account during simulation.

The performance of the antenna is compared with the existing three-element antenna designs and it is tabulated in Table 7.5. When compared to similar antennas available in the literature, the proposed antenna has maximum coverage for higher gain. The gain of the other antennas is restricted to much lower angular spread with very low radiation efficiencies in some instances. The proposed antenna provides higher gain with a wider angular coverage and electrically compact as required for WLAN antennas.

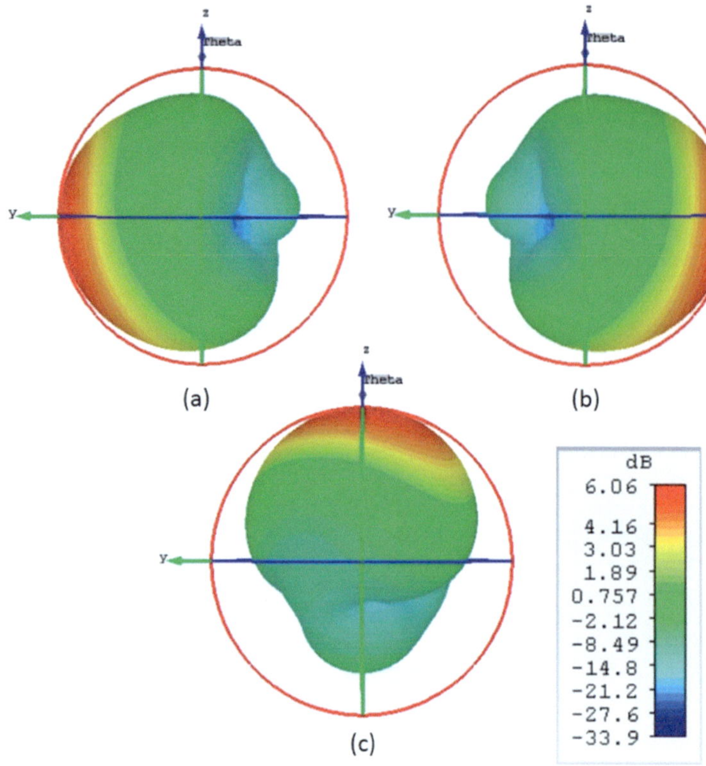


**Fig. 7.45** Three-dimensional gain plot at 5.2 GHz when **a** port 1 is excited, **b** port 2 is excited, **c** port 3 is excited [21]

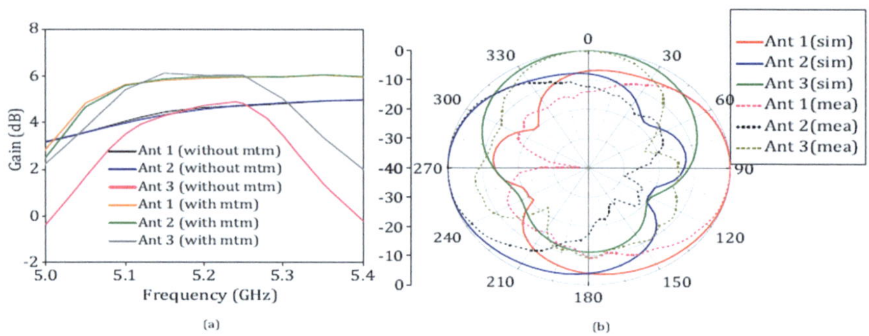


**Fig. 7.46** **a** Gain of the antenna without and with unit-cell structures, **b** angular coverage of the proposed antenna at 5.2 GHz [21]

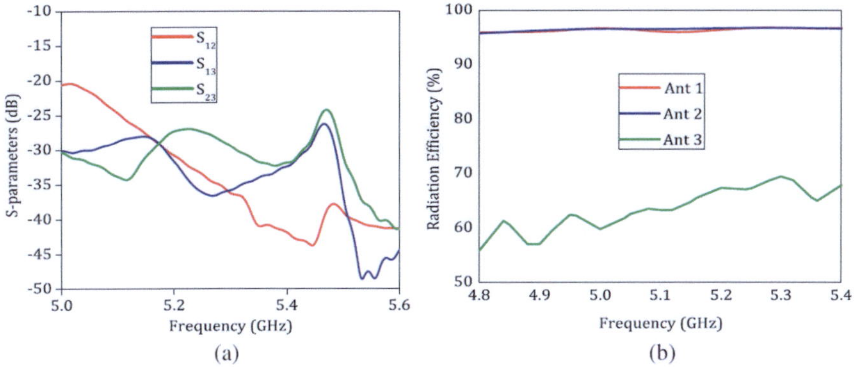


**Fig. 7.47** **a** Mutual coupling, **b** radiation efficiency [21]

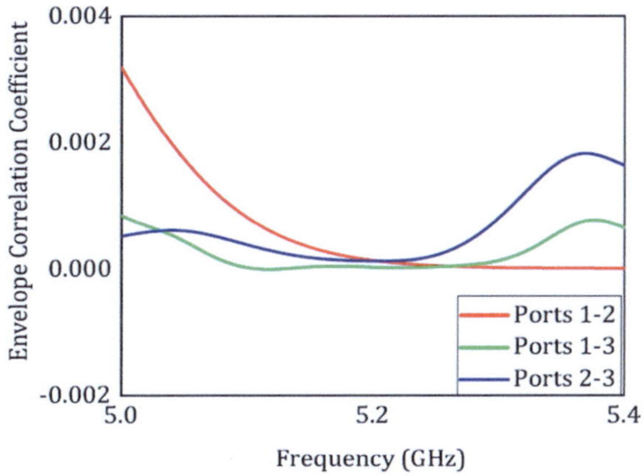


**Fig. 7.48** Simulated ECC for different antenna ports of the three-port antenna design

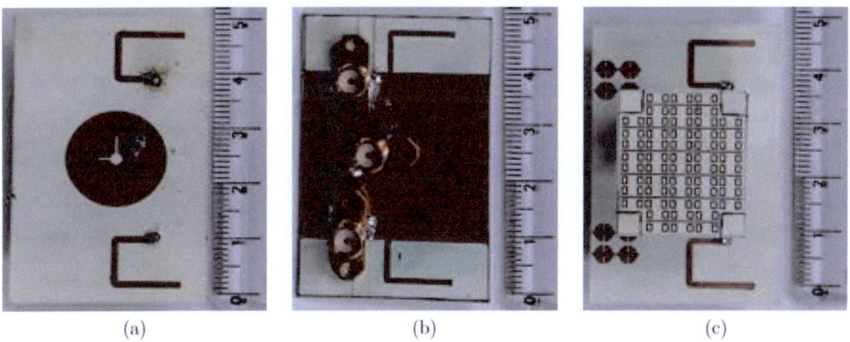


**Fig. 7.49** Photograph of the fabricated antenna **a** top view before unit-cell loading, **b** bottom view, **c** top view after the unit-cell loading [21]

**Table 7.5** Comparison of the proposed antenna structures

| Ref. | Freq. (GHz) | Operating range (GHz) | No. of bands covered | Size | Pattern diversity | Coverage | Isolation (dB) | Radiation efficiency (%) | Gain (dB) |
|---|---|---|---|---|---|---|---|---|---|
| [33] | 2.5 | 2.57–2.64 | 1 | $0.62\lambda \times 0.62\lambda \times 0.25\lambda$ | Omnidirectional, broadside | $120^0$ | 20 | 488,595 | 2.8 |
| [16] | 5.8 | 5.67–5.91 | 1 | $0.92\lambda \times 0.58\lambda$ | Broadside, endfire | 180° | 18 | 508,585 | 2.0 |
| [34] | 3.1 | 3.10–11.5 | UWB | $0.5\lambda \times 0.25\lambda$ | 0°, + 90°, − 90° | 120° | 18 | 70, 80, 85 | 2.5 |
| [35] | 2.6 | 2.57–2.64 | 1 | $1.04\lambda \times 1.04\lambda \times 0.1\lambda$ | + 20°, − 20° | 40° | – | 4146 | 6.0 |
| [36] | 2.44 | 2.42–2.5 | 1 | $0.8\lambda \times 0.48\lambda$ | + 90°, − 90° | 120° | 20 | 9451 | 0.3 |
| [37] | 5.8 | 5.5–6.3 | 1 | $0.6\lambda \times 0.5\lambda$ | Broadside, endfire | 120° | 13 | 9696 | 2.5 |
| [38] | 5.0 | 2.87–16.56 | UWB | $0.38\lambda \times 0.38\lambda$ | Conical patterns only | 30° | – | 70 | 7.0 |
| [39] | 5.8 | 2.54–15.36 | UWB | $0.32\lambda \times 0.32\lambda$ | Conical patterns only | 30° | – | 76 | 3.0 |
| PA | 5.2 | 5.12–5.27 | 1 | $0.56\lambda \times 0.9\lambda \times 0.18\lambda$ | Broadside, endfire | 330° | 26 | 969,667 | 6.0 |

*PA* Proposed antenna

## 7.6 Gain-Equalized Three-Port Antenna Using Fabry–Perot Cavity

In this section, the concept of Fabry–Perot cavity is used for gain enhancement in a three-port antenna. The gain at the three ports is also equalized to 10 dBi and the radiation beam of the antenna switches between 0°, + 25°, – 25°. The novelty of the antenna is in enhancing the gain using Fabry–Perot cavity concept in the oblique incidence rather than the broadside direction which is widely reported in the literature. The gain enhancement ensures an angular coverage of the antenna to 90° which is higher than other antenna designs reported in the literature.

### *7.6.1 Motivation*

The FPC designs available in the literature show that most antennas have their gain enhanced under normal angle of incidence. As illustrated in Fig. 7.50, the partially reflective surface (PRS) is placed over the antenna at a height of 0.5λ. The metamaterial superstrate is used like a mirror to reflect a portion of the incident wave and transmit the remaining part. The part between the ground plane and the PRS acts as the cavity where the wave undergoes continuous reflections. By adjusting the PRS structure design, the amount of transmission can be precisely controlled.

In the proposed design, the gain enhancement of Fabry–Perot cavity antenna at normal as well as oblique incidence is studied and designed. To achieve this, the PRS is placed at a tilted angle rather than horizontally above the radiating antenna. The tilted PRS enhances the gain of the antennas radiating in oblique incidences. This makes the concept of Fabry–Perot cavity antenna to be more versatile, and the concept can be used for different beam incidences.

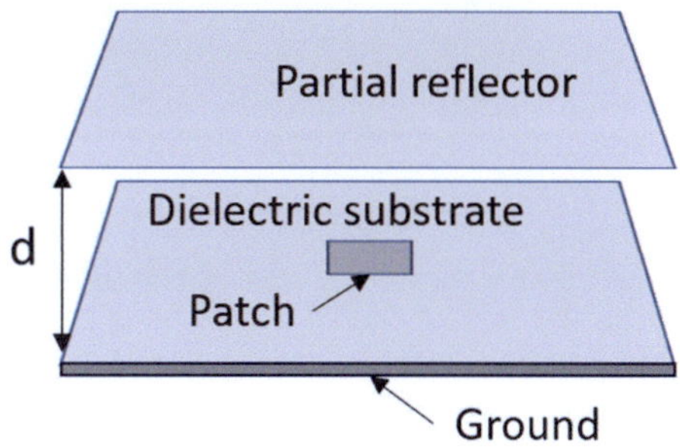


**Fig. 7.50** Fabry–Perot cavity design illustration

### 7.6.2 Antenna Design

Figure 7.51 shows the three-element patch antenna design. The antennas are fabricated on Rogers RO4350B substrate with dielectric constant $\varepsilon_r = 3.66$, thickness of 20 mil, and tan $\delta = 0.0037$. The ground plane area is $142 \times 60$ mm$^2$ ($2.4\lambda \times 1\lambda$), where $\lambda$ is free-space wavelength at 5.2 GHz. The large ground plane reduces the back radiation suffered by the antenna as the ground plane acts as a good reflector. The three antennas are coaxially fed with a port-to-port distance of $0.67\lambda$. This distance gives a maximum gain of the antenna with minimum mutual coupling between the ports. The coaxial feed allows elimination of the transmission line insertion loss, which reduces the aperture efficiency. As the transmission line cannot radiate, the PRS placed above the transmission line cannot contribute to the antenna gain enhancement. Hence, a higher aperture efficiency is achieved by using the coaxial feed.

The simulated and measured input reflection coefficients are shown in Fig. 7.52. The simulated impedance bandwidth for the antenna is 90 MHz (5.17–5.26 GHz). The half-power beamwidth (HPBW) of the antenna 1 is 36° and that of antenna 2 and 3 is 65°. The front-to-back ratio of antenna 1 is 18 dB and of antennas 2 and 3 is 20 dB due to an electrically large ground plane. The gain of antenna 1 as shown in Fig. 7.53 is 6 dBi and that of antennas 2 and 3 is 5.7 dBi along with a tilt of + 15°, − 15°. The patterns of antennas 2 and 3 are tilted due to the mutual coupling between the antennas at $0.6\lambda$ between the input ports.

A PRS superstrate for a Fabry–Perot cavity resonance is now designed to enhance the gain equally across the band and throughout the ports of the antenna.

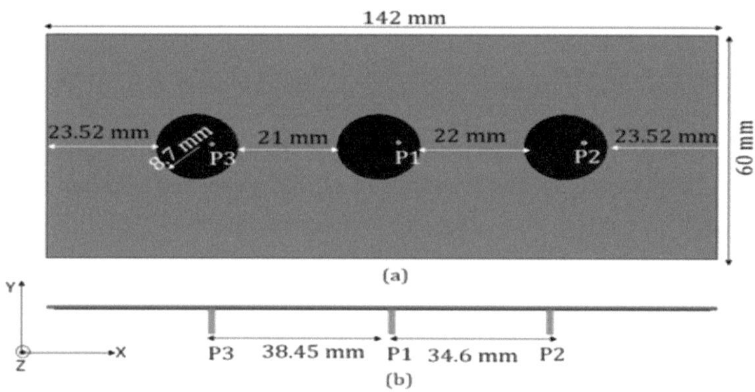


**Fig. 7.51** Schematic of three-element antenna **a** top view, **b** side view [40]

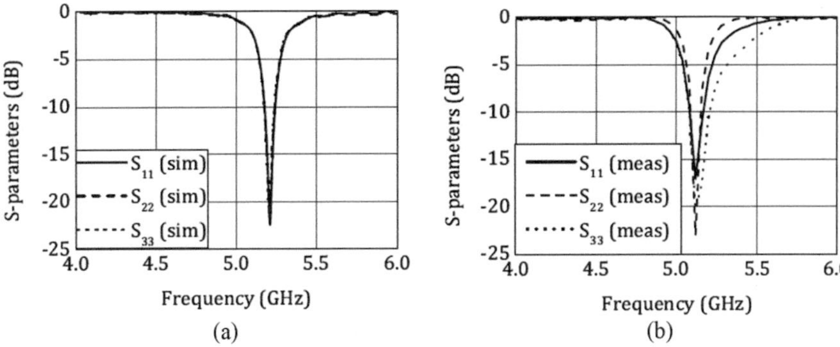


**Fig. 7.52** Input reflection coefficients of the three-patch antennas **a** simulated, **b** measured [40]

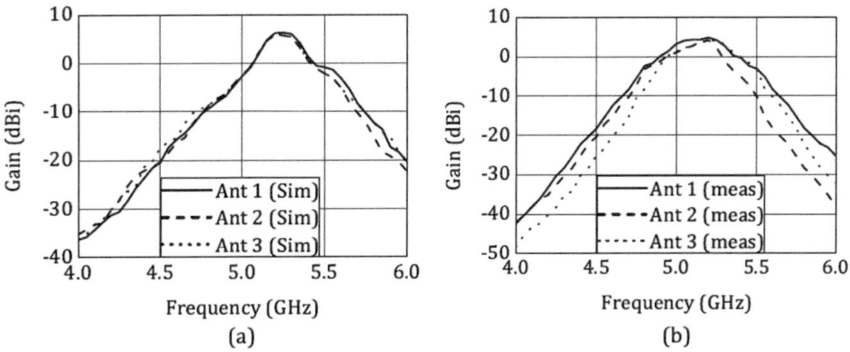


**Fig. 7.53** Gain of the three-patch antennas **a** simulated, **b** measured [40]

## *7.6.3 Metamaterial Design*

The unit-cell structure as shown in Fig. 7.54a is designed to improve the gain of the patch antennas. The *S*-parameters from Fig. 7.54b conveys that it is a partially reflective surface (PRS) with $S_{11} = -3.2$ dB and $S_{21} = -2.9$ dB at the operating frequency of 5.2 GHz. The substrate used is Rogers RO4350B with dielectric constant, $\varepsilon_r = 3.66$, 20 mil thickness, and $\tan\delta = 0.0037$. The substrate for the PRS was chosen to be the same as that of the antenna to ensure continuity and to have homogenous properties for the entire structure. The PEC and PMC boundary conditions were assigned in the *xz*- and *yz*-planes of the unit cell to extract the *S*-parameters. The two ports are assigned in the *z*-direction. The full-wave EM simulation software CST Microwave Studio is used to extract the parameters of the unit cell. A superstrate with periodic unit cells is then designed to be integrated above the patch. The PRS and the ground plane create a cavity where the EM waves undergo multiple reflections. The PRS is designed in such a way that a part of the reflections gets transmitted out of the cavity, which happens as the value of $S_{21}$ is $-2.9$ dB. The waves transmitted out of

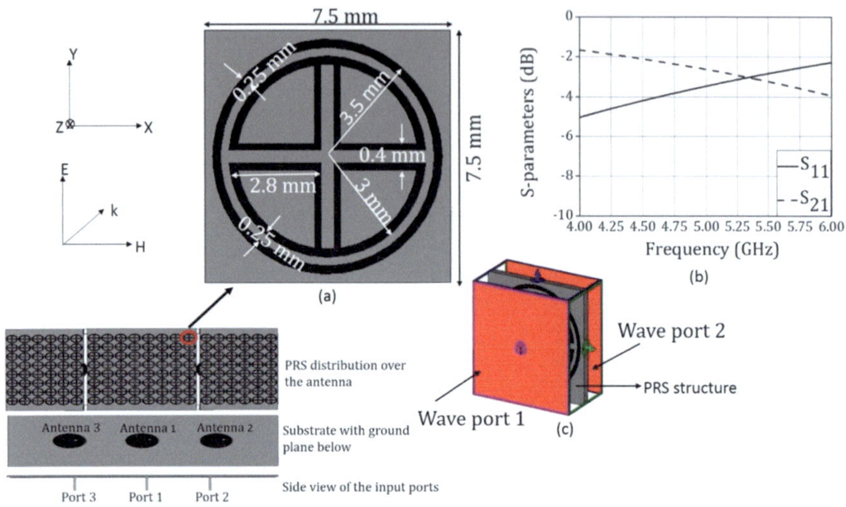


**Fig. 7.54** **a** Configuration of unit cell, **b** *S*-parameters of the unit cell, **c** simulation setup [40]

the cavity should sustain a constructive interference which then gets added up and boosts the gain of the antenna in a specific direction [41].

## 7.6.4 Antenna with Metamaterial Design

Figure 7.55 shows the proposed design consisting of three-patch antennas along with the shaped PRS integration. The antennas emit electromagnetic waves which undergo multiple reflections inside the cavity formed by the PRS and the ground plane. These multiple reflections result in gain enhancement of the antenna [42–44]. The PRS is placed at a height of approximately $\lambda/2$ above the patch for port 1 for boresight gain enhancement.

Initially, a PRS layer consisting of $20 \times 8$ unit cells was placed at the same height above the three antennas. In this case, the gain of patch antenna 1 was lesser by 2 dB compared to the other two patches. To equalize the gain between the patches, the height of the $8 \times 8$ unit cells that were contributing to the gain enhancement of the antenna 1 was varied. It was then noted that at a height of $0.68\lambda$, the antenna provided a gain of 10 dBi indicating 4.5-dB gain enhancement. A similar gain was obtained for antennas 2 and 3 for a PRS height of $0.6\lambda$. This gain enhancement was obtained while the antennas maintained the corresponding 15° beam tilt in the respective directions. In order to increase the beam tilt further, for an extended coverage by the antenna, the set of $8 \times 6$ local superstrates were tilted at different angles. In the literature of Fabry–Perot cavity antennas [45–47], the reported PRS designs are

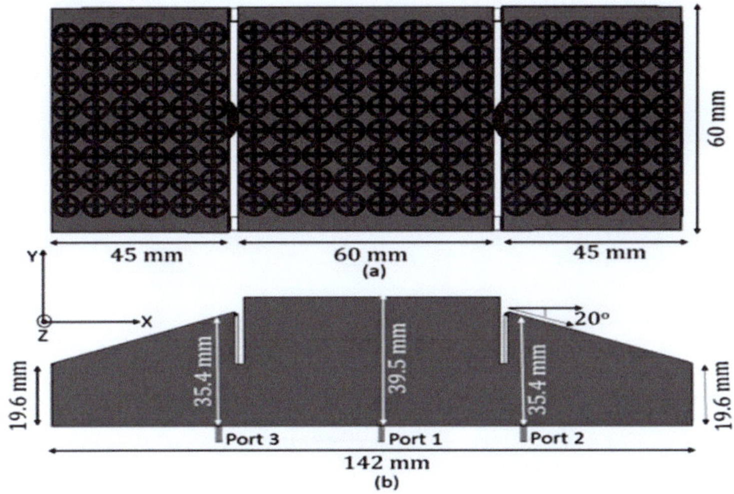


**Fig. 7.55** **a** Shaped PRS aperture arrangement, **b** side view of the arrangement [40]

planar in structure which is inadequate for beam tilting. Here, a structure that acts as a Fabry–Perot cavity with different tilt angles and different heights for the PRS is used.

The tilted superstrates ensure a higher phase correction to the EM waves than a planar structure for the beam shifted. For each case, the beam tilt obtained along with the resulting gain was noted down. From the observations, it was clear that at an angle of 20°, the antennas 2 and 3 of the FPC provided a beam tilt of 25° for a gain of 10 dB throughout the band. Further increase in angular tilt made the gain of the antennas to reduce. This happens as the height of the superstrate will become so low that proper resonance is not possible. A similar angular tilt given to the antennas by physically keeping them at + 25° and – 25° resulted in detuning of the antennas.

The electric field distribution before and after the PRS loading is shown in Fig. 7.56. The shaped PRS, especially the tilted PRS, re-orients the electric field of the leaky waves which subsequently results in the beam tilt. The PRS makes the phase correction wave front to be planar compared to the unloaded antenna which reduces the beamwidth and enhances the gain. For each port, the simulated and measured input reflection coefficients are shown in Fig. 7.57. The simulated impedance bandwidth of the proposed antenna is 90 MHz (5.16–5.25 GHz). This shows that the proposed antenna impedance bandwidth does not get altered after the superstrate integration. The height of the FPC is designed such that the resonance of the cavity can be merged to the patch antennas. Thus, the resonant frequency of the three antennas does not change post-PRS integration. This helps in achieving the design objective of higher gain at the impedance bandwidth of the three antennas. Another advantage is that the FPC separates the feed antennas and the PRS structures making it convenient to expand the design to further ports or multiple PRS layers.

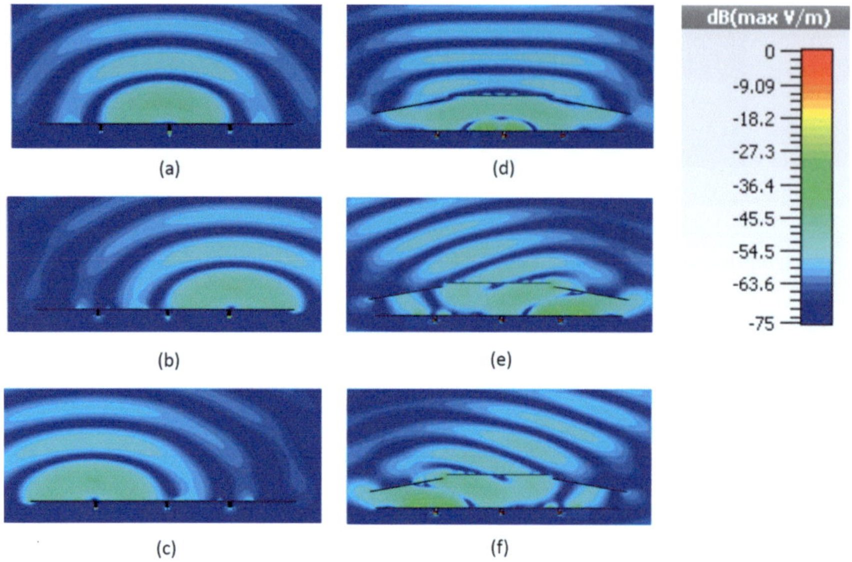


**Fig. 7.56** Electric field distribution for the three ports **a–c** before shaped PRS integration, **d–f** after shaped PRS integration [40]

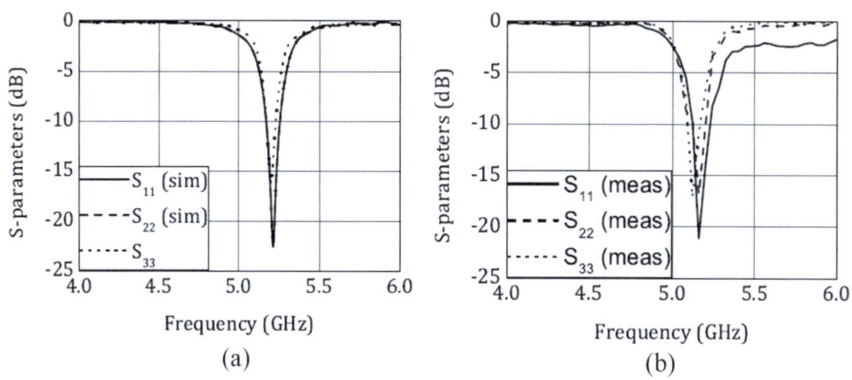


**Fig. 7.57** Input reflection coefficients of the three-patch antennas after shaped PRS aperture integration **a** simulated, **b** measured [40]

The radiation pattern of the antenna before and after the superstrate loading is shown in Fig. 7.58. As observed from the radiation patterns, there is a slight 10° tilt even before the PRS loading due to the strategic placement of the antennas at the required distance. After the PRS loading, the beam tilt increases to 25° in both directions along with an enhanced gain.

The proposed design and placement not only increase the gain but also the beam tilt of the antenna. The simulated half-power beamwidth of antenna 1 is 23° with a

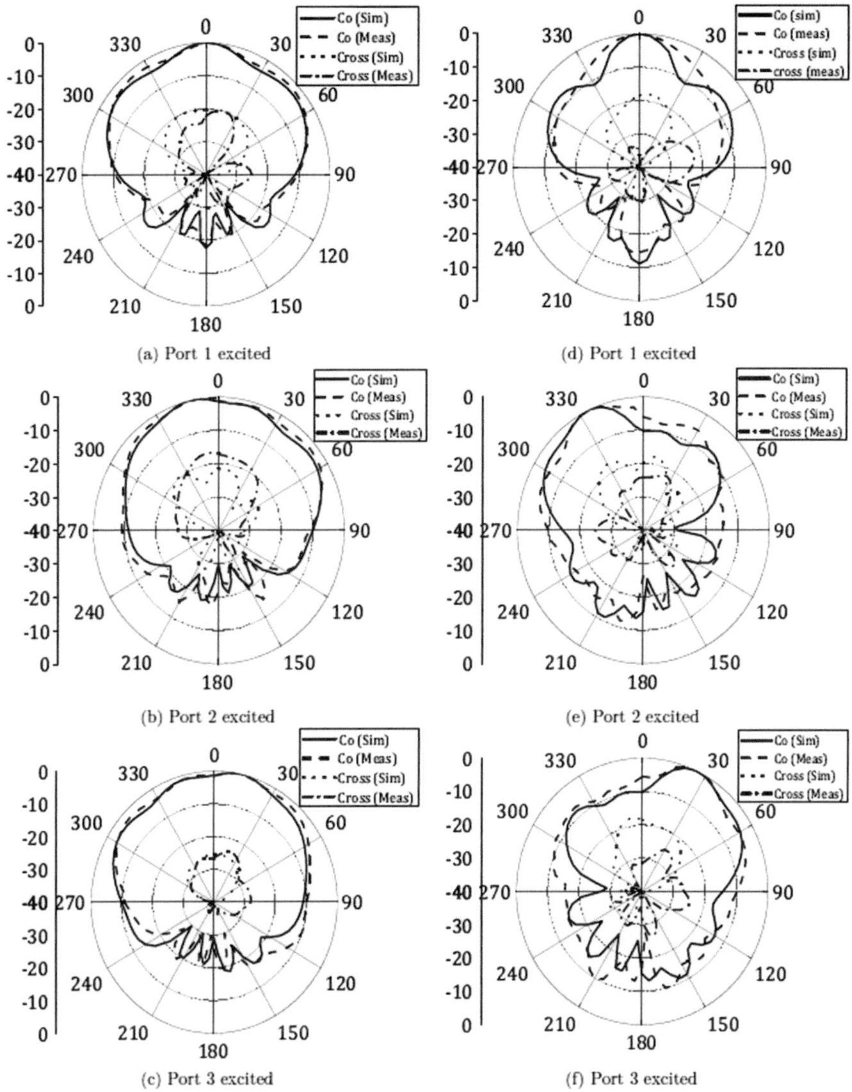


**Fig. 7.58** Radiation pattern at the operating frequency of 5.2 GHz **a–c** before unit-cell loading and **d–f** after unit-cell loading [40]

front-to-back ratio of 12 dB and that of antennas 2 and 3 the values are 30° and 16 dB. From the intersection of 3-dB values in the radiation pattern of the antenna, it is seen that the patterns are able to provide a coverage of 90° as shown in Fig. 7.59. The superstrate loading does not alter the polarization of the three antennas, and thus, it maintains the polarization integrity of the unloaded antennas.

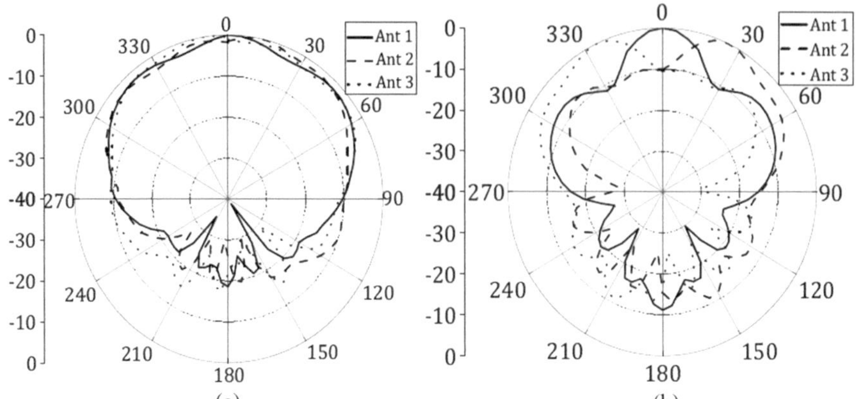


**Fig. 7.59** **a** Angular coverage before superstrate loading, **b** angular coverage post superstrate integration [40]

Figure 7.60 shows the gain of the three antennas after the shaped PRS integration. From the graph, it is seen that there is a gain enhancement of 4.5 dB with respect to the antenna after the shaped PRS integration. The PRS reflects a part of the electromagnetic wave, and the waves get trapped inside the cavity which forms a standing wave pattern. As the use of PRS reduces the radiation losses, the cavity should be properly designed to minimize the scattering and absorption losses. This continuous reflection of the waves inside the cavity is what causes the gain to enhance. In the proposed design, the PRS is designed and aligned for an equal gain enhancement over the three ports.

Figures 7.61 and 7.62 give the mutual coupling of the antennas between the ports before and after the PRS integration. The antennas maintain an isolation of 20 dB between the ports post-integration. The ECC graph for the operating bandwidth is

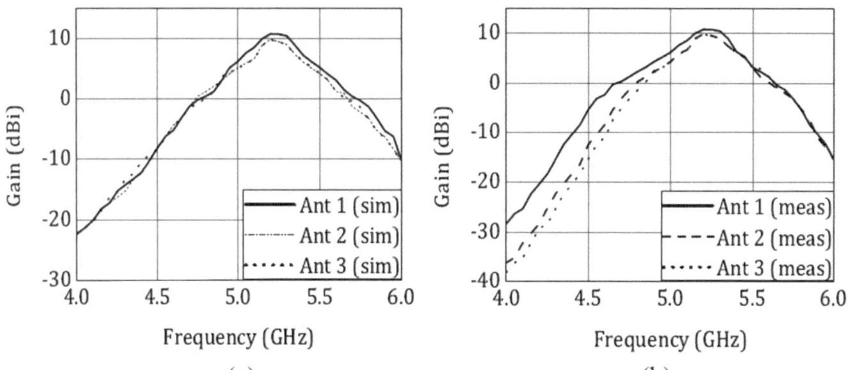


**Fig. 7.60** Gain of the antenna after PRS loading **a** simulated, **b** measured [40]

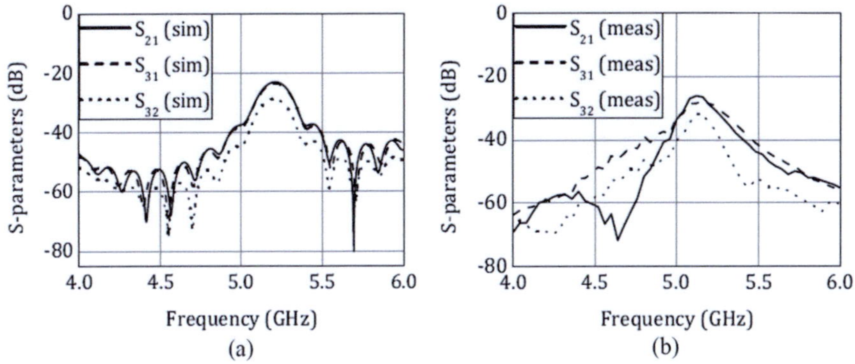


**Fig. 7.61** Mutual coupling of the antenna before PRS loading **a** simulated, **b** measured [40]

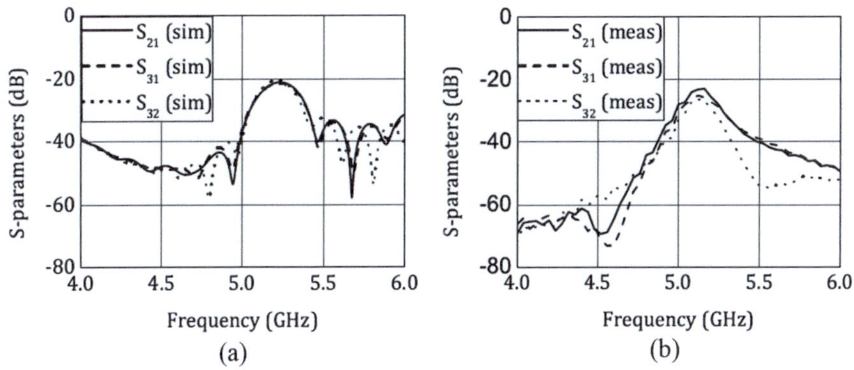


**Fig. 7.62** Mutual coupling of the antenna after PRS loading **a** simulation, **b** measurement [40]

shown in Fig. 7.63. All the values are below 0.005 which indicates a good diversity performance of the antenna. To validate the simulation results, a prototype of the proposed antenna has been fabricated as shown in Fig. 7.64. The antenna is fabricated using the photolithographic process while the scaffolding needed to keep the shaped PRS superstrate in place is 3D printed. The 3D printing was performed using a Raise 3D RxP2200 3D printer with the polymer polylactic acid (PLA) material which is used in commercial low budget 3D printing. Anritsu MS2028C was used to measure the S-parameter of the antenna and the far-field measurements are conducted inside an anechoic chamber. The ETS-Lindgren 3115 Model double-ridged waveguide horn antenna was used as the standard gain antenna for the measurement.

The proposed antenna with 10-dBi gain is an improvement over the commercial antenna for access points in [48] for a 90° coverage. The height of the antenna at the center is only 39.5 mm which is lower than the height of 55 mm for the commercial antenna.

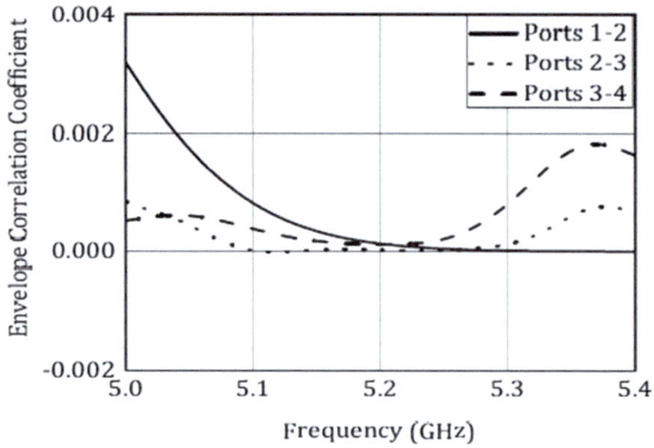


**Fig. 7.63** Simulated ECC for different antenna ports [40]

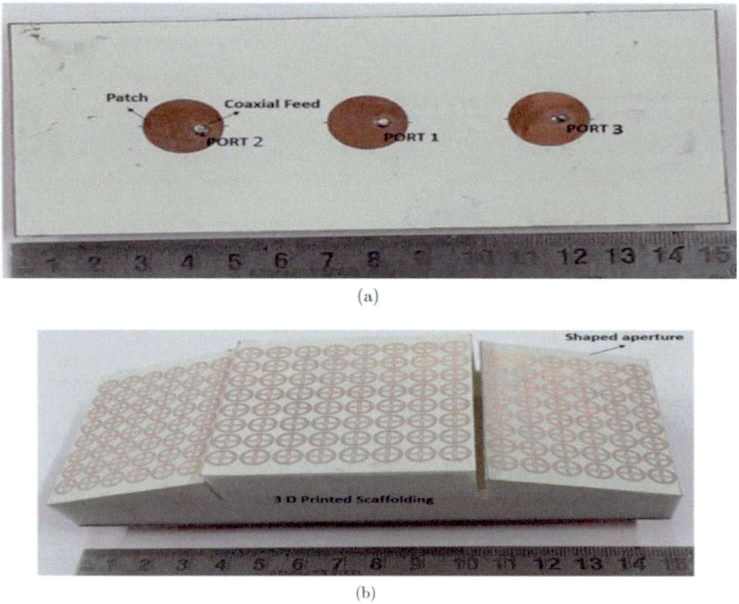


**Fig. 7.64** Photograph of the fabricated antenna **a** top view before unit-cell loading, **b** top view after unit-cell loading [40]

The performance of the proposed antenna is compared with the existing designs in the literature and is tabulated in Table 7.6. Antenna structures reported in [15, 16, 49] have very low gain with discontinuous angular coverage. Wang [50] reports a higher gain, but the angular coverage offered is very limited. Kim [30], Wang [51] show gain enhancement only in the broadside direction with low aperture efficiency. The antenna profile of [52] is very high. The structures reported in [53, 54] have low

**Table 7.6** Comparison of the three-port antenna with other antenna designs

| Ref. | Freq. (GHz) | Operating range (GHz) | Size | Isolation (dB) | Gain (dBi) | Patterns at (°) | Height of superstrate (mm) | Aperture efficiency (%) |
|---|---|---|---|---|---|---|---|---|
| [15] | 2.4 | 2.4–2.48 | 0.23λ × 0.23λ | 24 | 2.3 | 0°, + 120°, – 120° | – | – |
| [16] | 5.8 | 5.5–5.85 | 1λ × 0.56λ | 18 | 3.8 | 0°, + 90°, – 90° | – | – |
| [49] | 5.5 | 5.4–5.6 | 1.85λ × 1.85λ | – | 9.7 | 0°, + 10°, – 10° | 0.52λ | 27.33 |
| [50] | 2.45 | 2.3–3.9 | 0.29λ × 0.29λ | 18 | 3.5 | 0°, + 120°, – 120° | – | – |
| [30] | 5.8 | 5.71–5.87 | 2λ × 2λ | – | 6.9 | 0° | 0.39λ | 28.93 |
| [51] | 10 | 8.6–11.2 | 2.4λ × 2.4λ | – | 13.8 | 0° | 0.45λ | 40.9 |
| [52] | 10 | 8.6–11.4 | 4.3λ × 4.3λ | – | 20.7 | 0° | 2λ | 49.79 |
| [53] | 4.2 | 3.8–7.3 | 0.4λ × 0.3λ | – | 7.0 | 0° | 0.15λ | 48 |
| [54] | 5.8 | 5.75–5.82 | 1.17λ × 1.17λ | – | 12 | 0° | 0.44λ | 91.46 |
| [55] | 10 | 9.5–12.8 | 2λ × 2λ | – | 13.4 | 0° | 0.2λ | 43.5 |
| PA | 5.2 | 5.16–5.25 | 1λ × 2.4λ | 20 | 10 | 0°, + 25°, – 25° | 0.68λ | 68.5 |

*PA* Proposed antenna

aperture efficiency as well. The proposed antenna provides high gain for the angular coverage of 90° in the broadside direction which is higher than the existing designs.

## 7.7 Conclusion

In this chapter, gain-enhanced multiport antennas with pattern diversity are discussed. Gain enhancement along with gain equalization at the multiple antenna ports is achieved. Initially, a two-port and three-port antenna that uses a shared double-negative metamaterial structure is analyzed. The two-port metamaterialloaded antenna presents a gain improvement of 3.6 dB with the radiation beams at $0^0$ and $90^0$. The three-port metamaterial-loaded antenna presents a gain improvement of 3.4 dB with the beams radiating at 0°, + 45°, and – 45°. Another three-port antenna where the gain enhancement is achieved by loading a transmission-type and a reflectiontype metamaterial is discussed next. The antenna presents a wide angular coverage of $330^0$ at the operating frequency (5.2 GHz) with a gain of 6 dBi. A three-port pattern diversity Fabry–Perot cavity antenna with shaped aperture partially reflecting surface is proposed in the second design. The PRS is tilted by an angle of 20° above the two patches to obtain a 25° beam tilt with 10-dB gain for the three antennas. The PRS contributes a 4.6-dB gain enhancement to the three-patch antennas with a total beam coverage of 90°. The designs discussed in the chapter can be expanded further to multiple ports.

## References


1. Swapna, S., Karthikeya, G.S., Koul, S.K., Poddar, A.K., Rohde, U.: Gain enhancement of 3-D printed wideband cavity antenna for WLAN. In: IEEE International Symposium on Antennas and Propagation and USNC-URSI Radio Science Meeting, Portland, Oregon, USA, pp. 1195–1196, 23–28 July 2023
2. Garg, R.: Microstrip Antenna Design Handbook. Artech House, United Kingdom (2001)
3. Swapna, S., Karthikeya, G.S., Koul, S.K., Basu, A.: Gain enhancement using shared metamaterial aperture for WLAN antennas with pattern diversity. In: 2018 IEEE Indian Conference on Antennas and Propogation (InCAP), Hyderabad, India, pp. 1–4 (2018)
4. Ghosh, R.P., Patra, K., Gupta, B., Chowdhury, S.K.: Accurate formula to determine resonant frequency of double-sided printed dipole antenna. IETE J. Res. **64**(3), 331–336 (2018)
5. Zhou, Q., Dai, H.: Joint antenna selection and link adaptation for MIMO systems. IEEE Trans. Veh. Technol. **55**, 243–255 (2006)
6. Liu, L., Cheung, S.W., Yuk, T.I.: Compact mimo antenna for portable uwb applications with band-notched characteristic. IEEE Trans. Antennas Propag. **63**(5), 1917–1924 (2015)
7. Ren, J., Hu, W., Yin, Y., Fan, R.: Compact printed mimo antenna for uwb applications. IEEE Antennas Wirel. Propag. Lett. **13**, 1517–1520 (2014)
8. Das, G., Sharma, A., Gangwar, R.K.: Dielectric resonator-based two-element mimo antenna system with dual band characteristics. IET Microwaves Antennas Propag. **12**(5), 734–741 (2018)
9. Metamaterials: Physics and Engineering Explorations. Wiley, Germany (2006)
10. Luciano Tarricone, A.E.: Advances in information technologies for electromagnetics, 1st edn. Springer (2006)
11. Smith, D., Schurig, D., Rosenbluth, M., Schultz, S., Ramakrishna, S., Pendry, J.: Limitations on subdiffraction imaging with a negative refractive index slab. Appl. Phys. Lett. **82**, 1506–1508 (2003)
12. Ziolkowski, R.W., Heyman, E.: Wave propagation in media having negative permittivity and permeability. Phys. Rev. E **64**, 056625 (2001)
13. Zheng, Y.J., Gao, J., Zhou, Y.L., Cao, X.Y., Xu, L.M., Li, S.J., Yang, H.H.: Metamaterial-based patch antenna with wideband rcs reduction and gain enhancement using improved loading method. IET Microwaves Antennas Propag. **11**(9), 1183–1189 (2017)
14. Saha, C., et al.: Multifunctional Ultrawideband Antennas: Trends, Techniques and Applications. CRC Press, United States (2019)
15. Wang, H., Liu, L., Zhang, Z., Li, Y., Feng, Z.: Ultra-compact three-port mimo antenna with high isolation and directional radiation patterns. IEEE Antennas Wirel. Propag. Lett. **13**, 1545–1548 (2014)
16. Sharma, Y., Sarkar, D., Saurav, K., Srivastava, K.V.: Three-element mimo antenna system with pattern and polarization diversity for wlan applications. IEEE Antennas Wirel. Propag. Lett. **16**, 1163–1166 (2017)

17. Wang, K., Li, L., Eibert, T.F.: Compact three-port printed diversity array with combined antenna types. In: 2014 IEEE Antennas and Propagation Society International Symposium (APSURSI), pp. 488–489 (2014)
18. Mak, A.C., Rowell, C.R., Murch, R.D.: Isolation enhancement between two closely packed antennas. IEEE Antennas Wireless Propag. Lett. **56**(11), 3411–3419 (2008)
19. Sarkar, D., Singh, A., Saurav, K., Srivastava, K.V.: A four element quad-band MIMO antenna employing split ring resonator and interdigital capacitor. Microw. Antennas Propag. **9**(13), 1453–1460 (2015)
20. Zhang, S., Ying, Z., Xiong, J., He, S.: Ultrawideband MIMO/diversity antennas with a tree-like structure to enhance wideband isolation. IEEE Antennas Wireless Propag. Lett. **8**, 1279–1281 (2009)
21. Somanatha, S. P., Karthikeya, S.G., Koul, S.K., Basu, A.: Gain equalized three antenna pattern diversity module for WLAN access points. Prog Electromagnet Res **C96**, 215–227 (2019)
22. Bhartia, P., Bahl, I., Garg, R., Ittipiboon, A.: Microstrip Antenna Design Handbook (Artech House Antennas and Propagation Library), ser. Artech House Publishers, Artech House Antennas and Propagation Library (2000)
23. Hamid, M., Hamid, R.: Equivalent circuit of dipole antenna of arbitrary length. IEEE Trans. Antennas Propag. **45**(11), 1695–1696 (1997)
24. Ansarizadeh, M., Ghorbani, A., Abd-Alhameed, R.A.: An approach to equivalent circuit modeling of rectangular microstrip antennas. Prog Electromagnet Res B **8**, 77–86 (2008)
25. Pozar, D.M.: Microwave engineering, 3rd edn. Wiley, Hoboken, NJ (2005)
26. Balanis, C.A.: Antenna Theory: Analysis and Design. Wiley-Interscience (2005)
27. Latif, S.I., Shafai, L.: Pattern equalization of circular patch antennas using different substrate permittivities and ground plane sizes. IEEE Trans. Antennas Propag. **59**(10), 3502–3511 (2011)
28. Wu, C., Wu, K., Bi, Z., Litva, J.: Modelling of coaxial-fed microstrip patch antenna by finite difference time domain method. Electron. Lett. **27**(19), 1691–1692 (1991)
29. Waterhouse, R.: Printed Antennas for Wireless Communications, 1st edn., ser. RSP. Wiley (2007)
30. Kim, J.H., Ahn, C., Bang, J.: Antenna gain enhancement using a holey superstrate. IEEE Trans. Antennas Propag. **64**(3), 1164–1167 (2016)
31. Engheta, N.: An idea for thin subwavelength cavity resonators using metamaterials with negative permittivity and permeability. IEEE Antennas Wirel. Propag. Lett. **1**, 10–13 (2002)
32. Chen, K., Yang, Z., Feng, Y., Zhu, B., Zhao, J., Jiang, T.: Improving microwave antenna gain and bandwidth with phase compensation metasurface. AIP Adv. **5**(6), 1695–1696 (2015)
33. Saurav, K., Mallat, N.K., Antar, Y.M.M.: A three-port polarization and pattern diversity ring antenna. IEEE Antennas Wirel. Propag. Lett. **17**(7), 1324–1328 (2018)
34. Roshna, T.K., Deepak, U., Mohanan, P.: Compact uwb mimo antenna for tridirectional pattern diversity characteristics. IET Microwaves Antennas Propag. **11**(14), 2059–2065 (2017)
35. Naqvi, A.H., Lim, S.: A beam-steering antenna with a fluidically programmable metasurface. IEEE Trans. Antennas Propag. **67**(6), 3704–3711 (2019)
36. Yang, K., Bao, X., McEvoy, P., Ammann, M.J.: Pattern reconfigurable back-to back microstrip patch antenna. IET Microwaves Antennas Propag. **10**(13), 1390–1394 (2016)
37. Malik, J., Patnaik, A., Kartikeyan, M.V.: Novel printed mimo antenna with pattern and polarization diversity. IEEE Antennas Wirel. Propag. Lett. **14**, 739–742 (2015)
38. Saraswat, R.K., Kumar, M.: A frequency band reconfigurable uwb antenna for high gain applications. Prog Electromagnet Res B **64**, 29–45 (2015)
39. Saraswat, M.K.R.K.: Miniaturized slotted ground uwb antenna loaded with metamaterial for wlan and wimax applications. Prog Electromagnet Res B **65**, 65–80 (2016)
40. Somanatha, P.S., Karthikeya, S.G., Koul, S.K., Basu, A.: Three-port pattern diversity antenna module for 5.2 GHz ceiling-mounted WLAN access points, Prog Electromagnet Res C **98**, 57–67 (2020)
41. Costa, F., Bianchi, D., Monorchio, A., Manara, G.: Linear fabry-perot/leaky-wave antennas excited by multiple sources. IEEE Trans. Antennas Propag. **66**(10), 5150–5159 (2018)

42. Pan, W., Huang, C., Chen, P., Ma, X., Hu, C., Luo, X.: A low-rcs and high-gain partially reflecting surface antenna. IEEE Trans. Antennas Propag. **62**(2), 945–949 (2014)
43. Vaidya, A.R., Gupta, R.K., Mishra, S.K., Mukherjee, J.: Right-hand/left-hand circularly polarized high-gain antennas using partially reflective surfaces. IEEE Antennas Wirel. Propag. Lett. **13**, 431–434 (2014)
44. Ren, J., Jiang, W., Zhang, K., Gong, S.: A high-gain circularly polarized fabry–perot antenna with wideband low-rcs property. IEEE Antennas Wirel. Propag. Lett. **17**(5), 853–856 (2018)
45. Konstantinidis, K., Feresidis, A.P., Hall, P.S.: Multilayer partially reflective surfaces for broadband fabry-perot cavity antennas. IEEE Trans. Antennas Propag. **62**(7), 3474–3481 (2014)
46. Zheng, Y., Gao, J., Zhou, Y., Cao, X., Yang, H., Li, S., Li, T.: Wideband gain enhancement and rcs reduction of fabry–perot resonator antenna with chessboard arranged metamaterial superstrate. IEEE Trans. Antennas Propag. **66**(2), 590–599 (2018)
47. Guzmán-Quirós, R., Weily, A.R., Gómez-Tornero, J.L., Guo, Y.J.: A fabry–pérot antenna with two-dimensional electronic beam scanning. IEEE Trans. Antennas Propag. **64**(4), 1536–1541 (2016)
48. Cisco Aironet 2800 Series Access Points, Cisco, 2 (2019)
49. Xie, P., Wang, G., Li, H., Liang, J.: A dual-polarized two-dimensional beam-steering fabry–pérot cavity antenna with a reconfigurable partially reflecting surface. IEEE Antennas Wirel. Propag. Lett. **16**, 2370–2374 (2017)
50. Wang, H., Liu, L., Zhang, Z., Feng, Z.: Wideband tri-port mimo antenna with compact size and directional radiation pattern. Electron. Lett. **50**(18), 1261–1262 (2014)
51. Wang, N., Liu, Q., Wu, C., Talbi, L., Zeng, Q., Xu, J.: Wideband fabry-perot resonator antenna with two complementary fss layers. IEEE Trans. Antennas Propag. **62**(5), 2463–2471 (2014)
52. Li, H., Wang, G., Cai, T., Liang, J., Gao, X.: Phase- and amplitude-control metasurfaces for antenna main-lobe and sidelobe manipulations. IEEE Trans. Antennas Propag. **66**(10), 5121–5129 (2018)
53. Mao, C., Yang, Y., He, X., Zheng, J., Liu, T.: Design of high-gain dual-band dual circular-polarised antenna using reflective metasurface. Electron. Lett. **53**(22), 1448–1450 (2017)
54. Chen, Q., Zhang, H.: High-gain circularly polarized fabry–pérot patch array antenna with wideband low-radar-cross-section property. IEEE Access **7**, 8885–8889 (2019)
55. Aziz, R.S., Kim, T., Park, J., Ryu, Y., Park, S.: Em lens design using thin planar metasurfaces for high antenna gain and low sll applications. IET Microwaves Antennas Propag.Propag. **13**(7), 950–958 (2019)

# Chapter 8
# Multiport Antennas for WLAN

## 8.1 Introduction

The very first benefit of multielement antennas was recognized in 1920s and 1930s. It was observed that fluctuations in the received signal powers in a multipath environment, referred to as fading, were statistically independent at different antenna elements. When the power of a signal drops significantly, it is said to be in fade resulting in a loss of connection. To combat fading, engineers recommended selecting the antenna with the strongest signal at each time instant. This was the advent of selection combining receiver diversity antenna systems. By the late 1960s, coherent combining schemes like maximum ratio combining (MRC) and equal gain combining schemes were developed [1].

The foregoing linear combination schemes bestowed an extra gain, called array gain. In general, array gain is referred to an enhancement in average received signal-to-noise ratio (SNR) at the receiver as a result of coherent combining of the available signals. With further development of the multiport antenna concept, in late 1980s and early 1990s, the coherent combining was extended to cover transmitter antenna diversity, beamforming, and more significantly spatial multiplexing. The latter facilitated multiple data streams being transmitted in parallel while taking advantage of the space domain in addition to frequency and time domains [2]. This achievement revealed a remarkable advantage of multiport antennas in increasing the throughput of a wireless communication system, which is perhaps the primary reason for their worldwide popularity. The spatial multiplexing in its simplest form is a bit stream to be transmitted, can be demultiplexed into two half-rate substreams, and transmitted through different antennas simultaneously. Under a suitable channel condition, the spatial signatures of these signals at the receiver terminal are separable. Hence, the receiver having knowledge of the channel can differentiate between these two co-channel signals and extract the corresponding substreams accordingly. In addition to spatial multiplexing, the multiport antenna technology provides a further critical advantage. For instance, the frequency reuse in wireless channel is the main

 
S. K. Koul et al., *Antenna Systems for Modern Wireless Devices*, Signals and Communication Technology, https://doi.org/10.1007/978-981-97-3369-9_8

source of interference. When multiport antennas are used, the difference between the spatial signatures of signals with the same frequency makes it possible to reduce the interference between them too [3].

The concepts and the principles upon which the multiport antennas are designed and developed have been known for many years [4]. However, today there is a need to realize a number of radiation elements on a compact chassis. The presence of radiation elements in the nearby vicinity results in high signal correlation and mutual coupling [5]. The latter not only reduces the overall performance of the communication system significantly but also has noticable adverse effects on the battery reservoir of the mobile terminal. This has turned out to be a long-standing challenge in personal wireless communication industries. These crucial concerns remain beside the detuning and gain imbalance of the radiation elements which occur upon implementation of such a multiport system [6]. The multiport antennas especially on the user side should retain their efficient performance in different propagation circumstances. For instance, depending on the distribution of the incoming electromagnetic (EM) waves and their angle of arrival (AoA), polarization, Doppler frequencies, and correlation, different propagation scenarios bestow different potential levels for diversity and spatial multiplexing gains. This chapter describes the various multiport antenna designs that preserve their effective performance in different multipath scenarios.

## 8.2 Multiport Antenna Designs

Multiport antenna designs coordinate radiating antennas by creating and dynamically adjusting various radiation patterns without modifications to the shapes and locations of the radiating elements. Phased arrays are a special case of coordinated antenna arrays where the relative phases and amplitudes of a signal are adjusted across the array elements. The objective is typically to create radiation patterns with selectivity only in the desired directions leading to a large effective isotropic radiation power (EIRP) in the transmit mode and a large signal-to-noise plus interference ratio (SNIR) in the receive mode.

Conventional narrowband phased arrays require variable phase shifters and variable amplitude control elements for each array element. The placement of these controllable components varies across different architectures. Multiport antennas can be arranged in the form of true time delay elements instead of phase shifters to process signals. The error introduced by using phase shifters instead of multiple antenna elements depends on instantaneous bandwidth and the propagation delay difference between array elements, which is proportional to their physical separation. A major advantage of multiple port antennas over single radiating elements is the ability to create versatile radiation patterns. The most basic form of phased array enables creating a single electronically steerable narrow beam. An even more complete and interesting application of multiple port antennas is the creation of

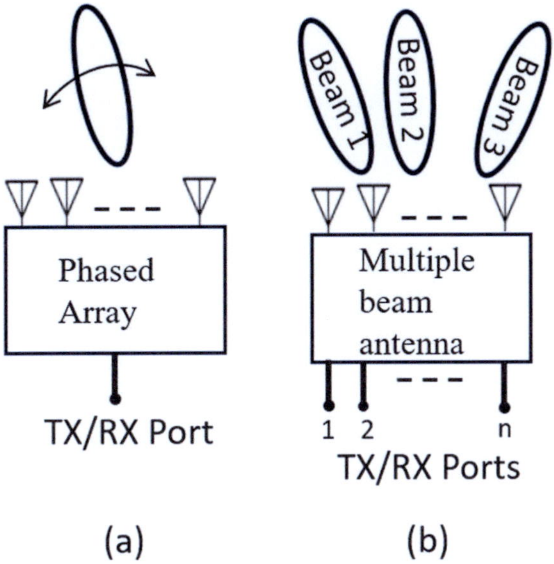


**Fig. 8.1** Two active array schemes. **a** Beam-scanning antenna array, **b** multiple beam antenna array

multiple independent beams as shown in Fig. 8.1. This, for instance, enables communicating independent data streams between the antenna and multiple spatially apart nodes.

The pattern and polarization diversity techniques are very effective to solve multi-path-fading effects in complex environments [7]. Comparing with classical antennas operating with a single radiation mode, the pattern diversity antennas, which are capable of radiating and receiving signals through different radiation modes simultaneously, have the advantage of performing higher effective gain while maintaining the same installation space. Several such antenna structures are proposed in the literature that mainly uses space diversity to achieve diversity gain. A four-port two-element antenna system using Planar Inverted-F Antennas (PIFAs) as antenna elements is reported in [8]. Each element of PIFA has two feeding ports connected to the common radiating plate using polarization and pattern diversities as the two feeding plates are at right angle to each other. In addition to polarization and pattern diversities, spatial diversity is also exploited between two PIFAs as the two antenna elements are positioned diagonally opposite ends of ground plane as shown in Fig. 8.2. Since the two ports under the same radiating element of each PIFA are placed very close to each other, there is a need of using an isolation technique for decreasing mutual coupling between two ports. For this purpose, a slot is cut in the ground plane beneath top-plate of each PIFA. These etched slots decrease the current flow between two feeds of the same PIFA through ground plane, which reduces the mutual coupling between them and make them work independently.

A compact four-port dual-band MIMO antenna with suppressed higher-order modes is presented in [9]. The complementary split-ring resonator (CSRR) loading is used on a square microstrip antenna to achieve simultaneous miniaturization and

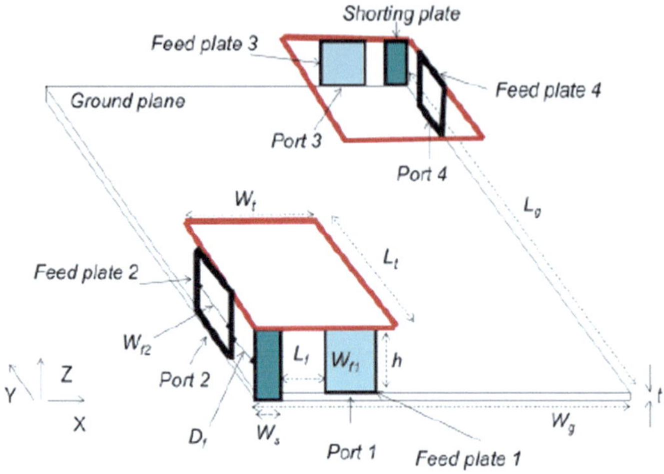


**Fig. 8.2** Structure of the four-port, two-element antenna [8]

dual-band response. By placing a single shorting post close to antenna center line, the higher-order modes of the four-port dual-band MIMO antenna are then suppressed, while maintaining satisfactory mutual coupling and impedance matching performance in the operating band. The four antenna elements loaded with CSRR to achieve dual-band operation and size reduction of the overall design is shown in Fig. 8.3.

Researchers in [10] proposed a technique that allows multiple antennas to be collocated on a common ground plane. T-slots are introduced between the radiating PIFA elements to minimize the coupling between adjacent elements. This allows those radiating elements to be positioned closer than half-a-wavelength on a shared ground plane. The technique allows direct and independent control of the gain of

**Fig. 8.3** Schematic diagrams of MIMO antenna design [9]

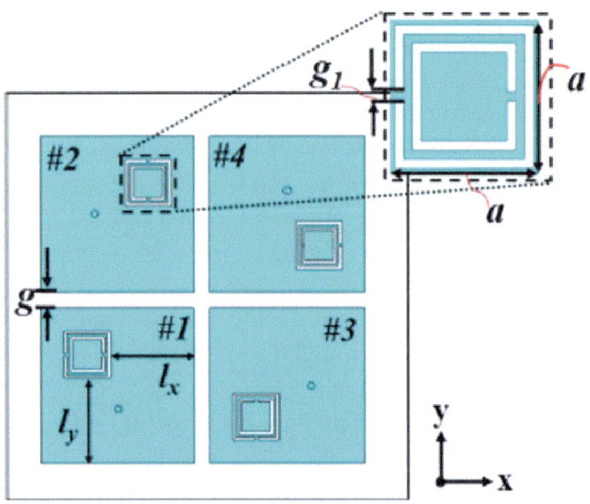

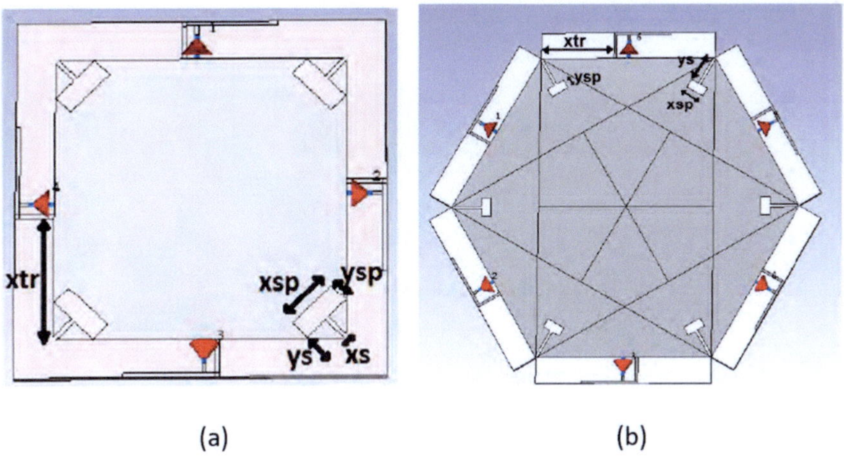


**Fig. 8.4** **a** Four-element antenna, **b** six-element antenna [10]

antenna element and number of elements in a MIMO system. Four- and six-element antenna developed using this technique is shown in Fig. 8.4a, b.

A four-port dual-polarized antenna is proposed in [11]. The four monopole radiators are deployed in a mirror image pattern. Antenna 2 is a mirror image of Antenna 1, whereas Antenna 3 and Antenna 4 are mirror images of Antenna 2 and Antenna 1. The incurred mutual coupling can be minimized by leaving sufficient space (spatial diversity) among the radiators. The ground planes of the four antenna elements must be interconnected. Therefore, a trapezium-shaped patch is loaded between Antenna 1 and Antenna 2, as well as between Antenna 3 and Antenna 4, so as to interconnect the ground planes, as shown in Fig. 8.5.

A new eight-port/four-element antenna system with miniaturized radiators and full radiation coverage is reported in [12]. The single-element is composed of square-loop resonators with dual-polarization diversity which have been fed using independently coupled T-shaped feed lines. Due to its configuration and low-mutual function, it can be considered self-isolated. The elements are highly miniaturized and operate at 3.6 GHz in the sub-6-GHz 5G spectrum supporting the frequency of 3.4–3.8 GHz. The antenna possesses several unique features such as good isolation, excellent radiation coverage, low ECC, and sufficient efficiencies. A common method in smartphone antenna design is to modify ground planes with slot and slit cuts. However, compared to the design with slots in the ground, it has limited radiation coverage which mainly covers the top side of the smartphone. Therefore, dual-polarized design with the modified ground plane is used as shown in Fig. 8.6.

A compact 2 × 2 MIMO antenna with an electrical dimension $0.28\lambda_0 \times 0.28\lambda_0$ mm$^2$ is proposed in [13]. The 5.03-GHz bandwidth and isolation are achieved with a tapered feed and a portion of the ground plane connected by stubs. The suggested antenna has a 15-dB average isolation over the operating band with a peak

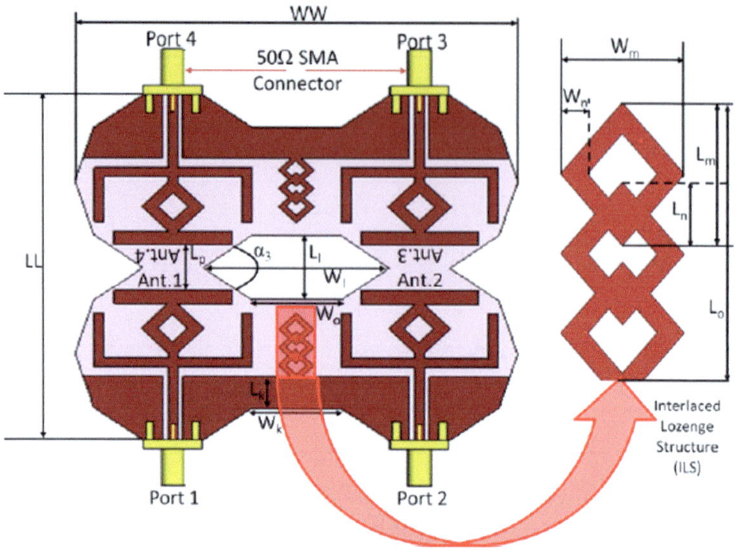


**Fig. 8.5** Geometry of the decagon-shaped flexible MIMO antenna [11]

**Fig. 8.6** Configuration of the MIMO smartphone antenna system with a full GND [12]

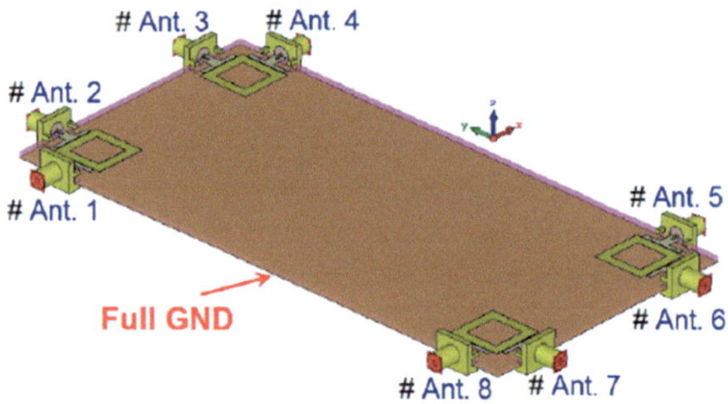


gain of 4.2 dBi. The proposed antenna, shown in Fig. 8.7, focuses on multiple wireless applications such as Bluetooth (2.400–2.483 GHz), ISM band (2.40–2.483 GHz, 5.725–5.850 GHz) WLAN/Wi-Fi, and 5G (sub-6 GHz) range. For bluetooth and ISM band applications, the proposed antenna offers wide impedance bandwidth with good return loss, radiation properties, and group delay < 1 nsec.

A novel wideband four-port diversity antenna that is capable of exciting four different radiation patterns is presented in [14]. The four-port diversity antenna consists of four magnetoelectric dipoles arranged in a ring configuration above a ground plane and a vertical electric monopole at the center of the ground plane. A matching network is used to combine the signals from the four magnetoelectric dipoles to produce two orthogonal broadside modes and one conical mode with horizontal polarization. The electric monopole is used to generate a conical beam with

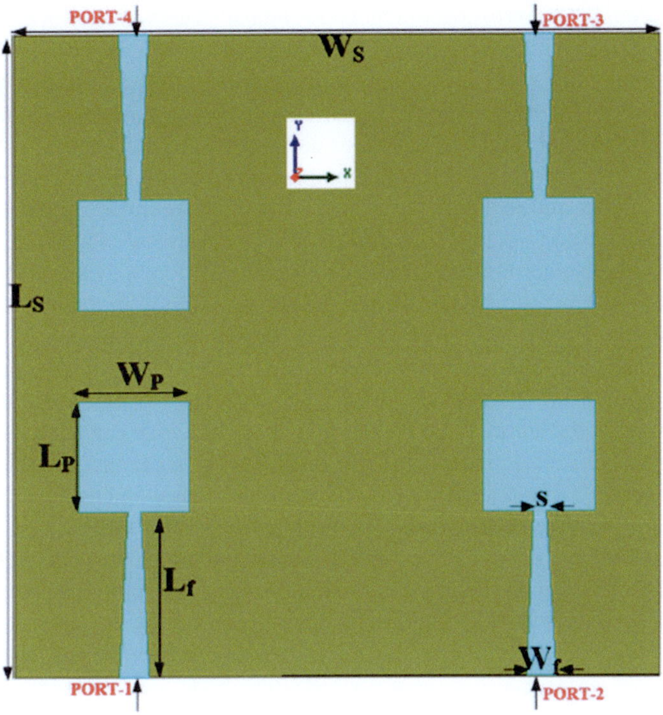


**Fig. 8.7** Proposed MIMO antenna [13]

vertical polarization in electric feed. The electric dipole is connected to the lower part of the antenna, which radiates like a folded magnetic current. The excitation of the four ports can be controlled by a switch or excited simultaneously.

## 8.3 Multibeam Antenna Designs

Multibeam antennas have emerged as a promising concept for applications where selective beam coverage is required. Multibeam antennas offer several advantages, especially in scattering environments, such as reducing the multipath fading, minimizing the RF interference, and providing simultaneous coverage of multiple areas. Multibeam antennas are adopted across a wide range of application spectrums, including surveillance systems, wireless and satellite-based communications, smart antennas, Doppler radio guides, and radio altimeters [15].

To serve a number of distributed users, multibeam antennas or beam-scanning antennas are highly demanded [16]. Multibeam antenna systems represent the key antenna technology for supporting a high data transmission rate, an improved signal-to-interference-plus-noise ratio and increased spectral efficiency [17]. It generates multiple radiation beams to cover predefined angular space which is an added advantage as compared to single-beam antennas. Compared to single-beam antennas, multibeam antennas offer more diversity, improved link quality, and higher geographical penetration, resulting in better signal coverage. The multibeam antennas can generate independent directive beams to cover a predefined angular range. They serve as the key hardware for enabling massive MIMO technology further improving the link reliability and reducing the outage probability. A single source or multiple sources can be used to create a multibeam antenna design. A single-source multibeam antenna can communicate simultaneously with different locations. In multiple source antennas, one independent beam gets activated at a time when each of the sources becomes active. An unrestrained expansion of the number of communication elements may cause interference among them. Conventional methods of designing multibeam antennas are mainly based on lenses, reflectors, and phased arrays. However, the lenses and reflectors are placed at a wide gap from the antennas increasing the overall antenna size [18]. Phased-array antennas generate multiple beams when multiple radiators are fed with different phases and amplitudes. But the high cost and complex feeding networks prohibit their wide applications.

Wireless link has a need of an omnidirectional antenna but with higher gains. Of course, there is no way of achieving both of them simultaneously. To solve this contradiction, Han et al. [19] proposed to design multibeam antennas, which divide the radiation pattern of omnidirectional antennas into several beams. The antenna, based on spoof surface plasmon polaritons (SSPPs) mode coupling, consists of a feed monopole and N double-side corrugated metallic strips circling around the feed. The fields of the feed are coupled into SSPP waves that propagate along the strips and radiate at the end, dividing the omnidirectional radiation pattern into several beams.

Multibeam radiation can be achieved leveraging the concept of metasurfaces, enabling the radiated wavefronts to be tailored in an all-electronic manner, making reconfigurable multibeam operation possible without the need for bulky feeding structures or complex phase shifting circuits. Two techniques are usually adopted. The first one is to use super-thin metamaterial substrates, which are usually placed above the antenna aperture to modulate the phase profile of transmitted waves. Anisotropic zero-index metamaterial lenses can realize multibeam radiations in predesigned directions [20]. Based on the effective medium theory, GRIN metamaterial lenses were proposed to enable high-directive beam shaping and beam steering in a broad frequency band [21]. The second technique is to use planar metasurface reflector, which modulates the phase profile of the reflected waves to form multiple beams.

A microstrip multibeam antenna based on metasurface is proposed in [22] as shown in Fig. 8.8. The metasurface is square with sides of $0.25\lambda g$. In order to realize multibeam radiation, four metasurfaces formed by square patches and Sub 3 are placed above the feeding network. According to the antenna radiation mechanism, the rectangular patch antenna can excite TM10 mode. The higher-order modes are also excited, and an electric field distribution like $TM_{10}$ mode will be generated on each patch. The entire $4 \times 4$ metasurface can be regarded as generating a quasi-$TM_{40}$

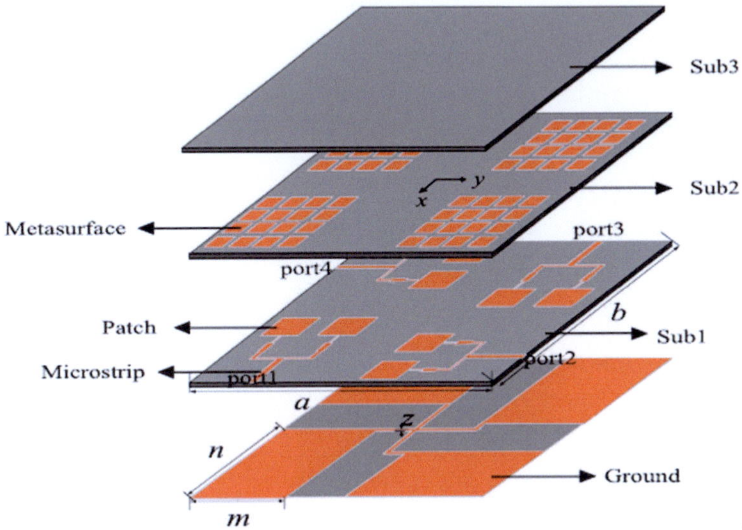


**Fig. 8.8** Multibeam antenna with metasurface [22]

electric field distribution. By adjusting the current distribution in every element of the metasurface, different patterns can be formed.

When ports 1–4 are excited, the antenna generates four different beams, and the results are shown in Fig. 8.9. It can be observed that the maximum radiation levels of the four oblique beams are ($\theta$, $\Phi$) = (32°, 0°), (32°, 90°), (32°, 180°), and (32°, 270°). Steering the beam in the elevation plane is realized by arranging metasurface above the feeding network. Furthermore, placing the substrate above the metasurface improves the beam performance. By exciting different ports, four beams pointing to different directions can be obtained. The operating frequency of the antenna is 5.82 GHz, and the gain can reach 9.3 dBi. At the same time, beam is steered in azimuth plane by adjusting the feeding network.

A hybrid metasurface using split-ring resonator unit cell and U-shaped unit cell for split beam at 3.5-GHz band is presented in [23]. The basic antenna reported in this work is a microstrip patch antenna (MPA), which is used to test the proposed metasurface and test its split property. The main purpose of splitting the beam in the proposed design compared to previous designs is to avoid the interference that comes from obstacles and buildings. In addition to this, the split beam technique helps in increasing the traffic capacity and reducing the size of the whole network. Therefore, a metasurface (MS) is proposed to direct surface current and to have high-gain and multibeam properties. To achieve this feature, a square split-ring resonator (SSRR) and U-shaped unit-cell metasurface is implemented as a superstrate to a single square patch antenna. The manipulation of this hybrid metasurface configuration can create opposite current flow on the unit cell and thus split the beam. The multibeam generations based on anisotropic zero-index metamaterials (AZIMs) are investigated in [24]. By rotating the principal axes of uniaxially or biaxially anisotropic media with

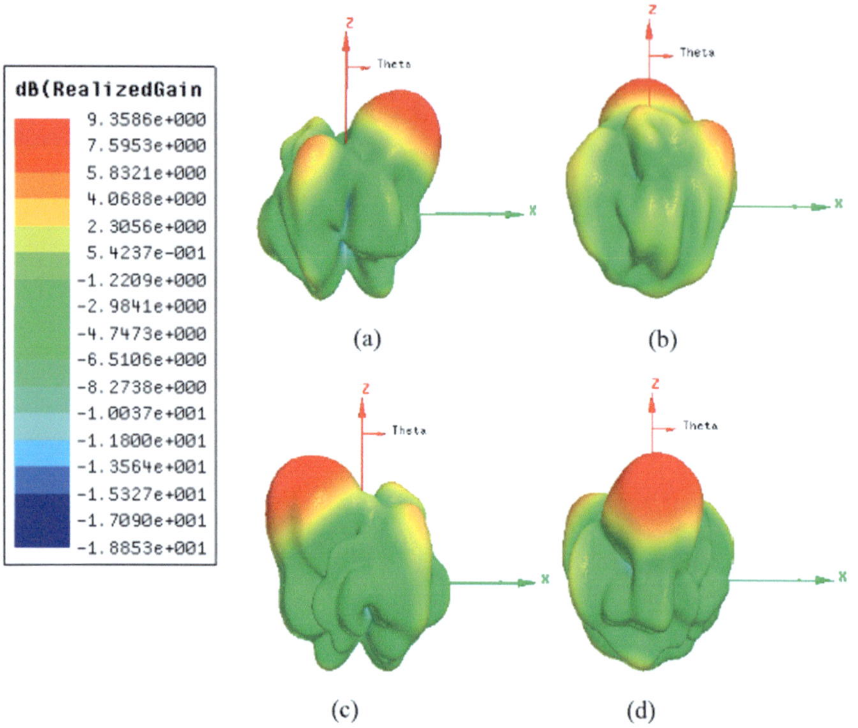


**Fig. 8.9** Radiation patterns of the antenna. **a** port 1 **b** port 2 **c** port 3, **d** port 4 [22]

one component of the permittivity or permeability tensor approaching zero, we can realize the high-gain beam generation toward an arbitrary direction under the excitation of a line source. By combining AZIMs with different orientations of principal axes, we can obtain multiple beams in the desired directions. This method is quite easy to implement and may find potential applications in antenna designs or other wireless techniques.

Researchers in [25] demonstrated that shorting pins can be used to control the radiation pattern of microstrip antenna. As shown in Fig. 8.10, by arranging shorted parasitic patches around a square patch, the corresponding current distribution is optimized, and radiation beam is divided into four parts. In addition, for single-source antenna, shorting pins along the edge of square patch can also adjust the operating frequency, gain and pitch angle.

After four parasitic patches shorted with a shorting pin are added, strong induced current is generated in the parasitic patches. As can be seen in Fig. 8.11, not in its fundamental mode, square patch works in its $TM_{20}$ and $TM_{02}$ mode. As a result, the beam is separated into four parts with a gain of 7.26 dBi and the pitch angle of each beam is 47°.

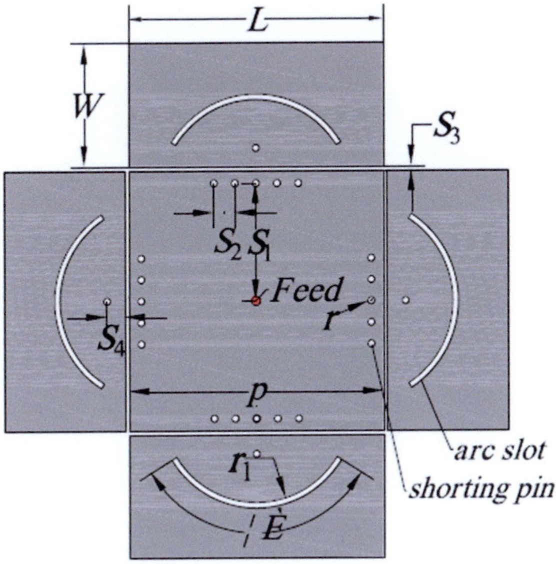


**Fig. 8.10** Single-source antenna geometry structure [25]

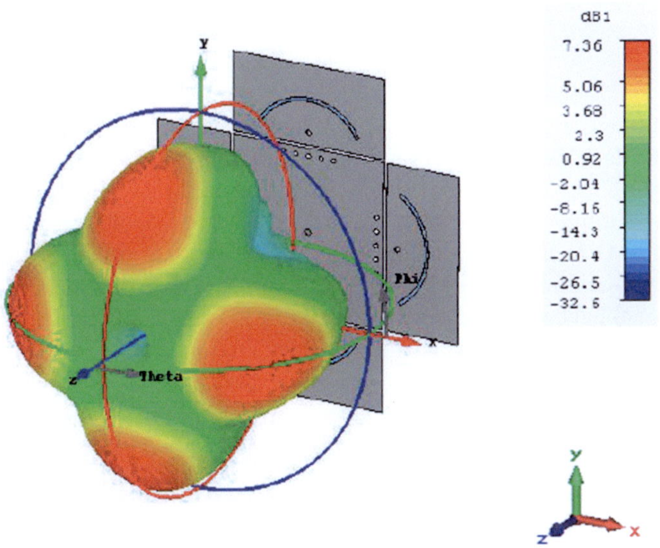


**Fig. 8.11** Radiation patterns of the single-source antenna at 5.3 GHz with shorting pins [25]

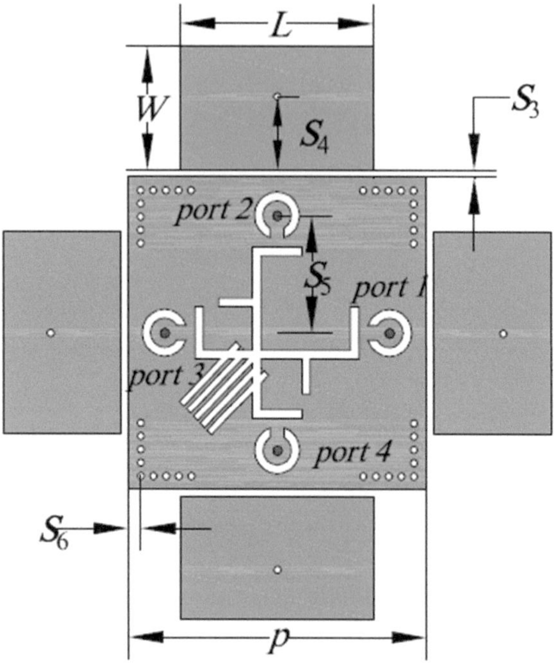


**Fig. 8.12** Multisource multibeam antenna structure [25]

Unlike single-source multibeam antennas, the multisource multibeam antenna consists of four ports, called ports 1–4, and the geometry of its patch layer is shown in Fig. 8.12. In addition, in order to improve the isolation between the four ports, an asymmetric gap is introduced in the square patch. For the multisource case, the principle of forming multiple beams is similar to that of single source. When port 1 is switched ON, the current mainly concentrates near the parasitic patch closing to port 1 and its direction is parallel to the short side of the parasitic patch. As a result, a beam having a maximum level at $(\theta, \Phi) = (0, 30^\circ)$ is formed. Compared to the normal multibeam antenna, the presented antennas are compact, low-profile, and suitable for various miniaturized communication systems.

## 8.4 Multiport Antenna Design No. 1 with Gain Equalization

In this section, a three-port 3D-printed series-fed antenna array with pattern diversity is presented. 1 mm air gap in the substrate ensures that the antenna module has higher gain and radiation efficiency than a conventional substrate antenna. The superstrate design consists of parasitic patches that enhance the gain in the broadside and oblique incidences. Therefore, the proposed superstrate design is more economical and versatile as it has the ability for enhancing gain for different beam angles and obtains equal gain at three ports. The three-port antenna switches between + 18°, 0°, and – 18° with 12 dBi gain.

### *8.4.1 Micromachined Patch Antenna Design*

First, the characteristics of a standard non-micromachined patch antenna at 5.8 GHz are studied. The substrate used is PLA with a dielectric constant of 2.4 ± 0.01 and a loss tangent of 0.015. The substrate thickness is 2 mm. The ground plane dimensions of the antenna are 0.7λ × 0.7λ, and the size of the patch antenna is 0.28λ × 0.28λ. From the substrate used for the standard patch antenna design, a part of the substrate is removed to introduce an air cavity as shown in Fig. 8.13. From the 2 mm substrate block, 1 mm substrate is removed to introduce an air cavity. This ensures higher radiation efficiency with adequate structural stability for the antenna. The introduction of the air cavity caused a detuning of the antenna, due to the variation of the substrate dielectric constant, and the patch antenna is again optimized for 5.8-GHz resonant frequency. The new dimension of the patch antenna is 0.35λ × 0.35λ. The simulated reflection coefficient of the patch is presented in Fig. 8.14. The radiation effectiveness and gain of the antenna are shown in Fig. 8.15. The data clearly show that adding an air gap between the substrate and ground plane results in gains that are improved by 2 dB and a radiation efficiency increment of more than 10%.

The proposed patch antenna is compared to various patch antennas in the literature in Table 8.1. The fused filament method is utilized to create the dielectric portion of the antenna in [27] and an ultrasonic metal wire mesh embedding technique is employed to fabricate the conducting portion. However, the antenna only has a 5.5 dBi

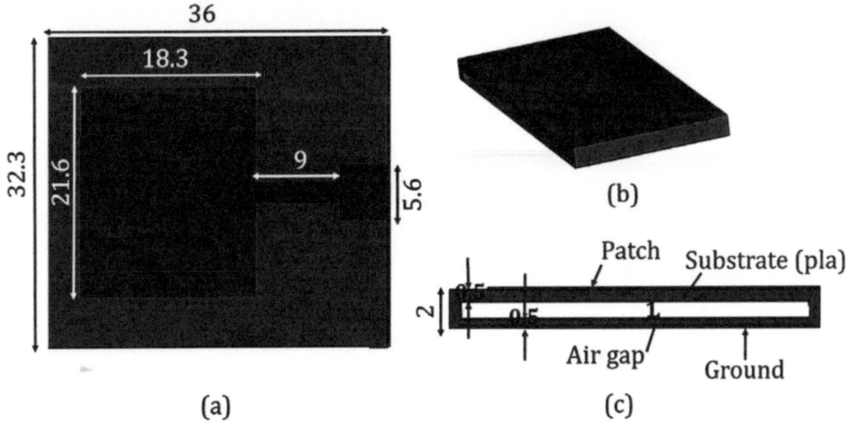


**Fig. 8.13** **a** Top view schematic of the proposed patch, **b** isometric cut-view, and **c** front view withal dimensions are in mm) [26]. Reprinted with permission from IEEE

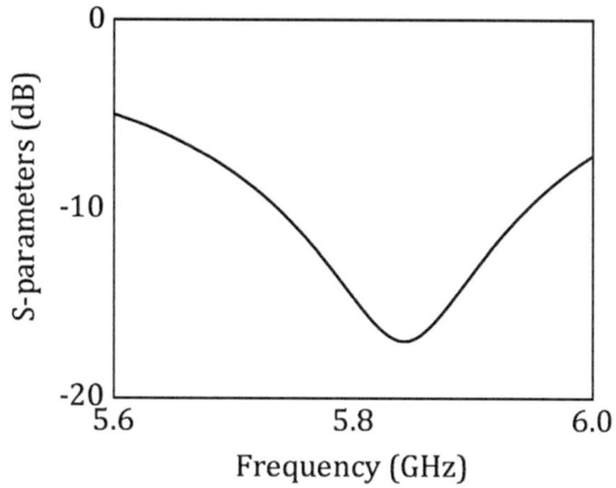


**Fig. 8.14** Return-loss characteristics of the micromachined patch antenna [26]. Reprinted with permission from IEEE

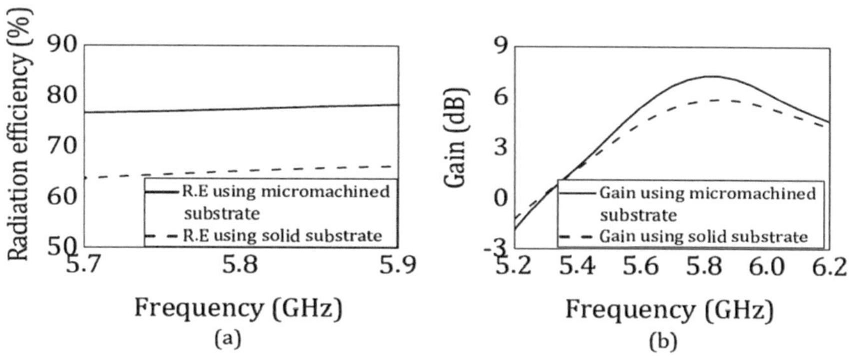


**Fig. 8.15** **a** Radiation efficiency **b** gain of the solid and micromachined patch design [26]. Reprinted with permission from IEEE

gain [28]. A square mesh-type antenna with resin slabs used to modify the permittivity is demonstrated. However, with a ground plane of dimension 2λ, the gain of the microstrip antenna is only 6.4 dBi. The suggested 3D-printed patch antenna features a simpler fabrication procedure and higher gain for a thinner substrate than these antennas. [29] illustrated a honeycomb substrate-designed antenna. Even though the honeycomb substrate lightens the overall weight, the patch radius of the honeycomb substrate is 1.6 times larger than the patch radius for a solid substrate, increasing the size of the antenna. The unique architecture of the honeycomb structure makes its fabrication challenging.

**Table 8.1** Comparison of the proposed additively manufactured patch antenna with other patch antennas

| Ref. | Resonant frequency (GHz) | Substrate thickness (mm) | Size | Gain (dBi) | Radiation efficiency (%) |
|---|---|---|---|---|---|
| [27] | 7.5 | 3.2 | $0.5\lambda \times 0.4\lambda$ | 5.5 | 84 |
| [28] | 2.45 | 4.5 | $2\lambda \times 2\lambda$ | 6.38 | 3.5 |
| [29] | 2.4 | 5.0 | $0.45\lambda \times 0.45\lambda$ | 8.0 | 81 |
| PA* | 5.8 | 2.0 | $0.6\lambda \times 0.7\lambda$ | 7.5 | 77 |

**PA* Presented antenna

### 8.4.2 Single Port Additively Manufactured Series-Fed Array

Using the proposed micromachined patch antenna, a five-element series-fed antenna array is designed. The schematic is shown in Fig. 8.16. The interelement antenna spacing is $0.25\lambda$. The substrate of the antenna has dimensions of $3.8\lambda \times 1.35\lambda$. It is observed in Fig. 8.17 that, as the array elements increase, the antenna gain increases, and the radiated antenna beam also gets tilted. Beyond five elements, there is no significant improvement in the beam tilt. For demonstrating the design, five-element antenna array is chosen. The radiation pattern of the array has $+ 18°$ beam tilt, and the antenna gain is 9.4 dBi.

A linear array with radiating and non-radiating edges with different feed networks consisting of a U-shaped feed network and a corner feed network is presented in [30]. The design of the feed networks is used for array size reduction and to obtain lower SLL. But only a single beam is radiated at the broadside and no flexibility about the beam in other angular directions is discussed in the design. In [31], a series-fed patch array with enhanced bandwidth and lower sidelobe level is proposed. The bandwidth is enhanced by a dual-band structure, and sidelobes are improved by the modulation

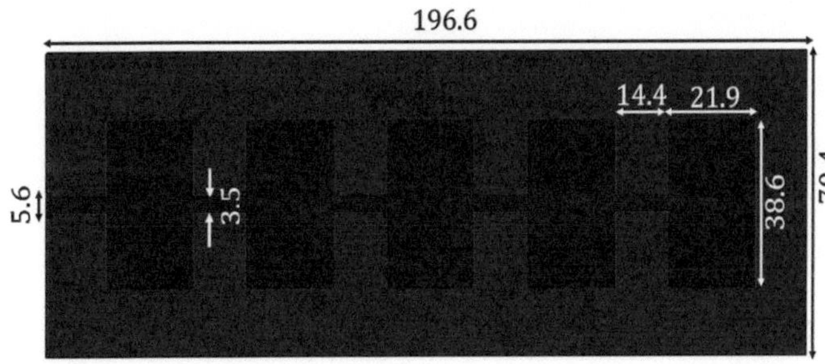


**Fig. 8.16** Schematic of the series-fed array (all dimensions are in mm) [26]. Reprinted with permission from Taylor and Francis

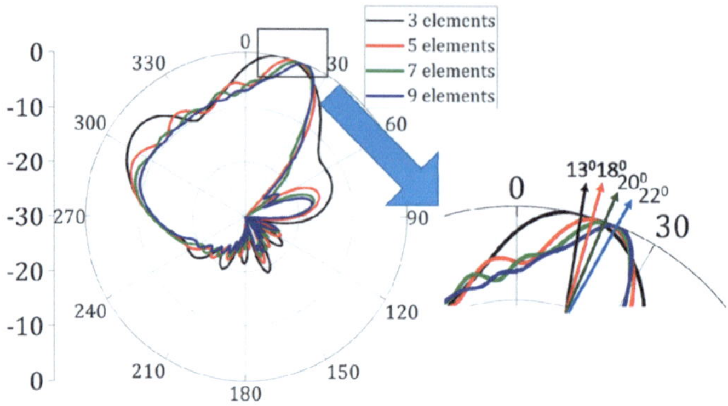


**Fig. 8.17** Beam tilt of the series-fed array as the number of elements is varied [26]. Reprinted with permission from Taylor and Francis

of the power transfer from one patch to the other. But it requires 14 elements for a gain of 14.2 dBi.

### *8.4.3 Two-Port Additively Manufactured Series-Fed Array*

The series-fed array is further extended to a two-port series-fed array as shown in Fig. 8.18. The distance between the two ports is 1.5λ. This topology allows the radiation pattern of the antenna to switch between + 18°, and − 18° with a gain of 9 dBi. The distance between the elements of the two arrays is only 2.8 mm, the nearest distance without any antenna detuning.

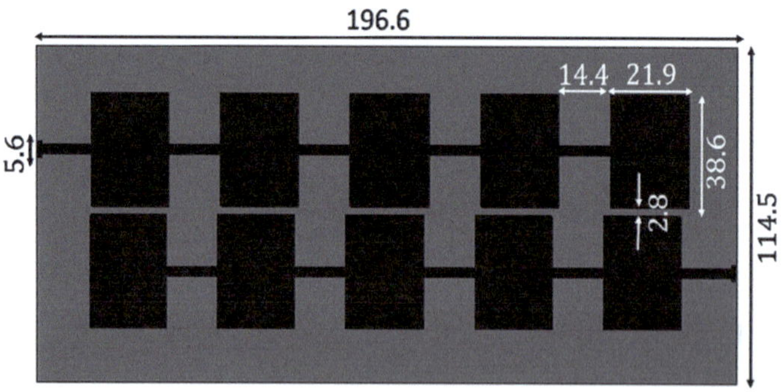


**Fig. 8.18** Schematic of the two-port series-fed array (all dimensions are in mm) [26]. Reprinted with permission from Taylor and Francis

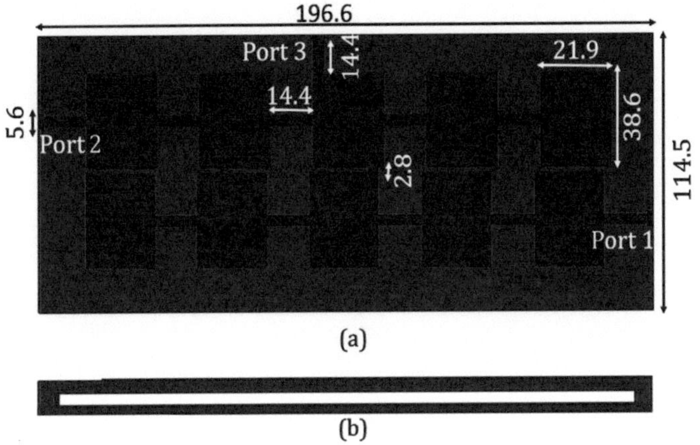


**Fig. 8.19** **a** Top view of the schematic of three-port series-fed antenna array and **b** front view with (all dimensions are in mm) [26]. Reprinted with permission from Taylor and Francis

### *8.4.4 Three-Port Additively Manufactured Series-Fed Array*

A third port is now introduced to the two-port series-fed array as shown in Fig. 8.19. To realize this, the third input port was moved along the length of the patch at the center element of the series-fed array. The input port is fixed at the edge of the patch length, as the antenna was detuning from the resonant frequency in other positions at the width of the patch antenna. Initially, when the input port was excited, the antenna has a beam tilt of 10°. At the middle of the series-fed array, a stub is attached between the two patches to reduce the beam tilt and move the beam to the broadside. The stub made sure that EM waves were coupled between the array and reduced the antenna's beam squint. Figure 8.20a displays the measured and simulated return-loss characteristics.

The simulated impedance bandwidth of the series-fed array is 510 MHz (5.36–5.87 GHz). Antennas 1 and 2 have an HPBW of 22°, and antenna 3 has an HPBW of 37°. Antennas 1 and 2 have a front-to-back ratio of 27 dB, and antenna 3 has a front-to-back ratio of 24 dB. Antennas 1 and 2 each have a gain of 9.4 dBi, and antenna 3 has a gain of 4.5 dBi as shown in Fig. 8.20b.

### *8.4.5 Gain Enhancement Using Parasitic Patch Superstrate*

To enhance the gain of the patch antenna, a superstrate with a parasitic patch is designed. Initially, the antenna is positioned horizontally, without any tilt. A parasitic patch of size 16.2 mm × 13.8 mm (0.75 times the patch size) placed at a height of 0.47λ from the patch provided a 2 dB gain enhancement. As the patch antenna was tilted for different inclinations, as shown in Fig. 8.21, some scanning losses were

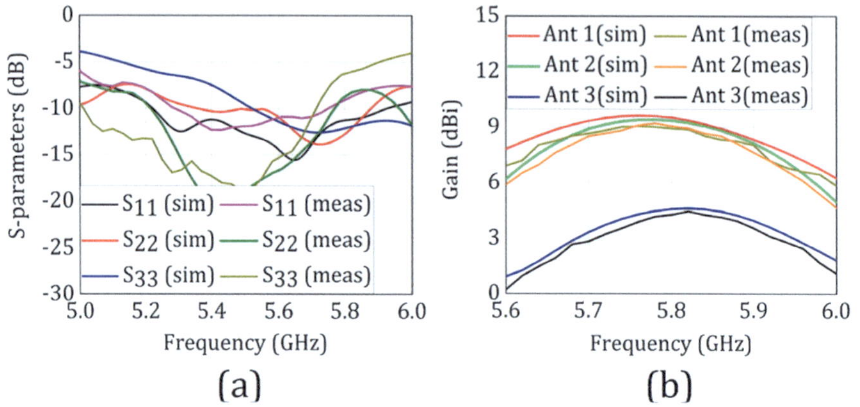


**Fig. 8.20** **a** Return-loss characteristics of the three-port series-fed array, **b** gain of the three-port series-fed array [26]. Reprinted with permission from Taylor and Francis

observed. However, with slight variations in the parasitic patch sizes, the scanning losses could be minimized. For 20° antenna inclination, a scaling factor of 0.67 times the patch size for the parasitic patch gave a scanning loss of 0.5 dB. These results indicate that for beam angles up to 20°, parasitic patches with varied scaling factors can be used for gain enhancement with minimal scanning loss.

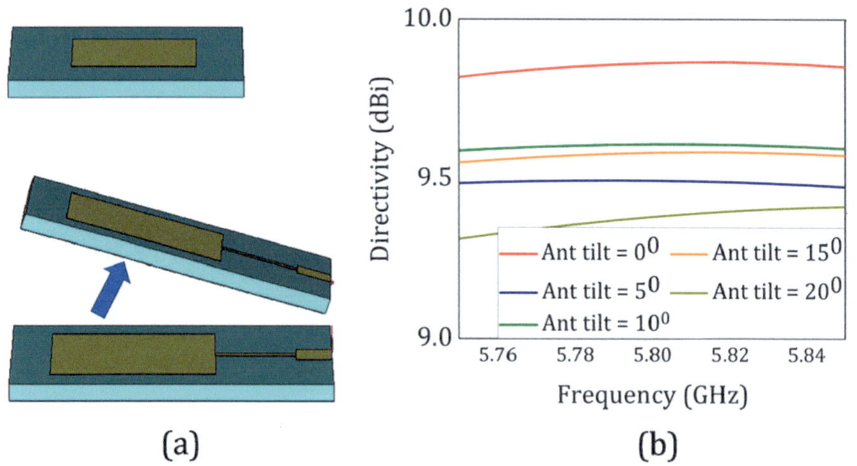


**Fig. 8.21** **a** Patch antenna with parasitic patch superstrate, **b** directivity variation of the patch for different antenna inclinations [26]. Reprinted with permission from Taylor and Francis

### 8.4.6 Three-Port Antenna with Gain Enhancement

In this section, a single superstrate that can enhance the gain of three antennas at different beam incidences is proposed. The researchers in [32] have explained the theory behind the gain enhancement using metasurface design in the broadside direction using an analytical model based on ray optics. In the proposed design, shown in Fig. 8.22, a single parasitic patch superstrate enhances the gain in the broadside and oblique incidence of EM waves. In literature, parasitic patches that enhance only the broadside gain are available. Increasing the gain and bandwidth of an array antenna using parasitic patches is presented in [33]. The design uses 16 array elements and 16 parasitic patches with a massive size of 1.6λ × 5λ for a gain of 19.8 dBi. Design using uniform and non-uniform parasitic patches above a circular patch is presented in [34]. However, the proposed antenna has a low gain of 6.34 dBi, and the gain increment is only in the broadside. A cavity-backed aperture antenna with a parasitic patch is proposed in [35]. The parasitic patch enhances the circular polarization of the antenna that gets deteriorated due to the cavity but offers a gain enhancement of only 1 dB.

For the proposed antenna, the parasitic patches are designed in non-micromachined PLA, the same substrate that was used for the three-port series-fed antenna arrays. The size of the substrate of the parasitic patch is the same as the size of the substrate of the antenna arrays. There are four factors of the parasitic patch superstrate that decide the antenna gain enhancement. First, the height of the parasitic patch superstrate; second, the scaling factors of the different parasitic patches; third, the placement of patches on the superstrate; and fourth, the thickness of the

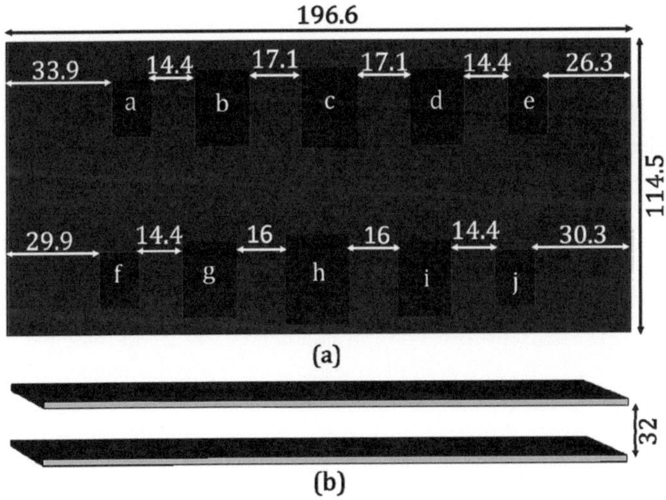


**Fig. 8.22** Schematic of three-port series-fed antenna array with parasitic patch superstrate. **a** Top view, **b** front view (all dimensions are in mm) [26]. Reprinted with permission from Taylor and Francis

**Table 8.2** Scaling factors of the parasitic patches

| Constant | Scaling factor | Constant | Scaling factor |
|---|---|---|---|
| *a* | 0.54 | *f* | 0.54 |
| *b* | 0.74 | *g* | 0.74 |
| *c* | 0.78 | *h* | 0.78 |
| *d* | 0.74 | *i* | 0.74 |
| *e* | 0.54 | *j* | 0.54 |

superstrate. To decide the height of the parasitic patch superstrate, the height of the superstrate was varied from 0 to 50 mm, and its observations were noted. It was observed that up to 30 mm height (0.5λ), the antenna excited by port 3 suffers from detuning. As the height is increased beyond 0.5λ the gain of all three antennas starts reducing. So, the ideal height of the superstrate is fixed at 0.5λ. Initially, the size of the parasitic patches considered for the superstrate design was the same size as the patches in the antenna array. But this resulted in gain variations in the three antennas by more than 2 dB. To equalize the gain in the three antennas, the parasitic patches in the substrate were scaled to varying dimensions. The scaling factor mentioned in Table 8.2 is how the patches were scaled over the superstrate. By slightly shifting the parasitic patches horizontally along the superstrate, the gain of the antennas was seen to be equalized. To decide the thickness of the substrate, the substrate thickness was varied from 0.5 to 8 mm, and the observations were noted. The gain of antenna 3 increases by 2 dB when the substrate thickness is varied from 0.5 to 2 mm. There is improvement in gains of antennas 1 and 2 also. Beyond 2 mm, gains of antennas 1 and 2 show some gain reduction, but the gain of antenna 3 still increases. Beyond 7 mm, the antenna gain of all three antennas starts to decrease. So, the substrate thickness was decided to be 2 mm.

Figure 8.23 shows the electric field intensity of the array before and after it is loaded with the parasitic patch superstrate. The antenna beam tilt is not affected much by the parasitic patch superstrate. The in-phase distribution of the antenna increases due to the presence of a parasitic patch superstrate, which enhances the antenna gain. This action is more pronounced in the third port resulting in a gain enhancement of 8 dB for that port.

The measured and simulated return-loss characteristics of the three ports are shown in Fig. 8.24a. The simulated impedance bandwidth for the antenna is 390 MHz (5.5–5.89 GHz). The half-power beamwidth (HPBW) of antennas 1 and 2 is 20°, and antenna 3 is 32°. The front-to-back ratio of antenna 1 is 22 dB, antenna 2 is 20 dB, and antenna 3 is 21 dB. The gain of antenna 1 is 11.8 dBi, antenna 2 is 12.2 dBi, and antenna 3 is 11.6 dBi with radiation patterns at − 18°, + 18°, 0°. The isolation characteristics between different ports of the antenna after the parasitic loading are shown in Fig. 8.25. Figure 8.26 shows the radiation pattern of the antenna module before and after the parasitic patch loading. Table 8.3 provides a comparison of the proposed design with the other antenna designs in the literature.

Figure 8.27 shows the proposed antenna prototype. The substrates for the antenna module are additively manufactured using a Raise 3D RxP2200® 3D printer which

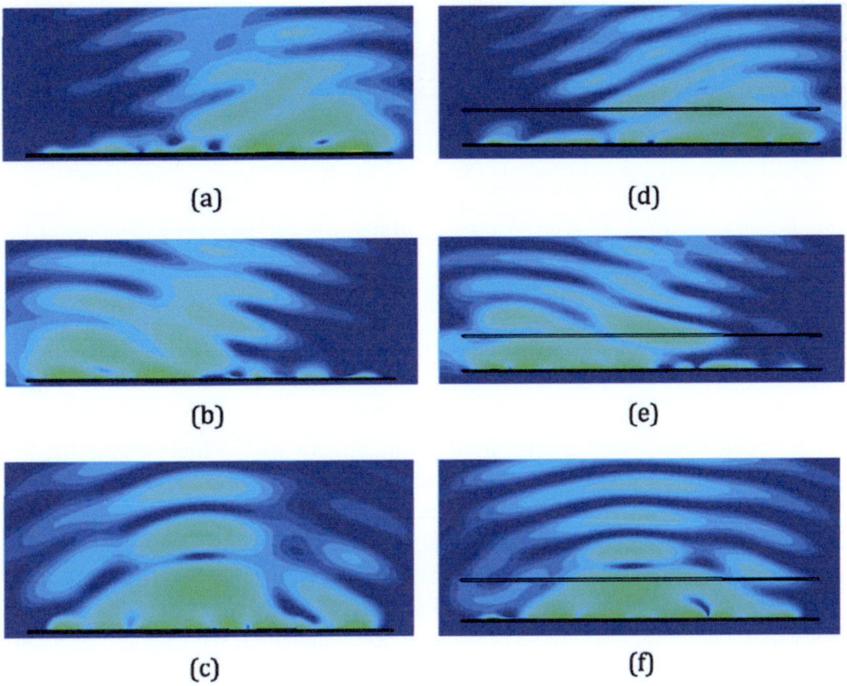


**Fig. 8.23** **a–c** Electric field distribution of three-port series-fed antenna array before parasitic patch loading, **d–f** after parasitic patch loading [26]. Reprinted with permission from Taylor and Francis

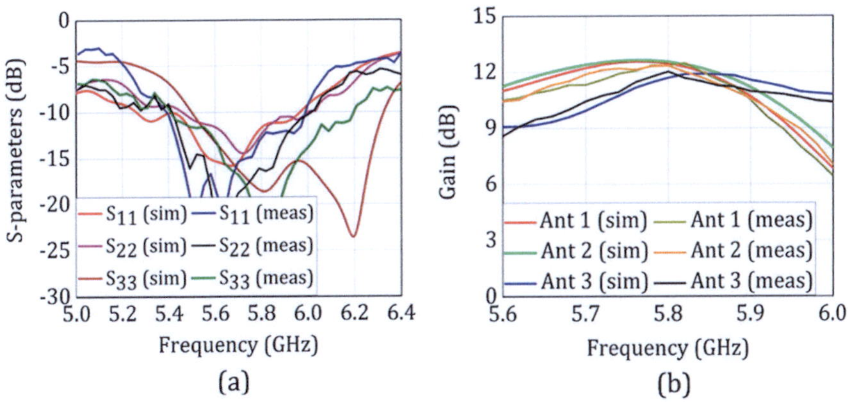


**Fig. 8.24** **a** Return-loss characteristics, **b** gain of the three-port series-fed array with parasitic patches [26]. Reprinted with permission from Taylor and Francis

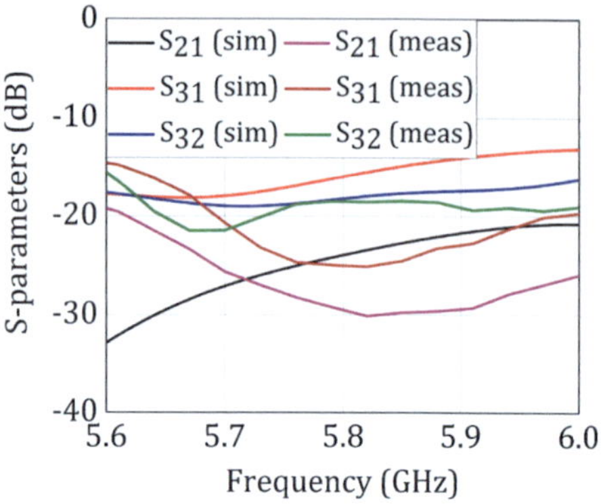


**Fig. 8.25** Mutual coupling between the antenna ports [26]. Reprinted with permission from Taylor and Francis

is a desirable low-budget printer used commercially with a high resolution. The total printing time for the substrates of the antenna and superstrate was 10 h, for an infill percentage of 25%. The design is fabricated through the photolithographic process using a copper tape of thickness 80 μm. The entire fabrication process is low-cost. Anritsu® MS2028C Vector Network Analyzer is used for return-loss measurement. The far-field measurements were performed in an anechoic chamber with ETS-Lindgren® double-ridged 3115 Model horn antenna as the standard antenna.

A 3D-printed three-port series-fed antenna array that utilizes a micromachined PLA substrate to enhance and equalize the gain across the ports at 5.8-GHz frequency is proposed. The gain equalization of 12 dBi is obtained using scaled parasitic patches. To enhance the gain and efficiency, a portion of the substrate is removed to introduce an air cavity, which is made possible through 3D printing.

The beam of the antenna scans the radiations at + 18°, 0°, and + 18° for a 12 dBi gain across the ports. The parasitic patches do not affect any other parameters of the antenna other than enhancing its gain, both in broadside and oblique wave incidence. There is good agreement between the simulation and measurement results.

## 8.5 Multiport Antenna Design No. II with Gain Equalization

In the proposed design, four different patch antennas with four input ports are placed in the same plane with a distance of 0.5 mm (0.008λ) between them. The important novelty here is the use of antenna elements, which have only 0.5 mm spacing between their edges, and the shift in their phase centers is taken advantage in beam tilting. When individual ports are excited one after another, the individual antenna beam radiates toward + $45^0$, + 15°, − 15° and − 45°. This performance of the antenna where the beam of the patch is tilted due to a shift in its phase center, instead of the normal broadside direction is not reported in the literature. Then an all-dielectric

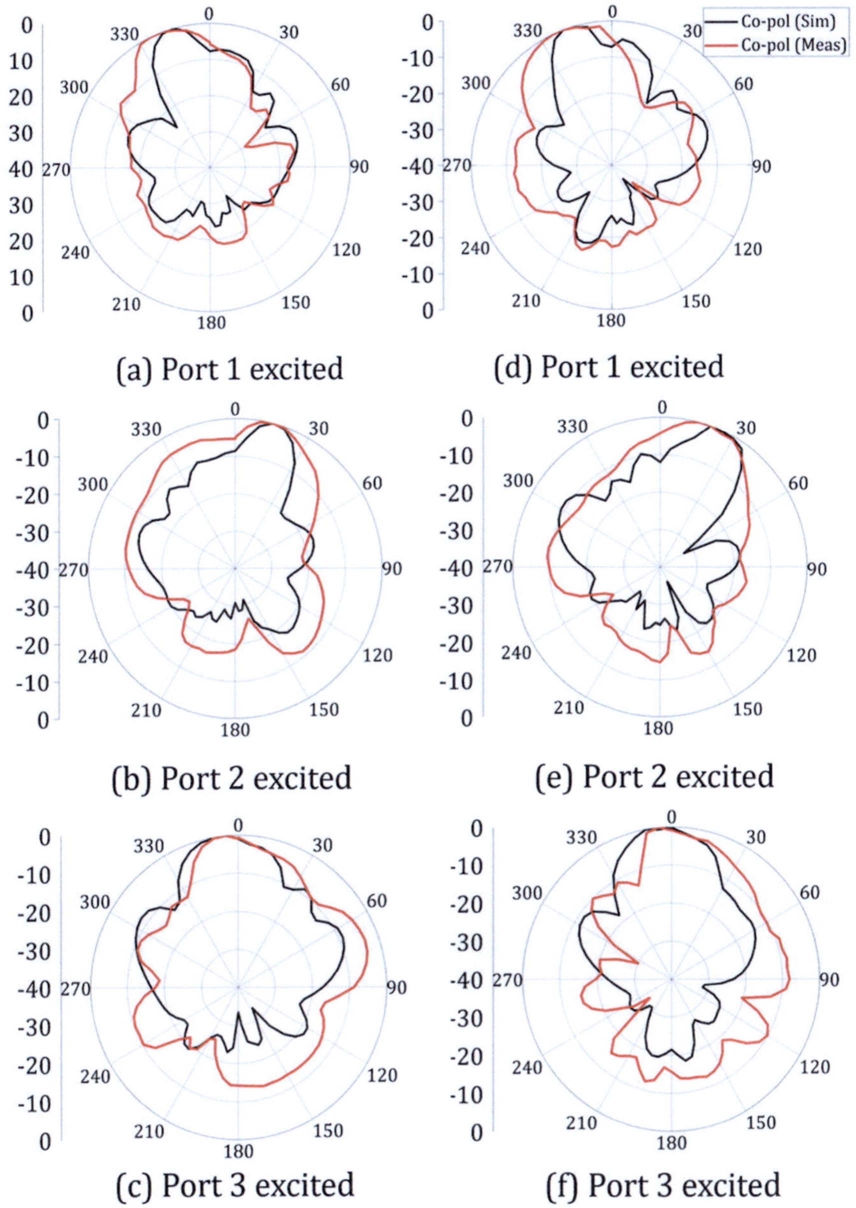


**Fig. 8.26** **a–c** Radiation pattern at 5.8 GHz before dielectric loading, **d–f** radiation pattern at 5.8 GHz after dielectric loading [26]. Reprinted with permission from Taylor and Francis

**Table 8.3** Comparison of the proposed antenna with other designs in the literature

| Ref. | Freq range (GHz) | Size | Ant profile | Feed antenna structure | Gain (dBi) | Scan loss (dB) | Patterns at (°) |
|---|---|---|---|---|---|---|---|
| [36] | 2.57–2.64 | 1.04λ × 1.04λ | 0.06λ | Planar | 5.7 | – | – 20°, + 20° |
| [37] | 5.3–5.9 | 10.1λ × 7.9λ | 3.46λ | Planar | 18.5 | 5.5 | + 30°, – 30° |
| [38] | 2.4–2.5 | 1λ × 1λ | 1.55λ | Non-planar | 9.2 | 3 | + 16°, – 15° |
| [39] | 5.7–5.8 | 1.8λ × 1.7λ | – | Planar | 12.5 | – | 0° |
| [40] | 9.7–10.1 | 5.2λ × 0.3λ | – | Non-planar | 10 | – | 0° |
| [41] | 8.1–8.5 | 4.8λ × 0.7λ | 0.05λ | Planar | 9.1 | 2 | – 31°, + 6° |
| PA | 5.50–5.87 | 3.8λ × 1.35λ | 0.6λ | Planar | 12 | 0.6 | 0°, + 18°, – 18° |

*PA* Proposed antenna

**Fig. 8.27** Photograph of the fabricated antenna [26]. Reprinted with permission from Taylor and Francis

3D-printed superstrate is stacked over the four-element pattern diversity antenna module.

Here, the superstrate has a novel design in which it increases the gain at ± 45°, ± 15° directions such that the final gain becomes equal in all four directions. This ensures a continuous coverage of the antenna for 120°. The all-dielectric superstrate designs, available in the literature, do not increase the gain at oblique incidences by a fair amount, whereas the proposed all-dielectric superstrate can increase the gain by 3.8 dB even at 45° incident angle. Other gain enhancement techniques, such as planar metamaterials, lenses or Fabry–Perot cavity, do not increase the gain of the antenna at 45° directions by more than 3 dB like the proposed design. As the topology of the antenna itself is capable of providing the required tilt in the antenna radiation, there is no requirement of any phase shifters. Absence of phase shifters reduces the complexity of the design. The goal of designing this antenna is to develop an antenna system applicable to WLAN access point devices. The height of the antenna, including the superstrate, is 47 mm, less than the height of WLAN antenna for access points, as mentioned in [42], which is 55 mm.

### *8.5.1 Four-Element Antenna Design*

In the proposed design, the diversity scheme provides four inputs at the antenna system controlled by an SP4T switch. Each antenna, selected by the control switch as shown in Fig. 8.28 will radiate its own distinctive beam at different directions. Even though the antenna elements are identical patch elements, each of them needs to be excited individually for the antenna system to radiate which then covers 120° at the broadside. Figure 8.29 shows the four-element patch antenna design. These are circular patch antennas with an individual radius of 8.7 mm (0.15λ) each. The four antennas are coaxially fed with a port-to-port distance of 17.9 mm (0.3λ) and 12.4 mm (0.22λ) among them.

The antennas are fabricated on Rogers RO4350B substrate with dielectric constant $\varepsilon_r = 3.66$, thickness of 20 mil, and $\tan \delta = 0.0037$. The ground plane has an area of 74 × 40 mm$^2$ (1.3λ × 0.7λ), where λ is the free-space wavelength at 5.2 GHz. The center-to-center distance between the patches is at 18 mm (0.3λ) from each other with a 0.5 mm distance between the patches. It was observed that when the separation between the patches is less than 0.5 mm, the four-element arrangement displays a unique behavior where the beam is tilted at + 45° when antenna 1 is excited, + 15° when antenna 2 is excited, – 15° when antenna 3 is excited, and – 45° when antenna 4 is excited. Hence, by exciting a particular antenna out of the four, we can get any one of the four possible beams. The patch antenna radiates in the broadside direction. As the patch antenna is moved away from this phase center of the substrate, the beam of the patch antenna begins to tilt away from the broadside direction. Furthermore, the

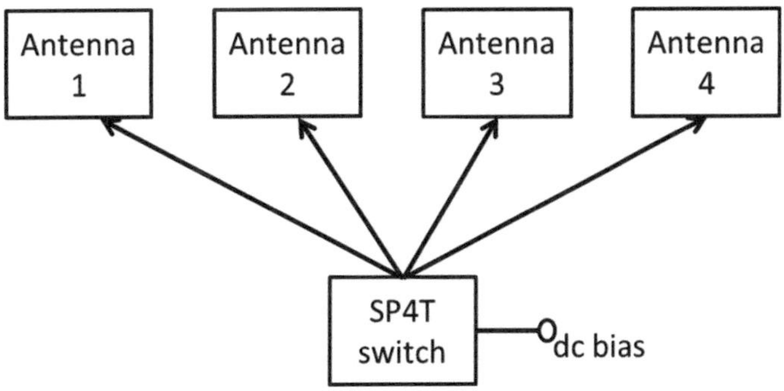


**Fig. 8.28** Antenna pattern diversity overview [43]. Reprinted with permission from Taylor and Francis

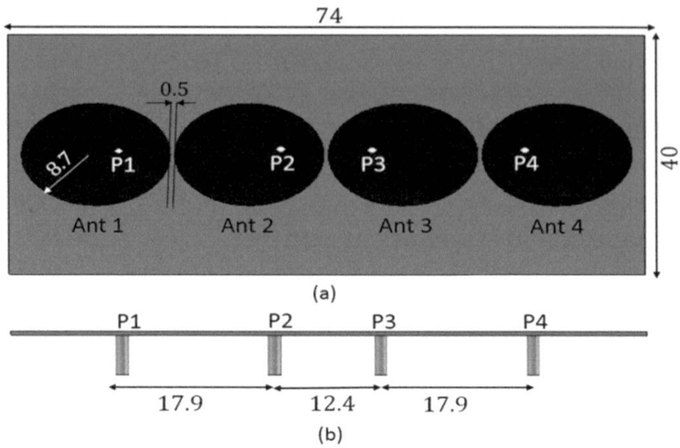


**Fig. 8.29** Schematic of four element antenna **a** Top view, **b** side view (all dimensions are in mm) [43]. Reprinted with permission from Taylor and Francis

antenna is away from the phase center, further will be the beam tilt. It was observed that beyond 45°, significant sidelobes get generated, and the patch antenna loses its directional radiation pattern. Then, the patch antenna was shifted from the phase center providing a beam tilt of 45°. Then another patch was placed appropriately for a beam tilt of 15°. The antennas 3 and 4 are mirrored designs of antennas 1 and 2 in order to obtain symmetric beam tilt across the antenna module. The separation of 0.5 mm was chosen because further reduction in distance between the antennas made them to be non-resonant at the designed frequency. For separations more than 1 mm, the beam-tilting behavior of the antenna disappears, and the patch antennas radiate in the broadside direction. This also proves that wide-angle beam scanning could be achieved with highly electrically compact elements. The antennas 2 and 3 experience coupling from two antennas from either sides, which reduces their gain to 3.3 dBi as compared to the gains of antennas 1 and 4 which has a gain of 5.6 dBi.

The radiation pattern of the antenna is tilted by extending the ground plane and by precise positioning of the antenna in the ground plane. When the antenna is placed symmetrically at the center of the ground plane, there is radiation at 0° broadside. But, as the ground plane gets extended to one side, the phase center of the antenna also gets shifted causing the radiation to tilt. When the patch is symmetric to the ground plane, it radiates at 0° broadside. When the ground plane is extended to right side, the tilt of the radiation increases. But, once the ground plane extension reaches a length of 1λ, the single-beam radiation of the antenna transforms into a dual beam radiation. When the ground plane extends further to a length of 1.5λ, the radiation transforms to a tri beam pattern and the pattern begins to get highly distorted.

The simulated and measured input reflection coefficients are shown in Fig. 8.30. The simulated impedance bandwidth for the antenna is 90 MHz (5.16–5.25 GHz). The half-power beamwidth (HPBW) of the antennas 1 and 4 is 50° and that of antennas 2 and 3 is 100°. The front-to-back ratio of antennas 1 and 4 is 10 dB and of antennas

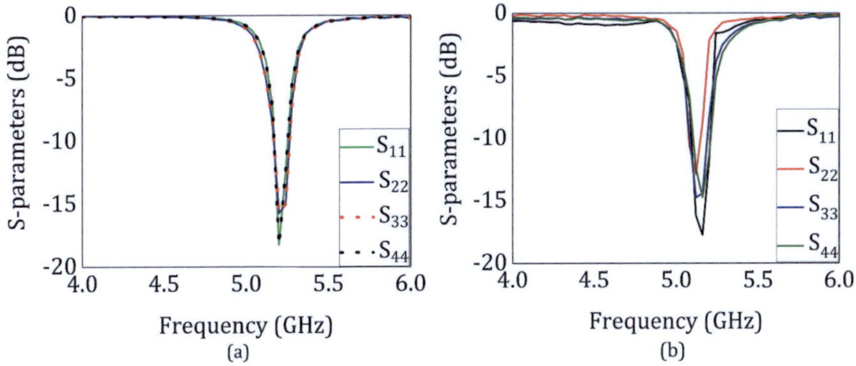


**Fig. 8.30** Input reflection coefficients for the four patch antennas. **a** Simulated, **b** measured [43]. Reprinted with permission from Taylor and Francis

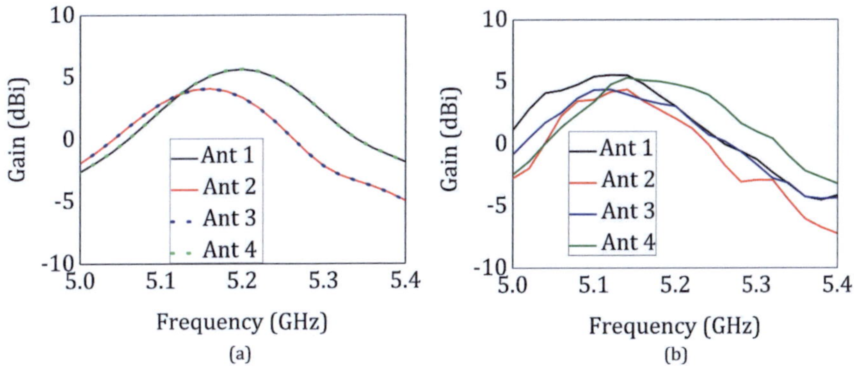


**Fig. 8.31** Gain of the four patch antennas. **a** Simulated, **b** measured [43]. Reprinted with permission from Taylor and Francis

2 and 3 is 18 dB due to the electrically large ground plane. The gain of antennas 1 and 4 is 5.6 dBi at 45° and that of antennas 2 and 3 is 3.3 dBi at 15°, as shown in Fig. 8.31. The patterns are tilted due to the amount by which the individual antennas are offset from the phase center of the antenna module. In order to increase the gain of the antennas, an all-dielectric 3D-printed superstrate is designed and stacked on top of the four-element antenna.

### 8.5.2 Four-Element Antenna Design with 3D-Printed Dielectric Loading

Figure 8.32 shows the proposed design, consisting of four patch antennas integrated with an all-dielectric superstrate. The dielectric superstrate is a 3D-printed block using polymer polylactic acid (PLA) material with dielectric constant of 2.58 and $\tan\delta = 0.015$. The researchers in [44, 45] proposed a gain enhancement design, using an all-dielectric superstrate, using a material with a dielectric constant of 10. The materials with high dielectric constant cannot be 3D printed at a cheap cost rather they use laser technique or milling followed by lamination process, which makes the manufacturing relatively expensive. Some designs of 3D-printed dielectric lenses, for gain enhancement using PLA, are reported in [46, 47]. The limitations in all the above cases are that the gain enhancement obtained is only in broadside direction and no discussion on gain enhancement for inclined incidence is provided. The dielectric superstrate has a dimension of 74 mm × 40 mm × 40 mm (1.3λ × 0.7λ × 0.7λ), and it is placed at a distance of 7 mm (0.12λ) above the antenna. This is much closer electrically when compared to the Fabry–Perot cavity antennas where the superstrate is placed 0.5λ above the antennas [48]. Due to closer placement of the superstrate, it is not affected by the air gap between them. The superstrate is designed such that the gain of the four patch antennas is increased and also equalized without compromising on the gain and pattern characteristics irrespective of the beam direction.

To design such a superstrate, first the characteristics of a solid PLA block are studied. The PLA block resulted in detuning of the antennas. Then, holes are drilled across the PLA block to change the material properties of the superstrate. This results in the variation of the dielectric constant of the superstrate. The characteristics of

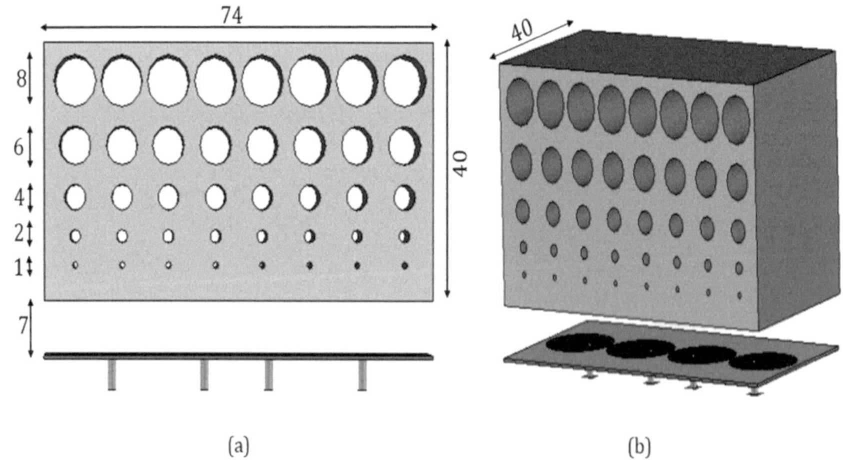


**Fig. 8.32** Schematic of the four-element antenna with dielectric superstrate. **a** Front view, **b** side view (dimensions in mm) [43]. Reprinted with permission from Taylor and Francis

different dimensions of holes are studied and it was observed that holes, with the same dimension throughout the PLA block, also cause detuning of the antenna. After some iterations, the superstrate with the radius of the holes increasing progressively was found to provide maximum gain without detuning or altering the beam tilting. By drilling holes, air cavity is being introduced in the superstrate which lowers the dielectric constant of the material. As the waves travel from the antenna, they go through a medium of decreasing dielectric constant which enhances the gain. The gain enhancement for 15° vertical incident beam is 5.4 dB compared to 3.8 dB for EM wave incident at 45°. But, this gain enhancement equalizes the gain of the four antennas as the gains of antennas 2 and 3 were already lower than the gains of antennas 1 and 4. The dielectric block, used to increase the gain of the antenna resulting in equal gain at different angles, is designed by gradually changing the dielectric constant of the PLA substrate. The dielectric constant is varied by drilling holes throughout the block with the radius of the holes increasing away from the antenna, as shown in Fig. 8.33.

As the radius is increased, the dielectric constant of the PLA reduces [44]. The dielectric constant of the PLA is 2.4. If we consider a PLA block and drill a hole with radius 2 mm, the dielectric constant reduces to 1.95. If we increase the radius of the hole to 4 mm, the dielectric constant reduces to 1.45. In the proposed design, each

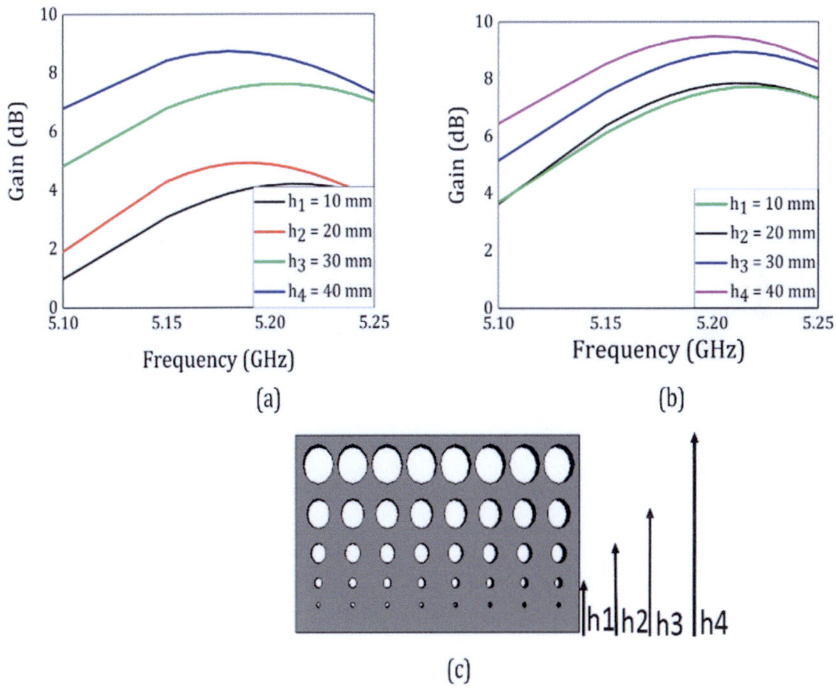


**Fig. 8.33** Gain enhancement of the antenna by the addition of different layers of the dielectric block **a** at 45° **b** at 15° [43]. Reprinted with permission from Taylor and Francis

such layer contributes certain gain to the loaded antenna. But through experiments, it was seen that the design, where the radius of the holes is gradually increasing, gave the maximum gain enhancement at tilted angles. Finally, the dielectric was designed in such a way to give a gain enhancement of 3.8 dB for antennas 1 and 4 and 5.4 dB for antennas 2 and 3 such that the maximum gain at different angles is equal. The aforesaid inferences can be better understood from the electric field distribution of antenna module, as shown in Fig. 8.34. The EM waves have a curved wave front when there is no dielectric loading, as shown in Fig. 8.34a, b for antennas 1 and antenna 2. The EM wave, once it passes through the dielectric superstrate, becomes much more planar and flatter, as in Fig. 8.34c, d. The electric field distribution is similar for antennas 3 and 4. The dielectric block does not affect the beam tilt that was seen prior to the superstrate loading. The phase correction, due to the dielectric loading, makes the waves to be more focused on the tilted beam angles, which then aids in gain enhancement in those angular wave positions.

Table 8.4 shows the proposed design along with the other gain enhancement techniques. For the transmission-type unit cell, we considered the design proposed in [49]. The unit cells are placed on top of the antenna module and the gain was checked at different heights. Again, enhancement at 15° was observed while not much gain enhancement is observed at 45° due to the reduction in the return loss of antennas 1 and 4. For antenna 2, the unit cells interfere with the angle of incidence and deflect the beam to − 5° instead of the expected + 15° even though this case shows gain improvement. For Fabry–Perot cavity the design proposed in [50] was

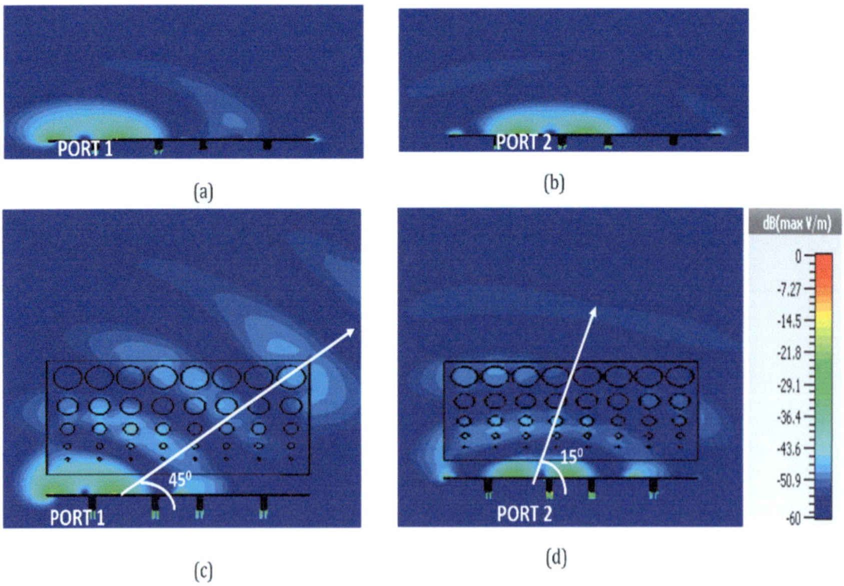


**Fig. 8.34** Electric field distribution for the two ports. **a**, **b** before dielectric loading and **c**, **d** after dielectric loading [43]. Reprinted with permission from Taylor and Francis

used. The unit cells are placed at a height of 0.5λ, and the observations are like that of transmission-type unit cells. The maximum gain enhancement was obtained for 15° which is closer to the broadside direction when the unit cells are at a height of 0.5λ. But, there is an increase in the relative height for the radiations from antennas 1 and 4; the unitcells are unable to enhance the gain for such instances. A metasurface lens as in [51] was also considered. The lens design also improved gain only in the broadside and failed to improve gain at 45° beam direction. By taking into consideration size and the required gain enhancement performance, the proposed design of all-dielectric metamaterial structure can be seen performing better than the other available design techniques at different angular incidences of the antenna. For each port, the simulated and measured input reflection coefficients are as shown in Fig. 8.35. The simulated impedance bandwidth of the proposed antenna is 80 MHz (5.16–5.24 GHz). This shows that the impedance bandwidth of the proposed antenna does not get altered significantly after the superstrate integration.

The antennas are linearly polarized which are unaffected after the superstrate loading. indicating high polarization purity across the ports. The simulated half-power beamwidth of antennas 1 and 4 is 40° with a front-to-back ratio of 16 dB and that of antennas 2 and 3 is 40° and 17 dB. The gain of the antennas after superstrate loading is 8.7 and 9.4 dBi, as shown in Fig. 8.36. The radiation pattern of the antenna before and after the superstrate loading is shown in Fig. 8.37. From the radiation patterns it is observed that the radiation from the antenna becomes more directional

**Table 8.4** Comparison of the proposed design with other reported design techniques

| Ref. | Gain enhancement technique used | Gain at 15° (dBi) | Gain at 45° (dBi) |
|---|---|---|---|
| [49] | Transmission-type metamaterial | 9.7 | 5.6 |
| [50] | Fabry–perot cavity | 9.5 | 5.2 |
| [51] | Metasurface lens | 9.8 | 5.1 |
| Proposed design | All-dielectric 3D-printed superstrate | 9.4 | 8.7 |

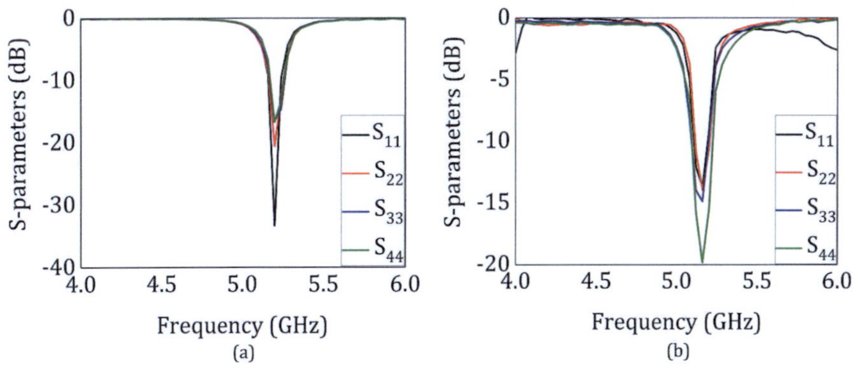


**Fig. 8.35** Input reflection coefficients of the four patch antennas after dielectric loading. **a** Simulated, **b** measured [43]. Reprinted with permission from Taylor and Francis

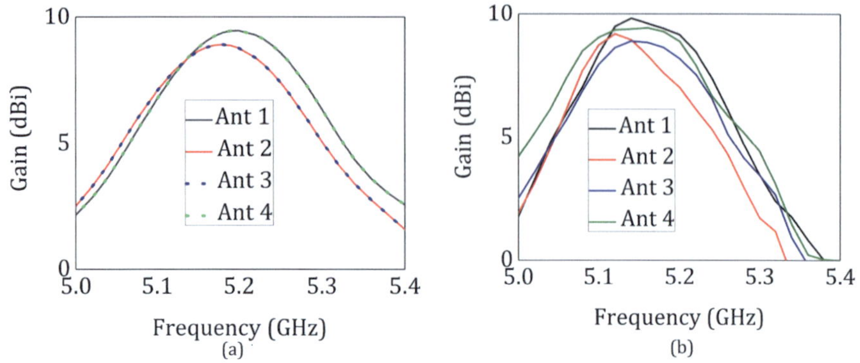


**Fig. 8.36** Gain of the four patch antennas after dielectric loading. **a** Simulated, **b** measured [43]. Reprinted with permission from Taylor and Francis

following the superstrate loading. This arises due to higher gain contributed by the all-dielectric metamaterial resulting in reduction in half-power beamwidth of the antennas. Due to the superstrate loading, the half-power beamwidth of antennas 1 and 4 decreases by $10^0$ and of antennas 2 and 3 by 60°.

While designing a planar antenna system, maintaining good isolation within a limited space is always a challenging task. The mutual couplings for the antenna system in the proposed design are less than − 13 dB throughout the band for adjacent ports with port-to-port distance of 0.22λ. For non-adjacent ports, the mutual coupling further reduces, as shown in Fig. 8.38. The mutual coupling of the antenna does not affect post the dielectric loading of the antenna. Compared to results presented in [52] with mutual coupling of 8 dB across the ports, the proposed design can achieve reasonably good port isolation. The measured radiation efficiency of the antenna is shown in Fig. 8.39. The envelope correlation coefficient is a standard parameter in MIMO antennas that measures the correlation of two radiation patterns. A lower correlation value is preferred which will result in higher data throughput of the antenna. Table 8.5 shows the ECC values at the resonant frequency of the antenna. All the values are below 0.01, which indicates a good diversity performance of the antenna. The ECC graph for the operating bandwidth is shown in Fig. 8.40.

A prototype of the proposed antenna has been fabricated as to validate the simulation results. The antenna is fabricated by the photolithographic process while the all-dielectric block is 3D printed. The 3D printing was performed using a Raise 3D RxP2200® 3D printer which is used in commercial low-budget 3D printing. Foam like material was used to maintain the height between the 3D-printed dielectric and the four-port antenna. This material did not affect the performance of the antennas during measurement. AnritsuMS2028C® VNA was used to measure the reflection coefficient of the antenna and the far-field measurements are conducted inside an anechoic chamber.

The ETS-Lindgren® 3115 Model double-ridged waveguide horn antenna was used as the standard gain antenna for the measurement. The setup used for the

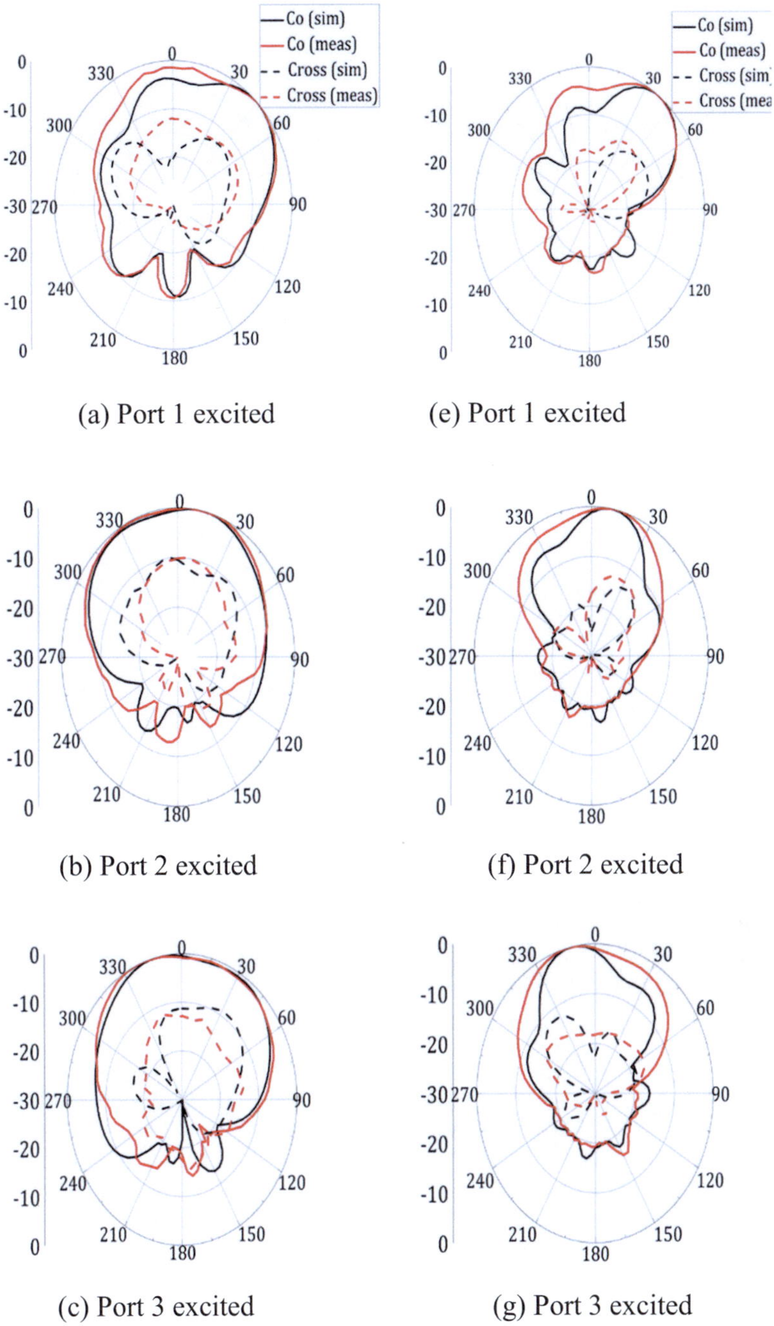


**Fig. 8.37** Radiation patterns at the operating frequency of 5.2 GHz. **a–d** Before dielectric loading, **e–h** after dielectric loading [43]. Reprinted with permission from Taylor and Francis

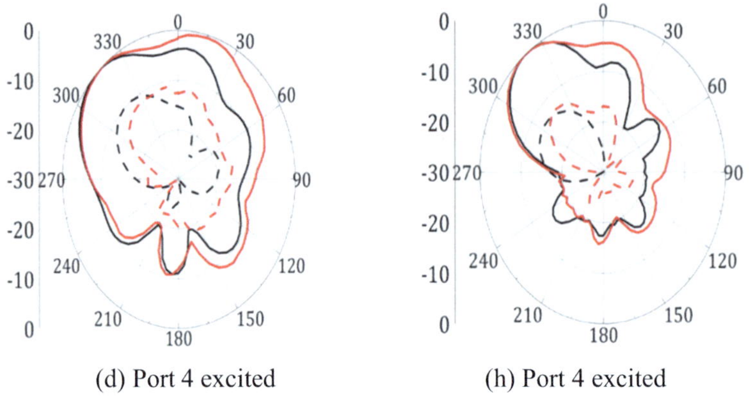


**Fig. 8.37** (continued)

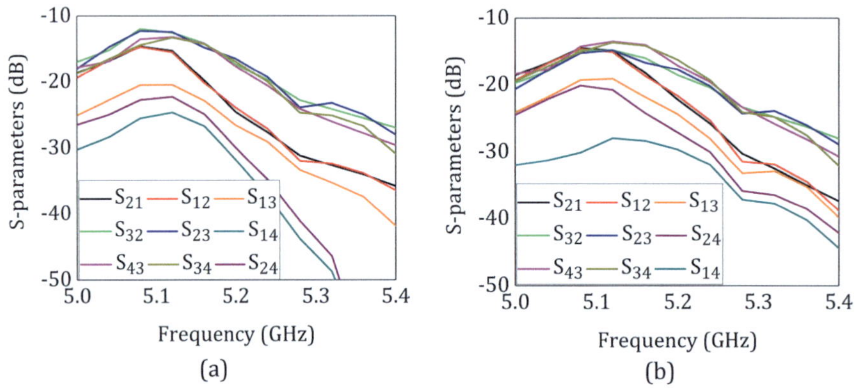


**Fig. 8.38** Mutual coupling of the antenna **a** Before dielectric loading, **b** after dielectric loading [43]. Reprinted with permission from Taylor and Francis

measurement of the antenna is depicted in Fig. 8.41a, and the fabricated antenna is shown in Fig. 8.41b. The proposed antenna can be connected to an off-the-shelf SP4T switch to select the preferred antenna. The performance of the proposed antenna is compared with the existing designs in the literature and same is tabulated in Table 8.6. The novelty of the proposed antenna is based on two aspects present in the antenna design. First, the antenna topology is such that it introduces beam tilting of ± 15° and ± 45° because the individual antenna elements have different phase centers with respect to the ground plane. Second, the 3D-printed all-dielectric metamaterial enhances gain at ± 45° and ± 15° such that the gain becomes equal throughout the band and across the 120° beam coverage.

The gain enhancement at oblique incidences using an all-dielectric metamaterial is not reported in the literature. The metamaterial structure enhances gain by 3.8 dB at 45° and 5.4 dB at 15°. The antenna proposed in [53] uses 6 diodes for a two-state

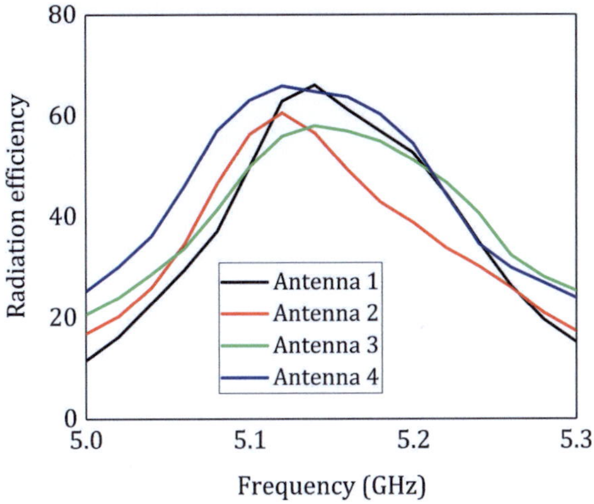


**Fig. 8.39** Radiation efficiency of the antenna [43]. Reprinted with permission from Taylor and Francis

**Table 8.5** ECC values at 5.2 GHz

| Ports | ECC |
|---|---|
| $Port_{12}$ | 0.00217 |
| $Port_{13}$ | 0.00031 |
| $Port_{14}$ | 0.00020 |
| $Port_{23}$ | 0.00281 |

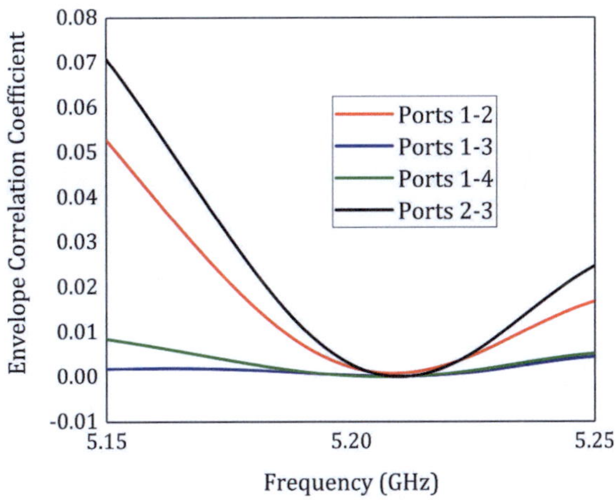


**Fig. 8.40** Simulated ECC for different antenna ports [43]. Reprinted with permission from Taylor and Francis

beam switching resulting in significant power loss across the diodes. The antenna also suffers from a 3 dB gain variation while scanning from 0° to 45°. The antenna proposed in [54] offers beam switching with very low gains of 1 and 3.8 dBi across the ports, which is not adequate for WLAN applications. The proposed antenna in [55] is

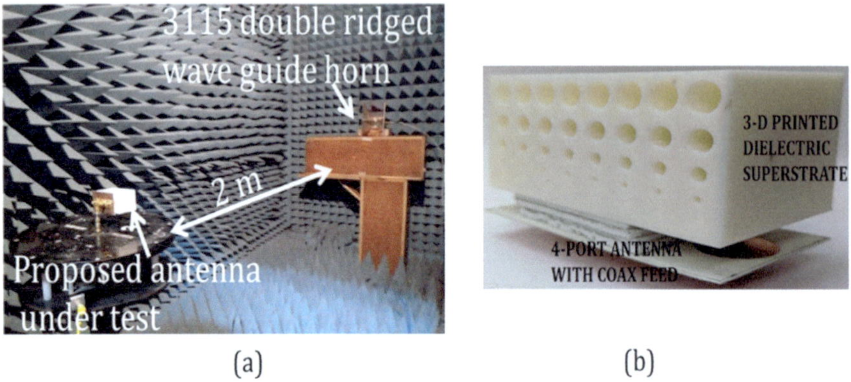


**Fig. 8.41** **a** Photograph of the measurement setup, **b** fabricated antenna [43]. Reprinted with permission from Taylor and Francis

electrically very large and tilts the beam only in one direction of the elevation plane. This design does not suit a ceiling mount access point design as we have to cover both sides of the elevation plane with the antenna radiations. The gain of the antenna in [56] also varies in different directions by significant amount compared to the proposed design where the gains of the antenna in different directions are nearly equal. [44] proposed a structure with all-dielectric metamaterials, but the gain enhancement is limited only in the broadside direction. Compared to the dielectric metamaterial in [44], the proposed design enhances gain at oblique incidences at $\pm$ 45° and $\pm$ 15° ensuring better coverage. Swapna [49] uses two different metamaterial structures at 0° and 90°, whereas in the proposed design, a single metamaterial block enhances gain in different angular direction. The antenna structure in [49] uses dipole antennas that radiate in the endfire direction to cover the beams at 90°, while the proposed antenna uses strategic placement of the patch antennas from the phase center to tilt the beams in various directions. Swapna [50] uses a Fabry–Perot cavity for beam-tilting and gain enhancement, but the beam tilt obtained is only up to 25°. The antennas proposed in [56, 57] are three-port antennas, but they offer limited coverage with very low gain. The antennas proposed in [58, 59] have high gain in the broadside, but the angular coverage is very limited with no radiation at 45°. The proposed antenna offers high gain with 120° coverage over the broadside direction.

## 8.6 Design Guidelines

In this chapter, we discussed some antenna designs whose objective was to enhance as well as equilize gain at different antenna ports. Similar to how the antenna gains were equalized, the concept can be extended to obtaining custom gain enhancements

**Table 8.6** Comparison of the proposed design with other designs

| Ref. | Operating range (GHz) | Percentage bandwidth (%) | Size | Total antenna profile | Isolation (dB) | Gain at 0° (dBi) | Gain at 15° (dBi) | Gain at 45° (dBi) | Gain enhancement technique used | Pattern at (°) | Radiation efficiency (%) |
|---|---|---|---|---|---|---|---|---|---|---|---|
| [53] | 2.25–2.8 | 2.4 | $2\lambda \times 0.9\lambda$ | $0.27\lambda$ | – | 7.7 | 10.5 | 7.7 | – | 13°, 32° | – |
| [54] | 5.5– 5.85 | 6.2 | $1\lambda \times 0.56\lambda$ | $0.033\lambda$ | 18 | 1.0 | 1.0 | 3.8 | – | 0°, ± 90° | 60 |
| [55] | 28.7–29.3 | 2.0 | $3.4\lambda \times 3\lambda$ | $0.83\lambda$ | – | 12.5 | 7.5 | 0 | 3D lens design | 0°, + 23°, + 27° | 82 |
| [44] | 5.65–5.87 | 3.8 | $1.95\lambda \times 1.95\lambda$ | $0.62\lambda$ | – | 6.9 | 5.0 | 0.0 | Holey superstrate | 0° | – |
| [49] | 5.12–5.27 | 2.9 | $0.56\lambda \times 0.9\lambda$ | $0.18\lambda$ | 26 | 6 | 6 | 6 | metamaterial structures | 0°, ± 90° | 67 |
| [50] | 5.16–5.25 | 1.7 | $1\lambda \times 2.4\lambda$ | $0.65\lambda$ | 20 | 10 | 7 | 7 | Fabry–perot cavity | 0°, ± 25° | 69 |
| [56] | 3.4–3.6 | 5.8 | $3\lambda \times 2.3\lambda$ | $0.01\lambda$ | 17 | 3.6 | 0 | 0 | – | 0° | 56.7 |
| [57] | 2.57–2.64 | 2.7 | $0.62\lambda \times 0.62\lambda$ | $0.25\lambda$ | 20 | 2.8 | 2.5 | 2.8 | – | 0°, ± 35° | 48 |
| [58] | 5.44–5.66 | 4.0 | $2.75\lambda \times 2.75\lambda$ | $0.5\lambda$ | – | 9.6 | 10 | 1.0 | Reconfigurable FSS structure | 0°, ± 22° | – |
| [59] | 5.5–5.7 | 3.8 | $3.1\lambda \times 3.1\lambda$ | $0.55\lambda$ | – | 12 | 11.7 | 0 | Reconfigurable PRS | 0°, ± 15° | – |
| PA | 5.16–5.24 | 1.5 | $1.3\lambda \times 0.7\lambda$ | $0.83\lambda$ | 13 | 8.6 | 8.7 | 9.4 | All dielectric superstrate | ± 45°, ± 15° | 67 |

*PA* Present design

depending on the needs and applications. Following guidelines can be followed for designing such antennas.

1. The single antenna design needs to be finalized first. The design parameters consist of size of the antenna, antenna gain, bandwidth, and radiation pattern. The antenna beam can be directed at broadside or any tilt as deemed necessary.
2. Identify the required number of antenna ports. A MIMO antenna design results in increased data rate and it is able to handle more data volume. So the number of ports is finalized based on the amount of data that the antenna system will be handling. For less data traffic, a smaller number of antenna ports are designed as higher number of ports can result in complex designs. But for higher data traffic, higher number of antenna ports are designed as the data handling would become much easier. Multiple ports also help to reduce the multipath fading during transmission and maintain the data integrity.
3. Once the number of antenna ports is finalized, the angle where the beam is to be radiated is considered. Based on the radiation pattern, the antennas connected to the ports are also redesigned. Radiation at the broadside direction needs a patch antenna, where a diploe antenna is preferred for radiation in the endfire direction. Similarly, the antenna design is revisited based on radiation pattern requirements and the beams are tilted at any direction.
4. Identify the suitable method for enhancing the antenna gain. It can be using various metamaterial structures, dielectric superstrates, 3D-printed superstrates, Fabry–Perot cavity antenna and so on.
5. The metamaterial and superstrate structures to be organized such that the required gain enhancement is obtained in different ports. These structures can be shared between the antennas and can be individual for each antenna or the structures can also be different for each antenna.
6. Finally, the antenna design is to be evaluated for the bandwidth, gain, radiation beam, and requirement of beam tilt being satisfied.

## 8.7 Conclusion

With the development and increased usage of MIMO antennas, multiport antennas have become crucial in wireless communication. Multiport and multibeam antenna designs are described in this chapter. Multibeam antennas can be obtained either from single antenna or from multiple antennas and both designs are described in the chapter. The design and implementation of multiport antennas with gain equalization is also discussed in this chapter. The first design is a three-port 3D-printed series-fed antenna array with pattern diversity. The second design presented is a four-port patch antenna with all dielectric superstrate for gain enhancement and equalization. Finally, some general guidelines for designing the multiport antennas are discussed.

## References


1. Molisch, A.F.: Wireless Communications. Wiley (2007)
2. Foschini, G., Gans, M.: On limits of wireless communications in a fading environment when using multiple antennas. Wireless Pers. Commun. **6**, 311–335 (1998)
3. Paulraj, A., Nabar, R., Gore, D.: Introduction to Space-Time Wireless Communications. Cambridge University Press (2006)
4. Karlsson, K., Carlsson, J.: Circuit based optimization of radiation characteristics of single and multiport antennas. Radio Eng. **18**(4), 438–444 (2009)
5. Morris, M., Jensen, M.: Improved network analysis of coupled antenna diversity performance. IEEE Trans. Wireless Commun. **4**(4), 1928–1934 (2005)
6. Sibille, A., Oestges, C., Zanella, A. (eds.): MIMO: From Theory to Implementation. Academic Press (November 2010)
7. Behdad, N., Sarabandi, K.: Dual-band reconfigurable antenna with a very wide tunability range. IEEE Trans. Antennas Propag. **54**(2), 409–416 (2006)
8. Chattha, H.T.: 4-Port 2-element MIMO antenna for 5G portable applications. IEEE Access **7**, 96516–96520 (2019)
9. Dash, J.C., Sarkar, D.: A four-port CSRR-loaded dual-band MIMO antenna with suppressed higher order modes. IEEE Access **10**, 30770–30778 (2022)
10. Kabiri, Y., Borja, A.L., Kelly, J.R., Xiao, P.: A technique for MIMO antenna design with flexible element number and pattern diversity. IEEE Access **7**, 86157–86167 (2019)
11. Kulkarni, J., et al.: Dual polarized, multiband four-port decagon shaped flexible MIMO antenna for next generation wireless applications. IEEE Access **10**, 128132–128150 (2022)
12. Parchin, N.O., et al.: Four-element/eight-port MIMO antenna system with diversity and desirable radiation for sub 6 GHz modern 5G smartphones. IEEE Access **10**, 133037–133051 (2022)
13. Mahto, S.K., Singh, A.K., Sinha, R., Alibakhshikenari, M., Khan, S., Pau, G.: High isolated four element MIMO antenna for ISM/LTE/5G (sub-6 GHz) applications. IEEE Access **11**, 82946–82959 (2023)
14. Wu, B., Luk, K.-M.: A 4-port diversity antenna with high isolation for mobile communications. IEEE Trans. Antennas Propag. **59**(5), 1660–1667 (2011)
15. Yurduseven, O., Smith, D.R.: Dual-polarization printed holographic multibeam metasurface antenna. IEEE Antennas Wirel. Propag. Lett. **16**, 2738–2741 (2017)
16. Song, L.-Z., Qin, P.-Y., Zhu, H., Du, J.: Wideband conformal transmitarrays for E-band multibeam applications. IEEE Trans. Antennas Propag. **70**(11), 10417–10425 (2022)
17. Hong, W., et al.: Multibeam antenna technologies for 5G wireless communications. IEEE Trans. Antennas Propag. **65**(12), 6231–6249 (2017)
18. Li, Y., Ge, L., Chen, M., Zhang, Z., Li, Z., Wang, J.: Multibeam 3-D-printed Luneburg lens fed by magnetoelectric dipole antennas for millimeter-wave MIMO applications. IEEE Trans. Antennas Propag. **67**(5), 2923–2933 (2019)
19. Han, Y., et al.: Multibeam antennas based on spoof surface plasmon polaritons mode coupling. IEEE Trans. Antennas Propag. **65**(3), 1187–1192 (2017)
20. Nguyen, V.C., Chen, L., Halterman, K.: Total transmission and total reflection by zero index metamaterials with defects. Phys. Rev. Lett. **105**(Art. no. 233908) (2010)
21. Xu, H.-X., Wang, G.-M., Tao, Z., Tie, T.J.: High-directivity emissions with flexible beam numbers and beam directions using gradient refractive-index fractal metamaterial. Sci. Rep. **4**(Art. no. 5744) (2014)
22. Zhang, J., Han, L., Chen, X., Yang, R., Zhang, W.: Multi-beam patch antenna based on metasurface. IEEE Access **8**, 37281–37286 (2020)
23. Fadhil, T.Z., Murad, N.A., Rahim, M.K.A., Hamid, M.R., Nur, L.O.: A beam-split metasurface antenna for 5G applications. IEEE Access **10**, 1162–1174 (2022)
24. Qiang, C., Wei, X.J., Tie, J.C.: Multi-beam generations at pre-designed directions based on anisotropic zero-index metamaterials. Appl. Phys. Lett. **99**, 131913 (2011)

25. Zhang, Z., Zhang, W., Liu, Y., Ma, R., Han, L.: Low-profile beamforming-network-avoiding multi-beam antenna based on parasitic patch and shorting-pin. IEEE Access **8**, 14749–14755 (2020)
26. Swapna, S., Karthikeya, G.S., Shiban, K.K., Ananjan, B.: A high gain multi-port series fed array realized with micromachined low-cost 3D-printed substrate. IETE Tech. Rev. (2023)
27. Whittaker, T., Zhang, S., Powell, A., Stevens, C.J., Vardaxoglou, J.Y.C., Whittow, W.: 3D printing materials and techniques for antennas and metamaterials: a survey of the latest advances. IEEE Antennas Propag. Mag. **65**(3), 10–20 (2023)
28. Wang, S., Zhang, X., Zhu, L., Wu, W.: Single-fed wide-beamwidth circularly polarized patch antenna using dual-function 3-D printed substrate. IEEE Antennas Wirel. Propag. Lett. **17**(4), 649–653 (2018)
29. McKerricher, G., Titterington, D., Shamim, A.: A fully inkjet-printed 3-D honeycomb-inspired patch antenna. IEEE Antennas Wirel. Propag. Lett. **15**, 544–547 (2016)
30. Chopra, R., Kumar, G.: Series- and corner-fed planar microstrip antenna arrays. IEEE Trans. Antennas Propag. **67**(9), 5982–5990 (2019)
31. Sacco, G., D'Atanasio, P., Pisa, S.: A wideband and low-sidelobe series-fed patch array at 5.8 GHz for radar applications. IEEE Antennas Wirel. Propag. Lett. **19**(1), 9–13 (2020)
32. Ratni, B., Lustrac, D., Piau, G.P., Burokur, S.N.: Modeling and design of metasurfaces for beam scanning. Appl. Phys. A **123**(50) (2016)
33. Khalily, M., Tafazolli, R., Xiao, P., Kishk, A.A.: Broadband mm-wave microstrip array antenna with improved radiation characteristics for different 5G applications. IEEE Trans. Antennas Propag. **66**(9), 4641–4647 (2018)
34. Feng, G., Chen, L., Xue, X., Shi, X.: Broadband surfacewave antenna with a novel nonuniform tapered metasurface. IEEE Antennas Wirel. Propag. Lett. **16**, 2902–2905 (2017)
35. Yang, W., Zhou, J.: Wideband circularly polarized cavity-backed aperture antenna with a parasitic square patch. IEEE Antennas Wirel. Propag. Lett. **13**, 197–200 (2014)
36. Naqvi, A.H., Lim, S.: A beam-steering antenna with a fluidically programmable metasurface. IEEE Trans. Antennas Propag. **67**(6), 3704–3711 (2019)
37. Yu, Y.-H., Zong, Z.-Y., Wu, W., Chen, Q., Fang, D.-G.: Dual-Polarized linear array with overlapping handover of subarray to produce continuous beam scanning for transmitarray antenna. IEEE Trans. Antennas Propag. **69**(2), 859–868 (2021)
38. Gu, C., et al.: 3-D coverage beam-scanning antenna using feed array and active frequency-selective surface. IEEE Trans. Antennas Propag. **65**(11), 5862–5870 (2017)
39. Chopra, R., Kumar, G.: Series-fed binomial microstrip arrays for extremely low sidelobe level. IEEE Trans. Antennas Propag. **67**(6), 4275–4279 (2019)
40. Ogurtsov, S., Koziel, S.: A conformal circularly polarized series-fed microstrip antenna array design. IEEE Trans. Antennas Propag. **68**(2), 873–881 (2020)
41. Xu, S., Guan, D., Liu, L., Zhang, Q., Xu, H., Yong, S.: A narrow-band circularly polarized leaky-wave antenna with open stopband suppressed. Int. J. RF Microwave Comput. Aided Eng. **31**(7) (2021)
42. Cisco Aironet 2800 Series Access Points, Cisco, 2 2019
43. Swapna, S., Karthikeya, G.S., Koul, S.K., Basu, A.: Compact pattern diversity antenna with 3-D printed all-dielectric superstrate for WLAN access points. IETE J. Res. (2021)
44. Kim, J.H., Ahn, C., Bang, J.: Antenna gain enhancement using a holey superstrate. IEEE Trans. Antennas Propagat. **64**(3), 1164–1167 (2016)
45. Assadi, M., Sebak, A.: Gain and bandwidth enhancement of 2×2 square dense dielectric patch antenna array using a holey superstrate. IEEE Antennas Wirel. Propag. Lett. **16**, 1808–1811 (2017)
46. Zhang, S., Arya, R.K., Pandey, S., Vardaxoglou, Y., Whittow, W., Mittra, R.: 3D-printed planar graded index lenses. IET Microwaves Antennas Propagat. **10**(13), 1411–1419 (2016)
47. Malik, B.T., Doychinov, V., Zaidi, S.A.R., Robertson, I.D., Somjit, N.: Antenna gain enhancement by using low infill 3D-printed dielectric lens antennas. IEEE Access **7**(7), 102, 467–102, 476 (2019)

48. Sultan, F., Mitu, S.S.I.: Superstrate-based beam scanning of a fabry-perot cavity antenna. IEEE Antennas Wirel. Propag. Lett. **15**, 1187–1190 (2015)
49. Swapna, S.P., Karthikeya, G.S., Koul, S.K., Basu, A.: Gain equalized three antenna pattern diversity module for WLAN access points. Progress Electromagnet. Res. C **96**, 215–227 (2019)
50. Swapna, S.P., Karthikeya, G.S., Koul, S.K., Basu, A.: Three-port pattern diversity antenna module for 5.2 GHz ceiling-mounted WLAN access points. Progress Electromagnet. Res. C **98**, 57–67 (2020)
51. Jo, E., Kim, D.: 3-D printer based lens design method for integrated lens antennas. IEEE Antennas Wirel. Propag. Lett. **17**(11), 2090–2093 (2018)
52. Varshney, G., Gotra, S., Chaturvedi, S., Pandey, V.S., Yaduvanshi, R.S.: Compact four-port MIMO dielectric resonator antenna with pattern diversity. IET Microwaves Antennas Propagat. **13**(12), 2193–2198 (2019)
53. Lin, Y., Liao, C., Hsieh, T., Liao, W.: A novel beam-switching array antenna using series-fed slots with pin diodes. IEEE Antennas Wirel. Propag. Lett. **16**, 1393–1396 (2016)
54. Sharma, Y., Sarkar, D., Saurav, K., Srivastava, K.V.: Three-element mimo antenna system with pattern and polarization diversity for wlan applications. IEEE Antennas Wirel. Propag. Lett. **16**, 1163–1166 (2017)
55. Kim, E., Ko, S., Lee, Y.J., Oh, J.: Millimeter-wave tiny lens antenna employing u-shaped filter arrays for 5G. IEEE Antennas Wireless Propagat. Lett. **17**(5), 845–848 (2018)
56. Chen, W.-S., Lin, R.-D.: Three-port MIMO antennas for laptop computers using an isolation element as a radiator. Int. J. RF Microwave Comput. Aided Eng. (2020)
57. Saurav, K., Mallat, N.K., Antar, Y.M.M.: A three-port polarization and pattern diversity ring antenna. IEEE Antennas Wirel. Propag. Lett. **17**(7), 1324–1328 (2018)
58. Ji, L., Zhang, Z., Liu, N.: A Two-dimensional beam-steering partially reflective surface (PRS) antenna using a reconfigurable FSS structure. IEEE Antennas Wirel. Propag. Lett. **18**(6), 1076–1080 (2019)
59. Ji, L., Guo, Y.J., Qin, P., Gong, S., Mittra, R.: A reconfigurable partially reflective surface (PRS) antenna for beam steering. IEEE Trans. Antennas Propag. **63**(6), 2387–2395 (2015)

# Chapter 9
# Additive Manufacturing in Antenna Development

## 9.1 Introduction

3D printing is a technology that creates three-dimensional products layer by layer additively. It is different from the conventional manufacturing technique where a product is created by removing parts from a larger piece of material. 3D printing consists of the additive construction of any solid arbitrarily shaped object directly from computer models without the need for specialized tooling [1], as shown in Fig. 9.1. To print the object, the 3D model of the structure is generated and then imported into a CAD program, which is then converted to a series of layered slices. Once the printer receives the data for each slice, a series of print heads deposit a thin layer of ultraviolet-curable polymer in the construction stage. Then the ultraviolet lamps on the print head immediately cure the materials when they are being deposited. After one layer is completed, the construction stage is lowered and the next slice is printed on top. Once the entire model is complete, the construction tray rises and the part can be removed. This technology has the advantage of being capable of producing structures with different levels of complexity, with various shapes, great flexibility, and with low manufacturing cost, when compared to other traditional techniques. Also, 3D printing does not require the use of high-cost consumable parts, such as drills used in subtractive techniques, which are susceptible to breakage. Product customization has been a challenge for manufacturers due to the high costs of producing custom-tailored products for end users. Customized functional products are currently becoming the trend in 3D printing as this technology can produce small quantities of customized products with relatively low costs. This is specifically useful in the biomedical field, whereby unique patient-customized products are typically required [2].

Often 3D printing is also referred to as Additive manufacturing (AM), rapid prototyping (RP), or solid-freeform (SFF). But additive manufacturing consists of the whole process of creating the product and other steps beyond 3D printing are almost always needed. Additive manufacturing entails using 3D printing and the

 
S. K. Koul et al., *Antenna Systems for Modern Wireless Devices*, Signals and Communication Technology, https://doi.org/10.1007/978-981-97-3369-9_9

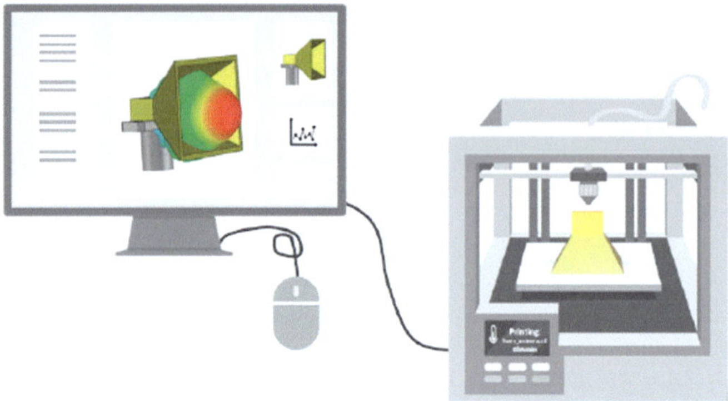

**Fig. 9.1** 3D printing technology [3]

product goes through several stages of post-processing and post-production before we get the finished product. However, 3D printing does simplify the whole process. It has received much attention recently with impressive demonstrations ranging from musical instruments, vehicles, housing components or even entire buildings [4]. It has been argued that 3D printing could be the future of manufacturing for creating traditional as well as futuristic designs.

### *9.1.1 3D Printing Methods*

The idea of 3D printing has been widely explored for a while now. Although Hideo Kodama was the first to describe the process of producing an object layer by layer, the first steps in the development of this technology were taken in the early 1980s. In 1986, the first 3D printing technique, stereolithography (SLA), was invented. This achievement is attributed to Charles Hull, who also founded 3D Systems Corporation and was the driving force behind the development of the SLA-1 printer. SLA technology was joined by selective laser sintering (SLS) technology as a trademarked 3D printing method in 1988. Carl Deckard came up with this method. Scott Crump patented the fused deposition modeling (FDM) a year later [3]. These methods are briefly explained in this section.

#### **9.1.1.1 Stereolithography**

Stereolithography (SLA) employs a UV light or electron beam to start a chain reaction on a layer of resin or monomer solution. The monomers, which are primarily acrylic or epoxy-based, are UV-active and, upon activation or radicalization, rapidly transform into polymer chains. Following polymerization, a pattern based on the intended structure is formed inside the resin layer. After printing is finished, the unreacted resin is removed. Some printed parts may undergo a post-processing procedure, such as heating or photo-curing, to achieve the desired mechanical performance. SLA prints high-quality parts at a resolution as low as 10 μm [5]. However, the printing process is relatively slow, expensive, and uses a limited range of materials for printing. SLA can be effectively used for the additive manufacturing of complex nanocomposites [6].

#### 9.1.1.2 Selective Laser Sintering

Selective laser sintering (SLS) is a type of powder bed fusion process where the powders in each layer are fused together with a laser beam or a binder. If the binder consists of polymers, metals, and alloy powders it is called SLS, and if the binder is made of certain metals such as steel and aluminum, the process is called selective laser melting (SLM). In the case of using a liquid binder, the method is referred to as three-dimensional printing. The laser can only be used for powders with a low melting/sintering temperature, whereas a liquid binder should otherwise be used. Laser scanning in SLS does not fully melt the powders and the elevated local temperature on the surface of the grains results in fusion of the powders at the molecular level. On the other hand, the powders are fully melted and fused together after laser scanning in SLM, which results in superior mechanical properties. A number of factors affect 3-DP, including the chemistry and rheology of the binder, the size and shape of the powder particles, the rate of deposition, the interaction between the powder and binder, and the post-processing methods [7]. In general, objects printed using binder deposition have a higher porosity than parts printed through laser sintering or melting, which are capable of printing dense pieces. This method is widely used in various industries for advanced applications such as scaffolds for tissue engineering, lattices, aerospace, and electronics [8].

Two filter prototypes in [9] are fabricated, one using metal-based SLM and the other by polymer-based SLA techniques. The first prototype shown in Fig. 9.2 was SLM-printed with an aluminum-copper-based alloy (92 weight percent (wt%) aluminum, 5 wt% copper, and 3 wt% others) in a powder form supplied in 15–53-μm particle size. The alloy was chosen due to its availability as a standard 3D printing material, as opposed to other aluminum alloys that are not normally used. The second prototype was SLA-printed with an Accura Xtreme resin with a printing resolution of 50 μm.

#### 9.1.1.3 Fused Deposition Modeling

Fused deposition modeling (FDM), also called fused filament fabrication (FFF) is the popular 3D printing method being used currently. In this method, a continuous filament of a thermoplastic polymer is used to 3D print layers of materials. The filament is heated at the nozzle to reach a semi-liquid state and then extruded on the platform or on top of previously printed layers. The thermoplasticity of the polymer

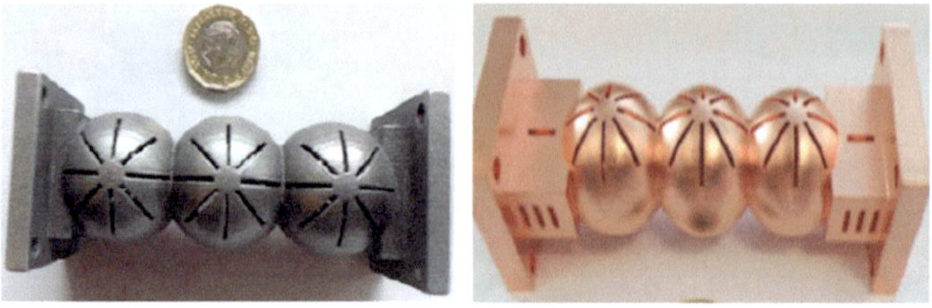

(a) (b)

**Fig. 9.2** **a** SLM-printed filter, **b** SLA-printed filter [9]

filament is an essential property for this method, which allows the filaments to fuse together during printing and then to solidify at room temperature after printing. The layer thickness, width, and orientation of filaments and air gap are the main processing parameters that affect the mechanical properties of printed parts. Low cost, high speed, and simplicity of the process are the main benefits of FDM. On the other hand, weak mechanical properties, layer-by-layer appearance, poor surface quality, and a limited number of thermoplastic materials are the main drawbacks of FDM. The development of fiber-reinforced composites using FDM has strengthened the mechanical properties of 3D-printed parts. However, fiber orientation, bonding between the fiber and matrix, and void formation are the main challenges that arise in 3D-printed composite parts [10].

### 9.1.2 3D Printing Materials

Throughout the years, the evolution of 3D printing techniques has been tremendous, with several additive manufacturing technologies being used nowadays. Some examples of these techniques are: Material jetting (MJ), binder jetting, direct metal laser sintering (DMLS), selective laser melting (SLM), electron beam melting (EBM), among others. One of the key decisions when dealing with 3D printing is the choice of material that will be used for each object's construction. There is a wide variety of materials with distinct properties (elastic, waterproof, resistant to impact, or high temperatures, among others), including polymers, ceramics, metals, concrete, and even biological tissues have been printed by various 3D printing technologies [11]. The choice of the most adequate material will always depend on the type of application. The most common materials used in FDM are acrylonitrile butadiene styrene (ABS) and polylactic acid (PLA). These are easy to print using basic off-the-shelf printers. The filaments are designed for their physical properties rather than their EM properties. The dielectric properties are measured with a range of techniques over the frequency range of 2–60 GHz: ABS had an $\varepsilon_r$ of ~2.45 and tan $\delta$ of ~0.005, and PLA had an $\varepsilon_r$ of ~2.55 and tan $\delta$ of ~0.009 [12]. There are also lower-loss thermoplastics available with similar properties, such as high-impact polystyrene, with an

$\varepsilon_r$ of ~2.49 and tan $\delta$ of ~8.53 × $10^{-4}$; polypropylene, with an $\varepsilon_r$ of ~2.03 and tan $\delta$ of ~1.45 × $10^{-4}$; and cyclo olefin copolymer, with an $\varepsilon_r$ of ~2.24 and tan $\delta$ of ~2.17 × $10^{-4}$, which makes these materials attractive for RF applications [13].

### 9.1.3 3D Printing in RF Engineering

In recent years, 3D printing has grown significantly and has become popular in the research and development community in many fields. This trend has been supported by the expiration of patents related to AM and 3-DP technology and the increasing availability of new AM machines and processes. The area of radio frequency (RF) and microwave engineering can greatly benefit from the use of AM. The growing interest of the RF industry in this technology has contributed to the development of RF devices, such as antenna elements. Although additive manufacturing provides significant flexibility in the production of antenna structures, an effort is currently being made to investigate different techniques and diverse materials to produce an antenna. New developments in printing equipment provide the capability to print 3D microwave and antenna structures using a variety of materials, which have been properly characterized. This is highly beneficial especially when the structures are too complicated and hard to fabricate by conventional methods. Particularly, 3D printing technology is highly suitable for antenna design at microwave frequencies because of the compatibility between antenna dimensions and 3D printer precision. In the microwave frequency range, various successful examples of 3D-printed antenna designs have been demonstrated, for instance, dielectric reflectarrays for sub-millimeter waves [14], wideband circular polarizer for millimeter waves [15], high-performance feed horns operating from Ku to V band [16], waveguide slot array [17], endfire antenna with designable permittivity [18], or dielectric resonator antenna utilizing different printing patterns [19]. These 3D-printed antennas and arrays have been shown to operate with performance similar or superior to antennas realized through conventional manufacturing techniques. In addition, their fabrication processes combine rapidity and simplicity.

The major advantage of using 3D printing technologies to fabricate microwave antennas includes rapid realization of designs without going through conventional processes such as machining and photolithography, ease of realizing complex geometry such as 3D conformal shapes, special tailored dielectric properties such as gradient-index structures, etc. One of the issues of 3D printing in microwave components and antennas is the limited printing resolution, which creates a surface roughness on printed devices and deteriorates the performance. The effect of surface roughness on the radiation characteristics and impedance matching of a dielectric rod antenna is studied in [20]. The surface roughness comes to exist due to the laying of melted material coming out of the tip on a congealed layer. The study shows that the roughness can deteriorate antenna performance like gain and impedance matching to a certain degree. The high-resolution tips for 3D printing are available; they lower the printing speed significantly and increase the cost of production. Although the existence of unavoidable surface roughness deteriorates the performance of the 3D-printed dielectric rod antenna, some advantages such as the ease of fabrication, more freedom on design, and the cost-effective process create motivation to 3D print some select types of high-gain antennas.

Since 3D printing tolerances are known to be worse than those of conventional fabrication methods; a study to evaluate the two methods is reported in [21]. It focusses on comparing manufacturing features of three-dimensional (3D) printing techniques versus conventional computer numerical control (CNC) milling in the context of gap-waveguide technology. The electrical performance of the 3D-printed antenna has been slightly worse due to several fabrication inconsistencies. In that sense, the current 3D printing accuracy is still not good enough to blindly trust on this technology for high-performance Ka-band antennas. Nonetheless, overall performance of the 3D-printed instance can be considered reasonably like those achieved by the accurately milled version. In regard to the price, the piece made of aluminum was roughly 3.5 times more expensive than the same piece in SLA 3D printing. The 3D-printed antenna was 50% lighter than the aluminum version, making it very appealing for exigent weight specifications. It is mentioned in [22] that the machined breadboard took five times as long as the 3D-printed piece. The used SLA 3D printing technique can be considered as a viable solution for low-cost manufacturing of gap-waveguide array antennas at Ka-band, becoming very appealing in terms of fabrication time, cost, and weight.

## 9.2 Classification of 3D-Printed Antennas

The use of 3D-printed structures in RF and antenna component design is a fast-growing research topic, as additive manufacturing provides many versatile methods of manufacturing components. 3D printing is ideal for the production of high-value components when conventional manufacturing techniques are not suitable. The fabricated electromagnetic structures range from large waveguides and filter components [23] to antennas at millimeter-wave frequencies [24]. While researchers address the challenges of 3D printing, such as roughness, plating, etc., its capability to model more complex geometries, with internal features, is being explored. Therefore, the rapidly developing trend of 3D printing technologies holds promise for cost-effective fabrications for various applications.

Depending on the printed geometry, 3D printing may be classified as—

1. Intricately shaped antennas.
2. Flexible substrates
3. Inhomogeneous substrate.

### 9.2.1 *Intricately Shaped Antennas*

The antennas discussed in this section are fabricated through 3D printing as the antenna structure is too complex and expensive to be fabricated through traditional methods.

Among the different available antenna topologies, the leaky-wave antenna can take benefits from 3D printing due to the usage of periodic structures to generate the leaky-wave modes. The antenna presented in [25] is a 3D-printed Archimedean spiral leaky-wave antenna at 18 GHz, using low-loss and low-cost dielectric filaments. To manufacture the antenna, a Creality CP-01 3D printer and ABS dielectric filaments provided by AVIENT is used. A relative permittivity of $\varepsilon_{r1} = 3$ is chosen for the grounded substrate, which is manufactured with the PREPERM ABS300. For the spiral, the ABS1000 filament is used with a thickness of 2 mm and a relative permittivity of $\varepsilon_{r2} = 10$. The manufactured Archimedean Spiral DLWA is shown in Fig. 9.3. For a circularly polarized antenna, the alignment between the grounded substrate and spiral is important. To ensure a proper alignment, the spiral and the substrate are manufactured in a single 3D printing process, using an offset setting that corresponds to the height h of the grounded substrate to change the filament used for the spiral deposition. The 3D printing technology gives a low-cost alternative for antenna topologies that have difficult shapes and are costly to manufacture using traditional methods and materials with the required permittivity values.

The phased-array antenna subsystem design reported in [26] uses two types of 3D printing technologies: (i) Polyjet, an industrial-level process with ultra-high precision; and (ii) low-cost, entry-level desktop FDM. A Polyjet 3D printer (Stratasys Objet500 Connex3) was used for fabricating the MPRWG parts of the phased-array antenna subsystem shown in Fig. 9.4. This printer is capable of printing large models at high resolution by using its polymer jetting technology. The Polyjet printer jets photopolymer droplets that solidify when cured by an internal ultraviolet (UV) light source. It utilizes wax-like support material that allows the 3D printer to print complex geometric features, such as curves and twists at the required resolution. The main

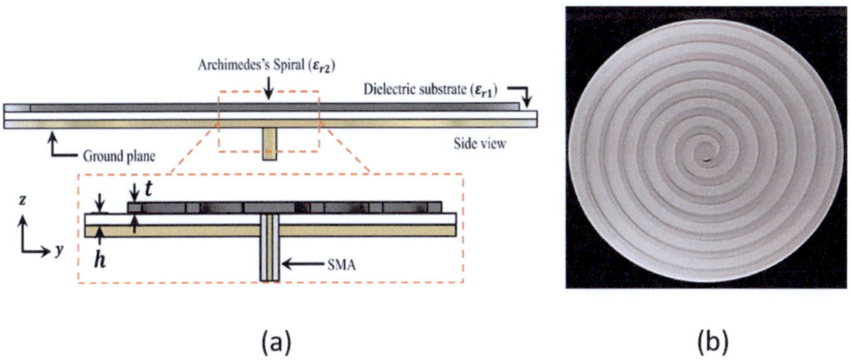


**Fig. 9.3** Polarized dielectric Archimedean spiral leaky-wave antenna. **a** Side view, **b** top view [25]

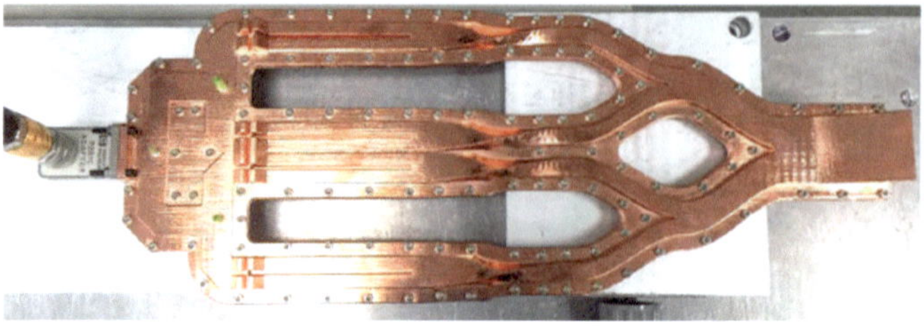

**Fig. 9.4** Phased-array antenna subsystem [26]

body of the phased-array antenna subsystem was printed in two pieces; the top and bottom parts were printed separately for ease of cleaning and metallization. A Polyjet resin, VeroWhite, was used for the main body. VeroWhite provides rigid 3D prints that offers high resolution comparable to injection-molded polymer counterparts. The support material was mechanically removed and cleaned with a high-pressure water jet. Fused deposition modeling 3D printing was used to manufacture the dielectric inserts for the tunable phase shifters, their mounting brackets and calibrated screw mechanisms. The dielectric inserts were fabricated using an Ultimaker 2 printer, with Formfutura's Premium ABS (White) filament. The brackets and screw mechanisms were fabricated using an Original Prusa i3 MK3, with polylactic acid (PLA) filaments. To reduce friction, petroleum jelly was applied to the screws.

An antenna subsystem which consists of an antenna and an orthogonal-mode transducer (OMT) which forms an essential subsystem for the polarimetric microwave radiometer system is reported in [27]. Conventionally, the antenna and the OMT are fabricated separately and then interconnected with the waveguide flange. This method is convenient for fabrication but results in an uncertainty factor to the reliability of the whole system which is unacceptable for instruments requiring extremely high reliability. Compared to the individual fabrication, an integrated fabrication of antenna and OMT certainly can increase the reliability and reduce the weight, which are preferred for spaceborne applications. However, it is usually difficult to achieve an integrated fabrication using conventional computerized numerical control (CNC) technology, due to its complex structure and small size. Fortunately, the rapid advancement of 3D printing technology has made it possible to achieve this kind of integrated fabrication. The advantage of using 3D printing can be seen from the comparison Table 9.1.

For the dielectric component structure, it is fabricated with the high-temperature resistant resin acrylonitrile butadiene styrene (ABS) using the fused deposition modeling (FDM) technology. The component is built with a dual extruder 3D printer and an inner support structure using the water-soluble polyvinyl alcohol is added during printing which is later removed. Finally, the structure is processed with electroless copper plating, and the component is plated with a thick layer of copper to form a metal surface with a certain thickness. The fabricated antenna is shown in Fig. 9.5, and the gain comparison of the two antenna prototypes is shown in Fig. 9.6.

**Table 9.1** Performance comparison between antenna/OMT subsystem fabricated by CNC and 3D printing

| Parameter | Return loss (dB) | Gain (dBi) | Fabrication time (hours) | Fabrication cost (US Dollars) | Weight (grams) | Integrative fabrication |
|---|---|---|---|---|---|---|
| Conventional fabrication | < − 10 for 11–13 GHz | > 14.9 for 11–13 GHz | ≈6 | 420 | 392 | No |
| 3D printing fabrication | < − 10 for 11–13 GHz | > 14.9 dBi for 11–13 GHz | ≈8 | 84 | 74 | Yes |

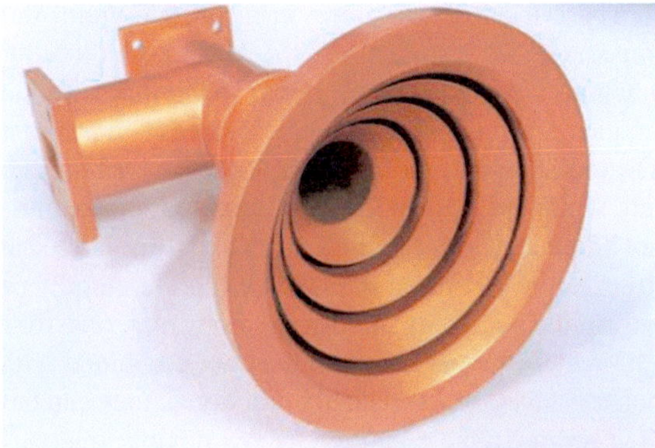

**Fig. 9.5** Photograph of the fabricated integrated antenna subsystem [27]

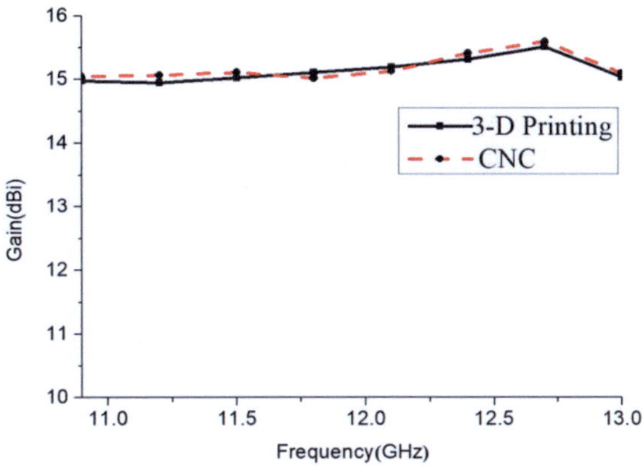


**Fig. 9.6** Gain comparison of antenna prototypes fabricated by CNC and 3D printing technology [27]

### 9.2.2 Conformable Substrates

Many complex mobile structures (aircrafts, ships, and automobiles) require conformal antennas for radio communication. This requirement can be particularly challenging for small platforms since a large antenna is needed at these frequencies. Concurrent requirements for greater bandwidth and multifunctionality imply an even greater need for conformality. Existing conformal antennas are still printed on rigid laminate substrates with curved shapes, making them expensive and cumbersome, if not impractical, to manufacture and, hence, not applicable for such platforms.

Further, an increasing demand for integration of antennas with radio-frequency (RF) front-end circuits makes use of such high-contrast substrates very attractive since they also allow for miniaturization.

The effects of curvature on resonance, gain, radiation pattern shape and input impedance were studied in [28] where both circumferentially and axially polarized antennas were considered. It was found that with increasing curvature, the resonant frequency increased and gain decreased with increasing curvature for both antennas. The radiation pattern for axially polarized antennas broadens with increasing curvature while the corresponding patterns for circumferentially polarized antennas does not broaden. It was concluded that axially polarized antennas exhibit more sensitivity to curvature compared with their circumferentially polarized counterparts. Polymers are rapidly becoming important among materials for microwave and electronic applications whether used in pure form or combined with ceramic powders. Though the polymers are not intrinsically functional, they can be doped and made functional and their soft or pliable nature enables flexible free-standing substrates in a variety of different shapes.

Three different fabrication procedures to print conformal antennas made of curved surfaces for devices such as wristbands and bracelets at the 2.4 and 5.5 GHz WLAN bands is described in [29]. The first technique is a multistep process consisting of 3D printing of the bracelet, deposition of a smoothing layer, aerosol jet printing of silver inks, flash curing, and electroplating with copper. The bracelet is printed using FFF method and the tracks of the antenna is deposited with Optomec's aerosol jet technology. It was then sent to an Ultimaker 3D printer using CURA software with print density set to 100%. White PLA was employed to make the bracelet. A layer of fast-acting adhesive (cyanoacrylates) was deposited to smooth the FFF surface and also increase the surface energy. Only then the metallic layers were added uniformly using Optomec's aerosol jetting process with Cabot CS-32 silver conductive ink. The thickness of the printed silver ink layer was about 20 μm. The printed dipole antenna was cured using a NovaCentrix PulseForge machine and then electroplated to further increase the conductivity. This added layer of copper was about 50 μm. The main advantages of this process are the high resolution of the printed antennas and the very smooth and thin metallic layer. The main disadvantages are the many tasks involved in the fabrication as the fabrication has to be done one after the other and the high cost of the equipment. The second technique involves printing the bracelet with thin grooves of 0.2 mm and then painting the metallic layers of the antennas using silver conductive ink by hand. Later a layer of copper of about 50 μm is deposited by electroplating to improve its conductivity. This process can produce three-dipole antennas on the same bracelet with various conformal shapes across the curved surface. The main disadvantage is the errors related to the painting of the antennas by hand. This process is more suitable for small-scale production and prototyping unless the painting of the metallic tracks is carried out using an automatic procedure. The third process employs a single machine combining FFF for the fabrication of the bracelet and a pneumatic dispenser for the addition of the metallic layer using silver ink. A flexible PLA substrate has been used so that the antennas are printed in a planar form and then folded to create the bracelet. The frame of the bracelet is printed on a rigid, durable yellow PLA plastic filament from 3D FilaPrint, which had an infill of 100%. The top layer of the bracelet was developed using a flexible PLA plastic filament from NinjaFlex. This has been proven to be the best manufacturing solution in terms of the number of tasks involved, the relative resolution of the printed tracks, and the fact that is an automatic procedure. These 3-DP techniques can be used for the fabrication of antennas and devices on bracelets customized to the user's wrist.

### 9.2.3 *Flexible Substrates*

Three-dimensional printing has found applications not only in the development of antenna structures that are rigid, but also in designing flexible substrates and wearable garments and devices. Antennas are needed if these wearables are smart, and wireless connected. Flexible antennas are necessary in many situations, but such antennas experience increased losses when the antenna structure is bent. Ideally, these antennas should be printed along the wearable structure. The flexible printable materials are characterized by the highest losses, and only rigid materials (such as PLA) have tan $\delta$ lower than 0.05. Consequently, it is difficult to maintain RF substrates that are both flexible and low loss. Due to the non-flexible and non-stretchable nature of ABS and PLA, they are not suitable candidates for such applications, including wearable applications, which enable devices integrated into clothes and affixed to skin. NinjaFlex® is a new 3D printable material, composed of thermoplastic and rubber. NinjaFlex® is a strong, lightweight, and flexible material, which makes it a good candidate for the fabrication of wearable radio frequency (RF) electronics and antennas. Previously, the use of fused deposition modeling (FDM) has been utilized to print NinjaFlex®, and the novel material has been successfully used as a substrate for wearable antennas [30]. However, in [30], copper tape has been used as the conductive antenna material. The use of copper tape limits the flexibility and the reliability of the wearable antenna. Authors in [31] presented an antenna design where two annular dipoles are wrapped on the surface of a printed spherical shaped acrylonitrile butadiene styrene (ABS) substrate of 0.76 mm thickness as shown in Fig. 9.7. The ring phase shifter generates the desired 90° phase shift between the annular dipoles and incorporates two layers. The top ring layer connects to the arms of the orthogonal annular dipoles, and the bottom ring connects to the other arms as shown in the Fig. 9.7. The top and bottom ring phase shifter has the same configuration as the top one, just rotated by 180° to form a sequential feeding network providing 0°, 90°, 180°, and 270° phases to the four arms. This also ensures 90° phase difference between the two annular dipoles. The substrate was 3D printed using PREFERM® ABS300 filaments on a Raise3D Pro2 printer. The spherical prototype substrate was printed in the form of a planar flexible sheet and subsequently rolled to form a spherical shape. Metallic patterns were printed using screen printing DuPont PE819 silver paste before the substrate was rolled. Fully printed prototypes were assembled and placed in an oven at 70 °C for two hours to cure the conductive paste. PREFERM R ABS300 filament has initial permittivity = 3 (from the manufacturer), which reduced to 2.68 after printing due to air gaps in the printed substrate. Thus, 3D printing can be used for designing larger and more complex folded dipoles antennas, which can provide larger impedance and radiation isotropy bandwidths.

Researchers in [32] manufactured a passive ultra-high-frequency (UHF) RFID tag on a 3D-printed substrate by using stretchable silver conductive paste (DuPont PE872) as the conductor. 3D direct write dispensing system is used to print the

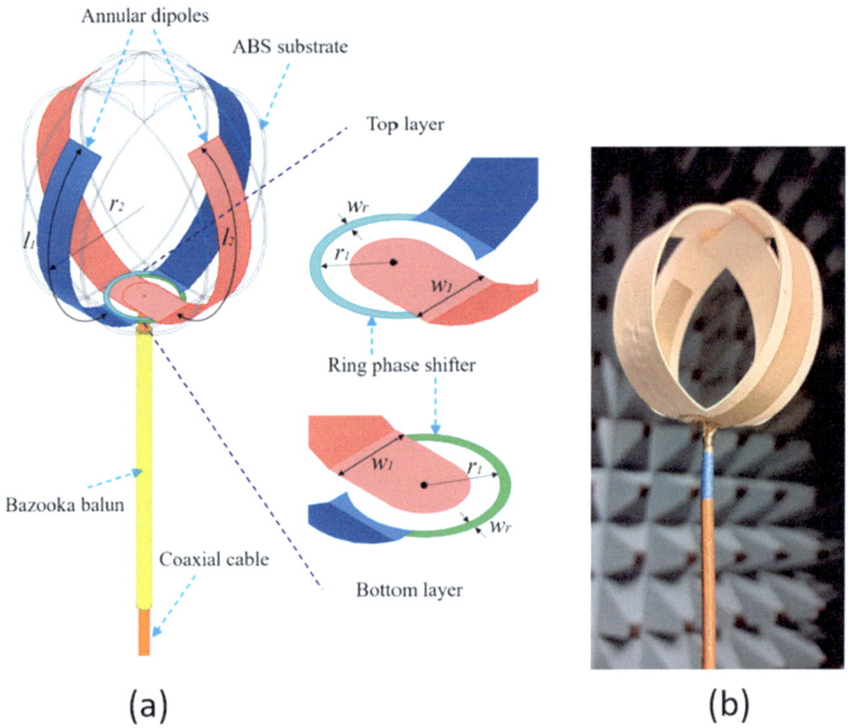


**Fig. 9.7** **a** Practical antenna with wideband radiation isotropy on a sphere, **b** fabricated antenna [31]

stretchable conductor on the flexible substrate using the Prenta Duo 3D printer. The stretchable silver conductive paste is brush-painted on the NinjaFlex 3D-printed substrate, and the structure is thermally cured at 110 °C. The manufactured tag can withstand harsh stretching up to a hundred times and maintain a peak read range of around 7 m, which means a 30% decrease from the initial performance.

Based on these results, the tag could be suitable for RFID applications where strain could be an important parameter to be considered.

### 9.2.4 *Inhomogeneous Substrates*

The relentlessly increasing demand for wirelessly enabled devices is continuously driving the requirement for more compact antennas without comprising the performance. Conventionally, antenna designers are restricted to use homogenous substrates with specific permittivities and thicknesses. Textured dielectrics have been used to reduce the size of the antenna while not adversely affecting the bandwidth or efficiency by using a lower permittivity in areas with high current densities and a higher permittivity elsewhere [33]. Researchers in [34] proposed further improvements that could be obtained by using a smooth variation between different dielectric regions. This could be achieved by controlling the volume ratio of the inclusions whereby the permittivity can be designed to have a continuous variation both horizontally and also vertically. A graded permittivity can be generated within the dielectric material through the use of small ($\ll 1$) inclusions within a host dielectric. The effective resultant permittivity is dependent on the properties of the host and inclusions as well as the size and spacing of these inclusions [35]. Analytical values of the effective permittivity with small repetitive air cavities can be calculated using the following equation [36]

$$k = k_1\left(1 - \frac{3f}{\frac{(2k_1+1)}{(k_1-1)} + f}\right), \tag{9.1}$$

where $k$ is the effective permittivity of the mixture, $k_1$ is the permittivity of the dielectric host material and $f$ is the volume fraction of air. The concept of using additive manufacturing as a method to construct heterogeneous substrates from a single building material via stereo-lithography is introduced in [37]. The additively manufactured heterogeneous substrates are produced from a single building material containing air cavities in a controlled configuration and geometry. Increasing the volume fraction of air inclusions within a dielectric reduced its effective permittivity. Therefore, a substrate with a locally graded permittivity can be fabricated from one material by varying the air cavity size or pitch. This creates new degrees of freedom for antenna engineers. By changing the aspect ratio of the air inclusions, the same process can be used to fabricate anisotropic substrates. An illustration of variable permittivity substrate is shown in Fig. 9.8.

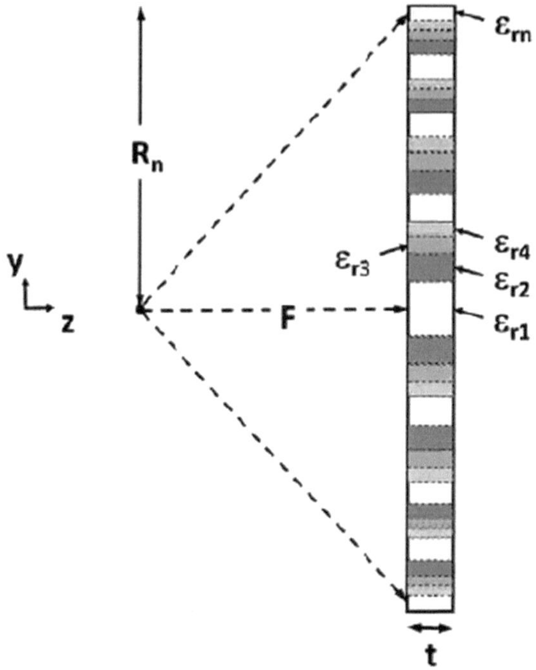


**Fig. 9.8** Illustration of graded permittivity [38]

Researchers in [39] presented a conical horn antenna with two synthetic dielectric materials loaded in the core realized by 3D printing. The dielectric loading is composed of an inner core and an outer shell that have permittivity of $\varepsilon_1$ and $\varepsilon_2$, respectively ($\varepsilon_1 > \varepsilon_2$) which results in exciting the hybrid $HE_{11}$ mode. It is challenging to fabricate a horn with a heterogeneous dielectric core using a single process traditional manufacturing technique. The two different dielectric materials with the desired profiles have been fabricated by using a single process AM technique. The air voids in the structures not only reduced the dielectric loss but also reduced the mass of the horn significantly. The polylactic acid (PLA) was used to fabricated both dielectric core and shell using the Ultimaker 2 + 3D printer. The antenna was then metalized using copper plating. This profiled horn offered wideband gain from 16.0 to 20.0 dBi in the frequency band from 9 to 15 GHz with aperture efficiency from 46 to 75%.

Instead of using different materials to obtain different dielectric constants, one can use different infill percentages while 3D printing. It is a parameter that we can set in the printer itself. If the infill percentage is 100%, it provides a solid block while printing. If the infill percentage is 50%, it will have 50% of dielectric material and 50% of air voids, and this will reduce the effective dielectric constant of the material. So by varying the infill percentages, one can vary the dielectric constant along the block with 3D printing. This feature can be exploited to fabricate dielectric substrates with the desired permittivity, and this property can prove to be very attractive in the

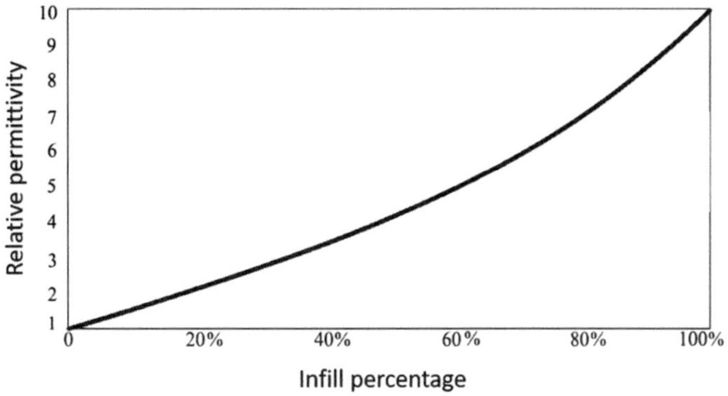


**Fig. 9.9** Variation of permittivity with infill percentage

design of printed components and antennas. Moreover, this manufacturing process permits to tune "on demand" the mechanical properties of the material, achieving the desired lightweight and flexibility properties. A technique to design a radiating structure made of multiple dielectric sections with various relative permittivities, ranging from 2 to 8.3 is reported in [18]. The effective permittivity of an elemental section can be controlled by varying the infill percentage in the printing process and its variation is shown in Fig. 9.9.

The dielectric block is 3D printed using a Flashforge Creator Pro 2016 3D printer with the filament from Premix Oy named PREPERM® 3D ABS DK 10.0 that has a bulk relative permittivity $\varepsilon_4 = 10.0$, loss tangent $\tan\delta = 0.003$, and density of 2.28 g/cm$^3$. The proposed antenna is designed to cover the X-band for communication and radar applications where endfire antennas are often required. The first dielectric section works as a surface-wave launcher, and its role is to couple the power from the coaxial feed into the dielectric material. In order to guide the surface waves propagating along the dielectric block and, thus, achieve endfire radiation and wide impedance bandwidth, the height, and permittivity of each section can be progressively reduced to smoothen the impedance transition to free space. In addition, to further increase the bandwidth and improve the antenna gain, the individual sections take the shape of isoceles trapezoids. The antenna is measured to have a gain greater than 9 dBi from 7.0 GHz onwards, and the maximum measured gain of 12.6 dBi is achieved at 12.4 GHz. The simulated radiation efficiency is above 98.5%, which is explained by the low-loss tangent of the 3D-printed dielectric material. A similar technique was used in [40] for a GRIN lens. The permittivity variation is realized through controlling the preset infill pattern and percentage in the 3D printer from 15 to 100%. The filament PREPERM based on ABS material loaded with ceramic particles to achieve a permittivity as high as 5 is used. For the infill percentage of 15%, permittivity of 1.5 is realized. The GRIN lens prototype shown in Fig. 9.10 is printed using Raise 3D Pro 3 3D printer. This work opens the possibility to simply

**Fig. 9.10** 3D-printed GRIN lens [40]

replace traditional metallic horn antennas with this new type of all-dielectric 3D-printed GRIN lens antenna. Measured aperture efficiency of 40% and directivity of 20 dB or higher is achieved at Ku-band for the presented antenna.

## 9.3 Metallization Techniques

Generally, 3D-printed antennas can be divided into three main categories in terms of materials used during manufacturing:

(1) Printed with dielectric materials and subsequently metallized.
(2) Produced through 3D metal printing
(3) dielectric antennas printed with dielectric materials.

This section focusses on various type of metalization techniques used in 3D printing antennas.

### *9.3.1 Conductive Spray Coating*

A compact, light, rigid, and low-cost three-dimensionally (3D) printed millimeter-wave antenna for a 28 GHz 5G communication system is presented in [41]. The plastic material used for fabrication is a transparent vero clear polyethylene, which enables fabrication of a solid, rigid, and light prototype. A 3D printer (Stratasys Objet30 Prime) is used to model the prototype and metalized using an electromagnetic interference called the radio frequency interference (EMI/RFI) conductive spray-paint technology. The commercial spray used for spray coating is RS EMI/RFI 400 mL shielding aerosol. The conductive spray paint has superior scratch resistance characteristics with very strong adhesive properties on a wide range of substrates. This conductive metal spray consists of metallic particles and acrylic resin along with compressed air. The spray coating simplifies the fabrication procedure since the paint can be used at room temperature, and the 3D-printed prototypes do not need any special pretreatment or post-treatment process before or after spraying. The paint delivers maximum conductivity after 24 h. This metalization method reduces the cost of fabrication and reduces the complexity of the fabrication process. However, the proposed spray-coating technique can be applied directly to open geometrical structures and it is not possible to apply the technique to complex and closed geometry structures unless they are disassembled. Another disadvantage of this metalization technique is that it is not possible to effectively control its thickness uniformity across the whole structure. Therefore, two coats are primarily applied to make sure that the interior sides of the slot and corrugations are painted, which makes the measured paint thickness on the surface of the prototype vary in the range of 45–65 $\mu$m, which is higher than the skin depth of copper $\sim$0.39 $\mu$m at 28.5 GHz. Hence, more expensive and complex plating techniques such as electroless plating are recommended as a reliable alternative plating method, if the thickness of the metalization layer is highly critical.

### *9.3.2 Electroless and Electroplating*

Two of the popular metalization techniques are electroless plating and electroplating. Electroless plating utilizes chemical reactions to bind metal particles onto the surface of an object. Electroless plating is a simpler plating process as it does not use electric current to bind the metal particles onto a surface. Since it does not use an electric current, electroless plating works with both conductive and non-conductive objects. It also produces thinner layers than electroplating. If a thick layer of metal is needed, electroplating should be considered instead. Electroplating is a plating process that involves the use of an electric current to bind the metal particles onto a surface. Because electroplating uses an electric current, it only works with conductive object. If the surface of an object is not conductive, the electric current would not flow through it, nor will the metal particles bind to it.

In [42], a 3D-printed spherical antenna under the operation of $TM_{101}$ and $TM_{211}$ resonant modes is proposed for its bandwidth enhancement. By symmetrically cutting five circumferential slots on the shell of a metallized spherical resonator, two resonant modes $TM_{101}$ and $TM_{211}$ can be excited simultaneously and merged with each other, resulting in wideband characteristics. A prototype is fabricated monolithically by incorporating SLA and electroless copper plating techniques. Measured results show

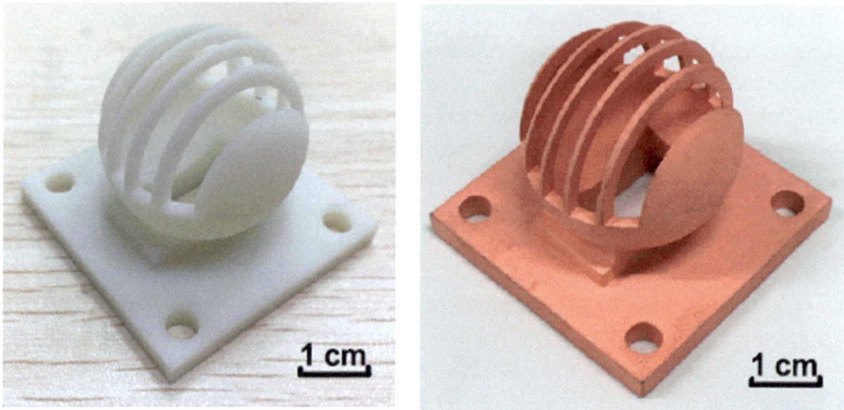


(a) (b)

**Fig. 9.11** Photographs of the antenna before and after plating [42]

that the antenna provides a wide bandwidth of 40.9% from 9.77 to 14.81 GHz and maintains stable radiation patterns over the operating band. Furthermore, due to the use of low-density resin as the building material, the weight of the antenna is dramatically reduced compared to the conventional metal ones. The final antenna prototype is shown in Fig. 9.11. The SLA printing was carried out in a vertical printing resolution of 50 μm. The ceramic-filled photosensitive resin Somos PerForm was used as the printing materials in this design due to high mechanical strength and high thermal handling capability. After the 3D printing process was completed, surface metallization process was carried out by electroless plating. In the metallization process, a 10 μm thick conductive layer of copper was plated on the resin model. The thickness of copper is 10× larger than the skin depth (~0.729 μm) calculated at 8 GHz. The sufficient copper thickness ensures good conductivity and reduces the loss from copper. The antenna is lightweight of 14.1 g due to the low density (1.61 g/cm$^3$ at 25 °C) of the utilized resin.

Researchers in [43] presented a design where an entire slot array, including all internal and external slots, and waveguide feed, was plated without disassembly. From an RF point of view, a plastic part plated with multiple skin depths of metal operates identically to a solid metal part. Initially it was plated at room temperature using electroless nickel plating. This step provides the conductive layer necessary to enable electroplating. Electroless plating followed by electro-plating has proven to offer good performance. The Ni electroless plating consists of four substeps: etching, activating, fixing, and deposition. Following the first three initial steps, the parts are immersed in the electroless Ni plating solution. The 3D-printed waveguides following the Ni plating have a smooth surface. After Ni plating, Cu is electroplated on each part. Electroplating is performed using copper-based electrolytic intended for thick deposits. Under appropriate conditions, this solution can deposit high-purity fine-grain films. The solution is electrolyzed prior to use. The antennas are

**Fig. 9.12** Metal-plated waveguide slot array [43]

plated with about 2 mils (50.8 microns) of copper externally, and inside there can be regions where the thickness is reduced to 10–20% of this thickness. That means there is a minimum of 5.08 microns of copper everywhere. At 5 GHz, that's 5.49 skin depths of copper, adequate for the metal-plated structure to function essentially the same way as a solid copper structure. The 3D-printed antennas are washed in fresh alcohol and cured with UV light. Parts that are properly cleaned and cured will result in more uniform deposition of nickel and subsequent copper over all surfaces. SLA 3D printing is followed for the presented design by thorough cleaning and curing. The resins used for SLA 3D printing of the slot array antennas presented here included DSM Somos Nanotool and Accura 25. Somos in particular uses a ceramic filler that makes it more rigid and stronger than some other choices, though it is also more brittle and dense, making parts heavier. The relatively low Coefficient of Thermal Expansion (C.T.E.) of these resins also enhances metal plating success. Platinum coated titanium wire, insoluble in the plating process is pulled through the waveguides to enhance fields in internal cavities.

This way, internal and external surfaces are both plated completely. A successful slot array created with this technique is shown in Fig. 9.12. The end caps are attached using silver epoxy after the array is plated.

### *9.3.3 Liquid Metal*

A miniaturized inverted-F antenna (IFA) operating at 885 MHz and suitable for wearable applications is designed in [44]. The NinjaFlex flexible plastic is used to realize the antenna through a 3D FDM printing process. The structure includes microchannels to delimit the shape of the conductive part of the antenna. The microchannels are filled with Galinstan liquid metal, an alloy of 68.5% Ga, 21.5% In, and 10% Sn is used to realize the radiating element. The effectiveness of the approach is validated for wearable applications. Compared to the use of sewed electro-textiles, which also allows the design of completely flexible antennas, the main advantage of this approach is the possibility of realizing 3D structures with larger thicknesses. The antenna provides a gain of 2.5 dBi and 70% efficiency. The antenna performances are maintained when the antenna is bent and placed over a human body phantom.

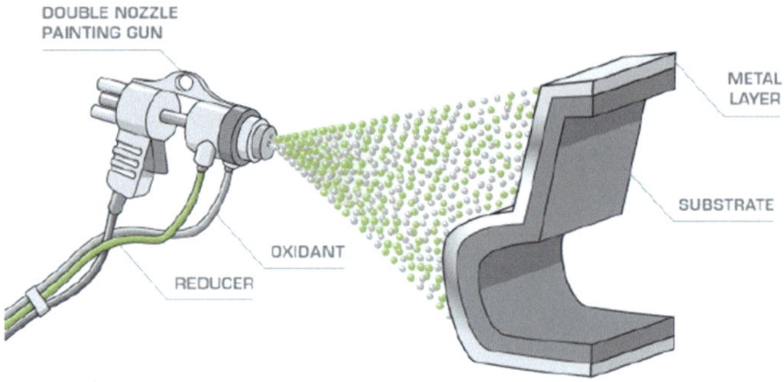


**Fig. 9.13** Principles of JMT metallization process [45]

## 9.3.4 *Jet Metal Process*

JMT, a low-cost metallization method alternative to electroless plating is reported in [45]. JMT is a direct plating method for silver particles developed by Jet Metal Technologies. The technique is used to coat a thin and a smooth silver film on top of the 3D-printed antennas. Firstly, the 3D-printed prototypes are cleaned by Isopropyl Alcohol in order to remove grease and stain. After cleaning, both antennas are flamed to get a good wettability on the whole surface and to increase the adhesion of the silver coating. Then, the prototypes are fixed on a holder that can be rotated in order to get a homogeneous coating. The silver is deposited to the surface of the antenna through continuous and simultaneous spraying of an oxidant agent (metallic ions $Ag^+$) and a reducing agent by using two double nozzle paint spray gun and compressed air at room temperature as shown in Fig. 9.13. Both antennas are coated with a film that has a thickness of 2.5 $\mu$m by using two spray guns simultaneously for 13 min, where the thickness of the silver layer can be easily controlled by controlling the spraying time.

## 9.3.5 *Vacuum Filling*

In [46], the fabrication and characterization of a 6 GHz microstrip patch array and a three-dimensional coaxial feed network embedded within a 3D-printed part is described. Internal cavities within the acrylic structure are filled with a gallium-based liquid metal alloy using a vacuum-driven process to form conducting elements. The insulating portions of the four-element array and integrated feeding network were printed using a ProJet 3500HDMax MultiJet printer. The empty cavities are metalized with EGaIn (an alloy of 75.5 and 24.5 wt% of gallium and indium, respectively) using a vacuum-filling process. These cavities ultimately form the radiating patch element, its ground plane, and the embedded coaxial transmission lines. To fill the internal cavities, a reservoir is attached at the inlet of the structure with an excess of EGaIn

(~30% more than the volume of the hollow cavities) and placed at a pressure of −30 in Hg relative to atmospheric pressure. After the air has been removed from the cavities, the pressure is returned to atmospheric pressure and the EGaIn is forced into the planar and curved microchannels. After the vacuum-filling procedure, the EGaIn reservoir is replaced by an SMA connector at the same inlet and sealed with a photocurable glue. This technique enables room-temperature rapid prototyping of intricate antenna array designs, combining the antenna elements and the feed structure as a single packaged structure. The array exhibits relatively low efficiency, which can be attributed to high dissipation in the printable dielectric material. However, this dielectric loss can be improved with the use of novel low-loss 3D printable materials capable of producing fine resolution and embedded cavities. In principle, the process can also be adapted to embed these intricate array structures into elastomers, enabling construction of flexible and reconfigurable fluidic antennas.

### *9.3.6 Wire-Mesh Embedding*

A 3D printable compact multilayer phased array operating at 3.5 GHz is reported in [47]. All the relevant parts of this multilayer phased array are 3D printed at room temperature using the thermal wire-mesh embedding for the conductive portion and the fused deposition modeling technique for the dielectric portion. Seamless integration of these two techniques allows robust and flexible RF components and potential RF systems to be fabricated with automation. The wire-mesh embedding procedure was to heat a wire mesh (0.1 mm wire diameter and 0.25 mm wire spacing) of a desired pattern using a resistive heater and simultaneously press the mesh into the thermoplastic–integrating plastic and metal structurally. After all necessary processes, the printed part was re-registered on the build plate of the 3D printer, and the build was resumed to print the subsequent substrate layers on top of the incomplete part. The additive process was repeated until the entire structure is finished. This is more advanced than the current conductive ink/paste traces on polymetric materials [48]. This work is different in that the manufactured metal conductivity using wire-mesh embedding is about ten times higher than conductive paste. The simulated array has a radiation efficiency of ~77% at scan angle of 0° at 3.5 GHz. The metallic layers of the phased-array system are made of copper with a thickness of 0.1 mm. The substrate layers are made of polycarbonate (PC) thermoplastic with a dielectric constant of 2.7 and a loss tangent of 0.005. A Stratasys Fortus 400mc material extrusion 3D printer with PC thermoplastic was employed to print the substrate.

## 9.4 3D Metal Printing

3D printing could be categorized as non-metallic and metallic printing based on the material used for printing. Compared with the non-metallic 3D printing that are firstly printed in split pieces and then metal plated, the SLM radio front end is printed using metal in one run. The metal plating process and post-assembly is not necessary, which simplifies the fabrication process. Compared with traditional injection molding and in-line laser cutting, the metallic 3D printing technology eliminates the cost of mold fabrication and has high material utilization, which results in a low cost and short turn-around time.

### 9.4.1 *Common Metal AM Processes*

The metal AM processes are classified based on the feedstock format and thermal source. Different existing feedstock formats include wire, filament, sheet, powder, and liquid. The thermal trajectory of each process and the amount of energy required are determined by the material compounds, chemistry, and transformation energetics.

1. Metal extrusion—A metal-filled filament can be printed using FDM printer machines [49]. In this method, the metal powder is infused in a standard ABS or PLA filament. The prints made using these filaments are not pure metal parts, but rather consist of metal particles in a polymer matrix, and are brittle and require care while handling the printed objects. The powder percentage of the polymer and metal particles can vary. The tensile strength of the filaments is reduced as the metal particle content increases. However, the thermal conductivity of the filaments increases and can make them suitable for fabricating circuits and electromagnetic parts.
2. Direct metal laser sintering—DMLS is used to produce metallic components where improved surface quality and lower levels of porosity are key advantages. The feed of the horn in Fig. 9.14 was DMLS 3D printed using selective metal sintering which sintered 316L stainless steel into a solid structure [50]. The assembled feed was metal printed in three parts: two waveguide splitters, and a feed bend. The horn itself was FDM printed from PLA and metallized using adhesive backed aluminum tape.
3. Directed energy deposition (DED)—In LENS, a focused, high powered laser beam is used to melt metallic powder [51] and the metal powder is applied only where powder is being added to the substrate at that moment. In LC, a laser heat source is applied to deposit a thin layer of preplaced materials on a substrate, which is moved using a controlled system [52]. The cladding material, which is in the form of powder, is deposited on the substrate's surface and then the laser beam is irradiated.

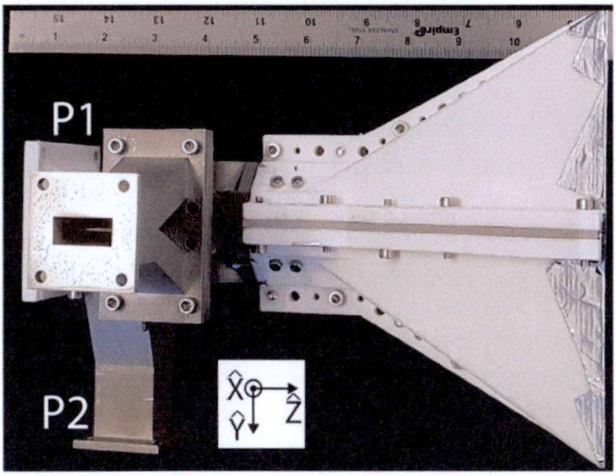


**Fig. 9.14** Metal 3D-printed waveguide feed with horn antenna [50]

4. Material jetting (MJ)—MJ technology uses photopolymers as the material and applies a UV light to selectively cure the printed layer. Nanoparticle jetting (NPJ) is one of the MJ technologies that enables the production of metal parts through the use of composite resins that are infused with metallic nanoparticles such as stainless steel.
5. Binder jetting (BJ)—BJ technology uses metal powder alongside a liquid-state binder to print the desired metal parts in layers. In this process, the binder droplets consolidate the powdered materials within and between sliced layers [53]. The BJ technology requires no support structures and prints much more accurate objects compared to metal powder bed fusion.

### *9.4.2 Antenna Designs Using Metal AM Technology*

The performance of Ka-band circuits and antennas manufactured using binder jetting metal printing is characterized in [54]. The parts are made of 316 stainless steel that is infiltrated with copper or bronze. A compound meshed parabolic reflector antenna system is designed to demonstrate the capabilities of BJ. This antenna has a feed network that includes the illuminating horn, rectangular waveguide section, and a filter, all in one piece. This monolithic approach eliminates the need for joints, which can be a source of failure, especially in satellites that are subjected to vibrations during deployment. Using a meshed surface for the reflector offers 40% reduction in weight, when compared with the solid version, while having minimal effect on the antenna gain, which is 24.56 dBi at 33.5 GHz. The antenna is fabricated with an ExOne Innovent metal BJ printer. The printing process starts by coating the printing bed with a uniform layer of powdered stainless steel. The print head then deposits liquid binder on the corresponding cross section of the part in the powder. The parts are then placed into a graphite crucible. The entire crucible is then filled with alumina powder and placed in a sintering oven. The high temperature burns off all the binder and causes the metal powder particles to fuse together. The liquid infiltrant metal with lower melting temperature is absorbed into the pores of the green part creating a composite material. For the proposed design, 316 SS powder is utilized for the BJ process and either bronze or copper is used as a second metal for the infiltration process. An alternative approach would be to perform electroplating instead of using the infiltration process, although other challenges such as plating uniformity can arise. The surface roughness of the components could also be improved by using a finer alumina powder for the infiltration process.

The use of metal AM for the development and fabrication of 3D fractal antennas at 2.4 GHz is presented in [55]. The 3D printing with metals gives mechanical strength, particularly in the joints between the various repetitions of the fractal elements. The antennas can be fabricated using a metal powder bed fusion process, and more specifically metal powder embinder printing. This metal technique was selected because of the mechanical robustness of the output and the possibility of printing highly detailed features. The technology prints the models by binding together layers of ultrafine grains of stainless steel powder in a precision inkjet printer. Once the printing is done, the part is sintered in an oven. The base metal used in this design is 316L stainless steel, a material with inferior conductivity. Consequently, a copper layer of approximately 50 μm was deposited on the antenna using an electroplating process. The alternative option of plating printed dielectrics would have mechanical weakness at the joints. The metal 3D-printed structures offer attractive electromagnetic features and are also able to reduce the amount of material use in AM processes. Fractal and complex antennas fabricated using metal 3D printing have applications in the defense and aerospace sectors where robustness as well as lightweight are key.

In [56], two different metals and related 3D printing technologies are studied to explore their potential for microwave applications using E-, D-, and H-band conical horn antennas. The first method is the binder jetting and sintering. The antenna is formed by binder-jetted 316L stainless steel powder, and then sintered to solidify. The second one is the Cu–15Sn by selective laser melting (SLM), which uses a laser beam to melt the Cu–15Sn powder layer by layer. The antennas have desirable performance up to 325 GHz, which proves their feasibility for mm- and sub-mm-wave applications. Different from the non-metallic 3D-printed horn antenna in two split metal-plated pieces, the metallic counterpart is printed in a whole metal piece in one run, which does not require assembling. Moreover, the screw thread in the flange of the non-metallic horn is easily worn out, which is not the case in the metallic horn. Compared with the traditional injection molding and micromachining implemented metallic horn, the 3D-printed horn has shorter turn-around time and lower cost, which is preferable for both industrial mass production and prototyping.

## 9.5 Latest Trends in 3D-Printed Antennas

3D printing is a flexible modern technology that can be integrated with many of the state-of-the-art technological advances. Research to utilize 3D printing more in the mm-wave and THz antenna systems are advancing. As the systems are getting more and more compact, higher antenna integration can be made possible using 3D printing. This section covers some of the latest areas where 3D printing is being used by antenna engineers.

1. ***Multilayer antenna array structure***

3D printing has the capability to integrate different antenna components and structural elements that may not be feasible using traditional manufacturing techniques. A 3D-printed wideband circularly polarized (CP) waveguide based 4 × 4 slotted multilayer antenna array at 20 GHz is proposed in [57]. In part 1 of the antenna shown in Fig. 9.15, there is a one-to- four-way corporate feeding circuit, which is mainly composed of a WR-51 waveguide port and several T-junction power dividers. There are 2 × 2 uniform coupling slots above the power dividers for transmitting electromagnetic waves. The back cavity is designed upon the slots for Part 1 to equally equalize the energy from the coupling slots and to excite the radiation slots in Part 2. The upper hexagonal aperture of Part 2 is designed such that the electromagnetic waves can be transmitted in a pair of degenerate modes to Part 3. This would eventually get synthesized into CP waves and radiated to the parasitic slots. The structure and dimensions of Part 3, including the simple air gap layer, are carefully designed to expand the AR bandwidth of the final antenna. Utilizing DMLS 3D printing technology not only mitigates the effects of manufacturing imperfections but also contributes to lower manufacturing costs. The design demonstrates the feasibility of the DMLS 3D printing technology in fabricating high-frequency microwave antennas.

3D printing enables the integration of electronics directly into the antenna structure. This can be particularly useful for incorporating feed networks, amplifiers, or signal processing components within the antenna itself. By integrating electronics, the overall size and complexity of the antenna system can be reduced while improving overall performance. Researchers in [58] used the 3D printing technology to integrate the 2D monopulse antenna array, along with its feed network and monopulse comparator as a whole structure to achieve higher integration level. The components of the entire array are compactly arranged and packaged by 3D printing technology to obtain high integration level. The printed array frame can also be viewed as a radome for protecting the array. The lens structure is printed with the array frame at the same time, which greatly simplifies the array fabrication process. With this integrated design, both the size and the losses can be reduced. The 3D printer ProJet MJP 2500, 3D System is used to fabricate the packaging frame with a printing precision of 0.018 mm. The printing material is RWT-Rigid White with relative dielectric permittivity 2.75 and loss tangent 0.02 measured at 10 GHz. By using the support material (wax) in the printing process, hollow and exquisite structures can be designed and fabricated [59].

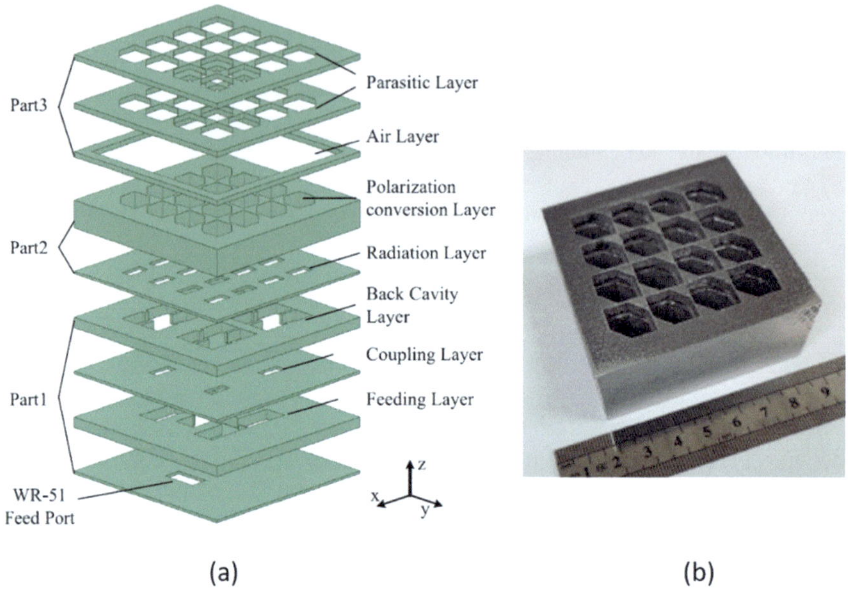


**Fig. 9.15** **a** Exploded view of the overall circularly polarized 4 × 4 slot antenna array **b** fabricated antenna array [57]

2. ***Less volume***

3D printing is now being explored for designing the fifth-/sixth-generation wireless network antennas. As 3D printing antenna designs can be optimized to any shape and size, it can be used to reduce the overall volume of the structure. Chen et al. [60] presented a novel compact mm-wave circular polarized 3D-printed DLA using an inset dielectric transition. It consists of two components: (1) a novel, low-profile transition is developed to solve the high-loss transmission between the air-filled waveguide and dielectric-filled waveguide; (2) a dielectric loaded antenna is excited with extra high-order modes to obtain narrower beamwidth and lower cross-polarization. The use of 3D printing technology enables seamless integration with a WR34 waveguide where the antenna is clipped to the waveguide for a smooth dielectric transition. The 3D-printed dielectric loaded design showed a 43% volume reduction and 90% weight reduction when compared with the metal antenna with similar performance. First, the solid dielectric in blue was printed by the Stratasys Objet30 3D printer as an integration. Then, the thin metal film was printed on the dielectric surface. Moreover, the weight of the dielectric antenna is only 4.17 g, which is less than the weight of a standard A4 size paper (typically 4.366 g). The compact transition and light-weight structure makes it superior when integrating with 5G/6G enabling antenna systems.

3. ***Machine learning***

Machine learning (ML)-based optimization methods is an alternative to traditional optimization methods, for complex antenna designs. This is an efficient methodology to tackle computational challenges, as it is capable of handling a large number of design parameters and is more efficient as well as informative. To demonstrate these advantages Sharma et al. [61] proposed two ML-based models for a quarter-wavelength monopole antenna surrounded by an optimized dielectric constant distribution on a finite ground plane. The dielectric profile is implemented as a $6 \times 6$ grid of 3D-printed dielectric blocks. By changing the filling ratio of each of the dielectric grids, effective dielectric distribution is controlled, as a result of which the radiation properties of the monopole antenna can be changed. The dielectric part of this design is 3D printed using a commercial stereolithographic apparatus (SLA)-based 3D printer, Asiga Max. This printer uses digital light processing technology, which is easy to use and allows for fast, precise, and efficient operation.

4. ***Structural flexibility***

A conventional antenna once fabricated does not allow any modifications to the structure. But a 3D-printed antenna design allows modifications in its structure also. There are some antenna designs where the structure was built using different smaller structures like lego bricks. So by removing any structure, the antenna parameters can be changed, and this allows more flexibility to the design. An assembled-type 3D FSS operating at 6.47 GHz is presented in [62]. This 3D FSS can be easily constructed using various FSS unit structures, and the reconfiguration can be implemented in a simple manner either by placing the structure or taking out the structure based on the design. The proposed structure has the advantage that the frequency characteristics can be easily changed as the FSS unit structure can be replaced structurally. Based on this, the cross-dipole, which is the basic structure of the FSS, is implemented on a quadrilateral pyramid. Consequently, a unit structure of the 3D FSS, which exhibits stability in polarization and incident angle of the incident wave, is designed. The cross-dipole-structured 3D FSSs are manufactured using 3D printing technology to achieve a high-speed and low-cost manufacturing process. To fabricate the proposed 3D FSS, a dielectric substrate is manufactured using a 3D printing method called fused filament fabrication method, which consists of melting plastic using high-temperature nozzles and stacking the melted plastic. The 3D printer used was Fortus 380 mc by Stratasys, and an ABS was used as the plastic material.

## 9.6 Enhancing Antenna Characteristics Through 3D Printing

3D-printed antennas and 3D-printed structures can be used to enhance various features of an antenna like gain, bandwidth, radiation efficiency.

1. ***Bandwidth enhancement***

Increasing the substrate thickness of a standard microstrip antenna is a simple way to increase its bandwidth, but leads to surface-wave generation and spurious radiation. Instead, a local modification of the vertical profile of a patch antenna can enhance the bandwidth as proposed in [63]. The substrate is planar, whereas the part of the substrate over which the patch is printed is made cylindrical through a 3D-printed substrate. The substrate thickness is maximum at the center and gradually decreases toward the edges of the patch. The bandwidth enhancement is achieved by exploiting the increase of the average substrate thickness in the region of the metal patch, whereas the remaining part of the dielectric substrate maintains a thin profile. Due to its peculiar curved profile, the substrate has been fabricated using a commercial 3D printer (Raise3D Pro2 Plus), using a PLA thermoplastic filament. After the printing procedure, a conductive tape has been glued to the cylindrical surface and to the bottom of the PLA substrate. This shows that non-flat antenna configurations can be employed to reduce the effect of the surface wave.

2. ***Beam scanning***

Customized 2D metasurfaces is a topic of considerable interest due to their ability to modify the phase of the propagating EM fields that makes them suitable for beam-steering antennas. However, such metasurfaces cannot be fully efficiently designed by using the commercial off-the shelf (COTS) materials that come with preset effective permittivity and permeability. Hence 3D printing allows greater flexibility to incorporate such metasurfaces. Such a metasurface design that uses metasurface for beam steering is presented in [64]. The meta-atoms shown in Fig. 9.16 is designed to attain different permittivities. It has a fixed-beam static base antenna at the bottom with a large radiating aperture and a pair of independently rotatable metal-dielectric composite metasurfaces (MDCMs) atop operating at 11 GHz, each separated by 4 mm or $0.15\lambda_0$ as shown in Figs. 9.17 and 9.18. The 3D printable dielectric material used with M1 is PREPERM® ABS 1000, which has a base permittivity ($\varepsilon_r$) of 10. The effective permittivity of the meta-atom is reduced below 10 by increasing the size of the air inclusion. It is noted that if traditional machining methods are used to introduce air inclusions, the drill bits available are of discrete sizes, and hence discrete effective permittivity values can only be achieved. To achieve continuous effective permittivity between 2.6 and 10, additive manufacturing can be used instead, which can provide accuracy at the sub-millimeter level to realize M1 with desired permittivity values. Different meta-atoms are then strategically arranged with a phase delay gradient to form MDCM, which has a constant phase in the $y$-axis and a progressive phase in the $x$-axis. MDCMs are designed such that they change the direction

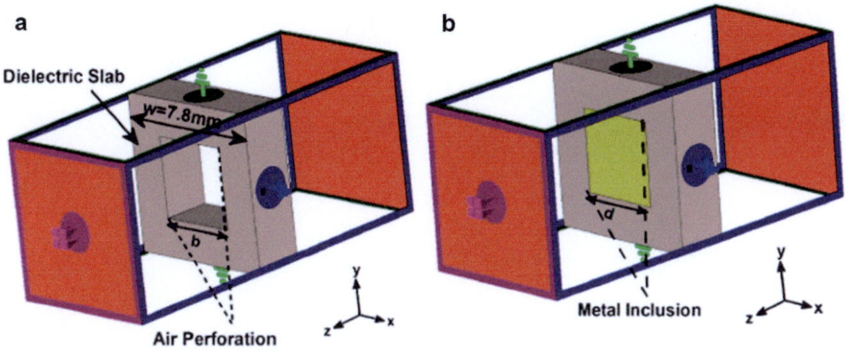


**Fig. 9.16** **a** M1: low-permittivity materials by introducing air perforation, and **b** M2: high-permittivity materials by introducing metal inclusion in the host dielectric [64]

of radiated field propagation from broadside to an offset angle $\delta$ by introducing a gradient in the phase of the electric fields. Thus, when used separately with the base antenna, each MDCM tilts the antenna beam by an angle of $\delta$. If a single MDCM is placed above the base antenna and rotated, its tilted beam with a fixed elevation angle can be moved to any azimuth angle between 0° and 360°. The combination of two MDCMs gives freedom to steer the beam in both elevation and azimuth planes in a conical region having an apex angle of at least $2\delta$.

3. ***Compact structure***

3D printing technology has been exploited in [65] to design a circularly polarized antenna for ETSI UHF RFID handheld readers. The antenna consists of four inverted-F monopoles with capacitive top loads, sharing a single ground plane. The upside-down mounted feed network implements the sequential rotation feeding technique to achieve a circularly polarized radiated field. The two PLA parts (i.e., the annular square for the antenna and the feeding network substrates) have been molded by using a Sharebot Next-Generation 3D printer having a 1.75 mm filament extruder, a 0.35 mm nozzle, and a $45 \times 45\ \text{cm}^2$ heatbed plate. The conductive parts (i.e., radiating elements, feed network, feeding lines, and ground plane) are realized by using 35 $\mu$m adhesive copper tape shaped through the cutting-plotter-based technique. Taking advantage from low losses and moldability of the 3D printing materials joint to a specific design strategy, the proposed antenna realizes an improved gain and an appreciable size reduction when compared with the state of the art.

4. ***Frequency tuning***

A 3D-printed tunable CP microstrip patch antenna is presented in [66] employing an L-slot in a square patch for CP operation. The use of the 3D printing has allowed the vertical mounting of varactors along the thickness of the substrate as well as the thicker substrate to increase the bandwidth of the antenna without stacking multiple substrates. This work has demonstrated the benefits of 3D printing to realize compact

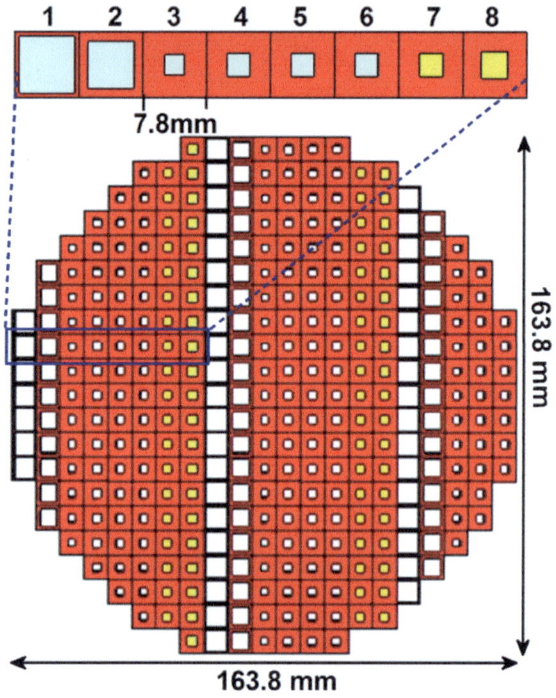


**Fig. 9.17** Front view of 3D printable synthetic MDCM model with a set of eight meta-atoms with varying geometries [64]

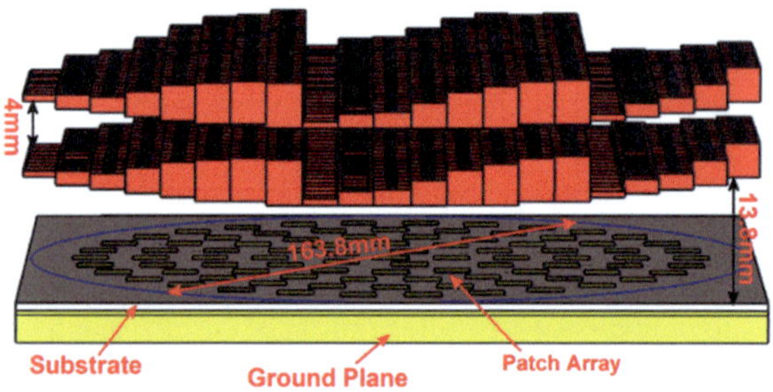


**Fig. 9.18** Beam-steering antenna using a pair of MDCMs with patch array antenna used as a primary feed source [64]

radio frequency components that were not possible by conventional fabrication techniques. The antenna can be tuned from 1.65 to 2.17 GHz while maintaining CP operation, thus providing a CP tunable range of 27.2%. The antenna substrate is 3D printed using the MakerBot Replicator2 printer. Four vertical posts are also realized in the air gap close to the edges of the patch for mounting of the varactors. Cylindrical posts are also made within the air gap to support the two layers of ABS. The two layers of ABS are 3D printed separately and then connected with the help of the cylindrical posts located at the four corners. Rectangle openings are left in the top ABS layer to assist in the mounting of the varactors. In order to connect the varactor leads across the patch at the top and the ground at the bottom, via holes are made within the top and bottom ABS layers. For metallization of the patch and the ground, an aluminum tape has been used, which has been milled to include the L-slot for the top patch using the LPKF ProtoMat D104 circuit board plotter. The via holes are metalized using the Circuitworks CW2400 conductive silver epoxy, which is also used to mount the varactors and the SMA connector as conventional soldering methods are not applicable for 3D-printed plastic because of high temperatures.

5. ***Gain enhancement***

A shortened horn antenna with high gain enabled by a 3D-printed gradient-index (GRIN) lens composed of high-permittivity zirconia ($ZrO_2$) is presented in [67]. The H-plane sectoral horn antenna is designed with length that is 1/3 of the optimal horn antenna and exhibits a low gain due to the high flaring rate of the horn. Gain is enhanced by adding a flat GRIN lens at the horn aperture as shown in Fig. 9.19. Since the lens is flat, the phase error is corrected by varying the effective permittivity of the material from the center to the edges along the $x$-axis. The permittivity along the $x$-axis is given by

$$\varepsilon_r(x) = \left[\sqrt{\varepsilon_{r,\max}} + \frac{R - \sqrt{x^2 + R^2}}{t}\right]^2, \tag{9.2}$$

where $\varepsilon_{r,\max} = \varepsilon_r(x = 0)$ is the largest permittivity that can be obtained with the composing material (i.e., its bulk permittivity) and thickness, $t$, of the lens alone is defined as

$$t = \frac{\sqrt{\left(\frac{a}{2}\right)^2 + R^2} - R}{\sqrt{\varepsilon_{r,\max}} - \sqrt{\varepsilon_{r,\min}}}, \tag{9.3}$$

where $\varepsilon_{r,\min}$ is the permittivity value at the edge ($x = a/2$).

To fabricate the GRIN lens, the continuously varying permittivity distribution of the lens is discretized, and the permittivity range is from the minimum of 3.6 to the bulk permittivity of 23. Then, the lens is implemented with appropriate permittivity

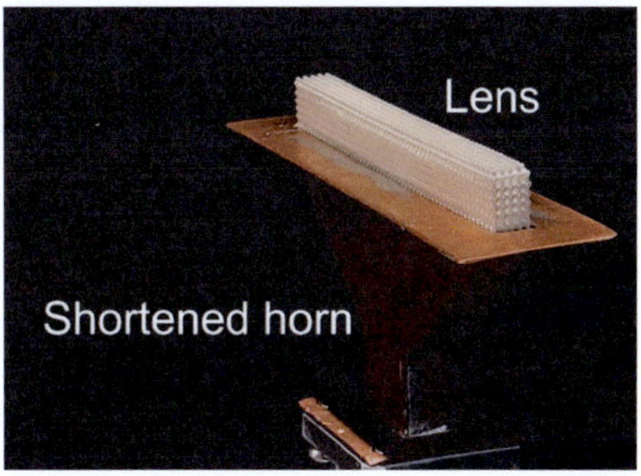


**Fig. 9.19** Shortened horn antenna with lens [67]

values using an additive manufacturing process. A single material (zirconia) is printed using an XJet Carmel 1400, and the required permittivity gradient is created by varying the fill of the printed material and air. The 3D-printed lens is composed of the three types of unit cells as illustrated in Fig. 9.20. The chamfered unit cell is used for a high-permittivity range of 17.2–23 (red dashed rectangle), the cubic unit cell is used for the mid-permittivity range of 5.9–16.2 (black and blue dashed rectangles), and the gapped unit cell (pink dashed rectangle) is used for the lowest permittivity value in the lens, $\varepsilon_r = 3.6$. The zirconia lens is 3D printed using a nanoparticle jetting (NPJ) technology.

6. ***High radiation efficiency***

A dielectric helical antenna based on 3D-printed zirconia ceramic is proposed in [68]. The antenna benefits from the low loss of zirconia ceramic, along with little dielectric loss and no conductor loss, resulting in high radiation efficiency. The proposed antenna makes full use of the advantage of 3D printing as without that the helical structure is difficult to realize in conventional craft. By using 3D printing, the purity of zirconia in helical structure can exceed 95% thus making the product white in color. The CERAMAKER 3D printer from 3DCeram SAS is used for manufacturing the zirconia ceramic helical structure. The radiation gain and efficiency of the fabricated

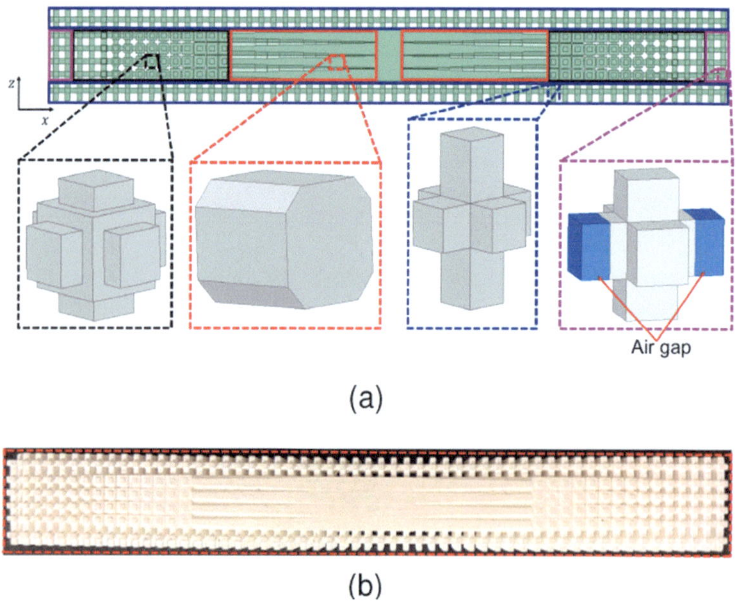


**Fig. 9.20** The continuously varying permittivity is quantized. **a** Schematic of the lens. A chamfered unit cell is used for high permittivity (red rectangle), a cubic unit cell is used for middle permittivity and ML (black and blue rectangle), and a gapped unit cell is used for low permittivity (pink rectangle), **b** photo of the printed lens [67]

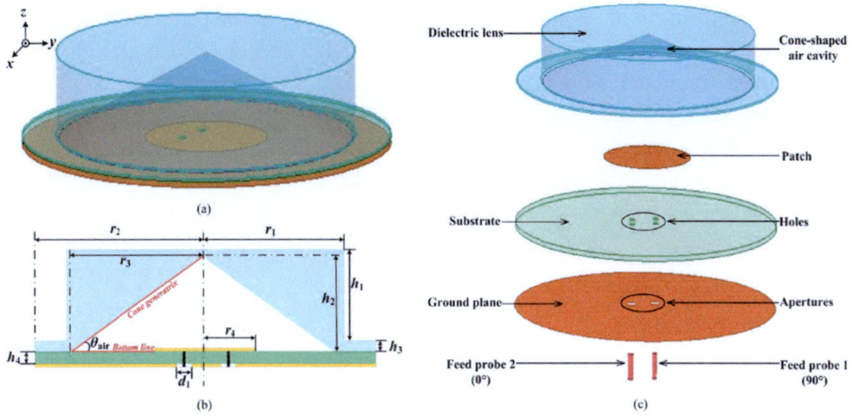


**Fig. 9.21** Geometry of the proposed CP antenna. **a** Perspective view, **b** side view, **c** explosive view [69]

antenna at 6 GHz are about 10 dBic and 95%, respectively, revealing its good radiation property.

7. ***Axial ratio beamwidth enhancement***

The design of a 3D-printed dielectric lens with a cone shaped cavity for broadening the 3-dB ARBW of CP patch antenna at 1.6 GHz as shown in Fig. 9.21 is reported in [69]. Due to the existence of lens, the CP wave radiated from the source antenna is refracted before radiating into free space. The proposed lens is fabricated by a 3D printer with thermoplastics nylon material which has a dielectric constant of 3.15 and a loss tangent of 0.02. The 3-dB ARBW of conventional patch is extended from 82° to 162° with the help of lens. The lens has been proven to be applicable for wide band CP antennas, finite ground plane and other types of CP antennas. The 3D printing can be used for printing lens with shapes necessary to improve the antenna parameters.

## 9.7 Conclusion

3D printing has immense potential to be used in the wireless electronics industry. This technology is used to develop RF antenna systems that were not possible using traditional methods. The chapter went through the classification of 3D printing based on the type of antennas being printed. The chapter also covered the metalization techniques used by the antenna designers once the dielectric part of the antenna is 3D printed. As in some cases direct 3D printing in metal is advantageous as discussed in this chapter. The chapter also described how the advantages offered by the 3D printing technology is being incorporated in the latest antenna designs to enhance the various features of the antenna. Some of the limitations of this technology are related to the printing of smaller dimensions, resolution of the printed products and surface roughness. As a result, 3D printing at higher frequencies is less common than at lower frequencies for now. However, it is anticipated that when 3D printing technology advances and more advanced 3D printers with better resolution become accessible, it will help to produce improved antenna designs in the future. As some of the designs explained in the further chapters of the book make use of 3D printing, this chapter can be used as a guide to understand the designs better.

## 9.8 Future Scope

As the field of wireless communication and additive manufacturing rapidly evolves, there is tremendous scope for future work in this important area. Some of the ideas are listed below:

- Investigating the design and fabrication of novel metamaterial antennas with functionalities like multiband operation, beam steering, and polarization control will push the boundaries of performance. Optimizing integration with existing antenna systems will be crucial for real-world applications, requiring exploration of hybrid structures and efficient feeding mechanisms.
- Investigating the potential for reconfigurable or tunable metamaterial antennas will enable dynamic control and adaptability, opening doors for new functionalities.
- Explore efficient methods for integrating 3D-printed ceramic antennas with existing systems, tackling challenges like feeding mechanisms and impedance matching. Miniaturization efforts will be crucial for applications like Internet of Things (IoT) and various other modern wireless devices.

To carry out research in these areas, the book can serve as an invaluable resource for researchers, students, and anyone interested in the future of additive manufacturing in antenna design.

## References

1. Redwood, B., Schoöffer, F., Garret, B.: The 3D Printing Handbook. 3D Hubs, Amsterdam, The Netherlands (2017)
2. Berman, B.: 3-D printing: the new industrial revolution. Bus. Horiz. **55** (2012). https://doi.org/10.1016/j.bushor.2011.11.003
3. Helena, D., Ramos, A., Varum, T., Matos, J.N.: Antenna design using modern additive manufacturing technology: a review. IEEE Access **8**, 177064–177083 (2020)
4. Liang, M., Shemelya, C., MacDonald, E., Wicker, R., Xin, H.: 3-d printed microwave patch antenna via fused deposition method and ultrasonic wire mesh embedding technique. IEEE Antennas Wirel. Propag. Lett. **14**, 1346–1349 (2015)
5. Wang, X., Jiang, M., Zhou, Z., Gou, J., Hui, D.: 3D printing of polymer matrix composites: a review and prospective. Compos. B Eng. **110**, 442–458 (2017)

6. Manapat, J.Z., Chen, Q., Ye, P., Advincula, R.C.: 3D printing of polymer nanocomposites via stereolithography. Macromol. Mater. Eng. **302**(9):1600553 (2017)
7. Utela, B., Storti, D., Anderson, R., Ganter, M.: A review of process development steps for new material systems in three dimensional printing (3DP). J. Manuf. Process. **10**(2), 96–104 (2008)
8. Mohamed, O.A., Masood, S.H., Bhowmik, J.L.: Optimization of fused deposition modeling process parameters: a review of current research and future prospects. Adv. Manuf. **3**(1), 42–53 (2015)
9. Zhang, F., et al.: 3-d printed slotted spherical resonator bandpass filters with spurious suppression. IEEE Access **7**, 128026–128034 (2019)
10. Parandoush, P., Lin, D.: A review on additive manufacturing of polymer-fiber composites. Compos. Struct. **182**, 36–53 (2017)
11. Ngo, T.D., Kashani, A., Imbalzano, G., Nguyen, K.T.Q., Hui, D.: Additive manufacturing (3D printing): a review of materials, methods, applications, and challenges. Compos. B Eng. **143**, 172–196 (2018)
12. Lee, C.-K., et al.: Evaluation of microwave characterization methods for additively manufactured materials. Designs **3**(4), Art. no. 47 (2019)
13. Stevens, C.J., Spanos, I., Vallechi, A., McGhee, J., Whittow, W.: 3D printing of functional metal and dielectric composite meta-atoms. Small **18**(10), 1–9 (2022)
14. Nayeri, P., et al.: 3D printed dielectric reflectarrays: Low-cost high gain antennas at sub-millimeter waves. IEEE Trans. Antennas Propag. **62**(4), 2000–2008 (2014)
15. Wang, K.X., Wong, H.: A wideband millimeter-wave circularly polarized antenna with 3-D printed polarizer. IEEE Trans. Antennas Propag. **65**(3), 1038–1046 (2017)
16. Addamo, G., et al.: 3-D printing of high-performance feed horns from Ku- to V-bands. IEEE Antennas Wireless Propag. Lett. **17**(11), 2036–2040 (2018)
17. Tak, J., Kantemur, A., Sharma, Y., Xin, H.: A 3-D-printed W-band slotted waveguide array antenna optimized using machine learning. IEEE Antennas Wireless Propag. Lett. **17**(11), 2008–2012 (2018)
18. Huang, J., Chen, S.J., Xue, Z., Withayachumnankul, W., Fumeaux, C.: Wideband endfire 3-d-printed dielectric antenna with designable permittivity. IEEE Antennas Wirel. Propag. Lett. **17**(11), 2085–2089 (2018)
19. Huang, J., Chen, S.J., Xue, Z., Withayachumnankul, W., Fumeaux, C.: Impact of infill pattern on 3D printed dielectric resonator antennas. In: Proceedings IEEE Asia–Pacific Conference Antennas Propagation (APCAP), Auckland, New Zealand, pp. 233–235, Aug 2018
20. Honari, M.M., Mirzavand, R., Saghlatoon, H., Mousavi, P.: Investigation of the 3d printing roughness effect on the performance of a dielectric rod antenna. IEEE Antennas Wirel. Propag. Lett. **17**(11), 2075–2079 (2018)
21. Ferrando-Rocher, M., Herranz-Herruzo, J.I., Valero-Nogueira, A., Bernardo-Clemente, B.: Performance assessment of gap-waveguide array antennas: CNC milling versus three-dimensional printing. IEEE Antennas Wirel. Propag. Lett. **17**(11), 2056–2060 (2018)
22. Tamayo-Dominguez, A., Fernandez-Gonzalez, J.-M., Sierra-Perez, M.: Groove gap waveguide in 3-D printed technology for low loss, weight, and cost distribution networks. IEEE Trans. Microw. Theory Techn. **65**(11), 4138–4147 (2017)
23. Balzer, J., Weidenbach, M., Busch, S., Koch, M.: 3D printed waveguides for 120 GHz. In: Proceedings 2016 German Microwave Conference, pp. 1–3 (2016)
24. Zhang, B., Zirath, H.: Metallic 3-D printed rectangular waveguides for millimeter-wave applications. IEEE Trans. Compon., Packag., Manuf. Technol. **6**(5), 796–804 (2016)
25. Lira-Valdés, T., Rajo-Iglesias, E., Pizarro, F.: 3-d-printed spiral leaky wave antenna with circular polarization. IEEE Open J. Antennas Propag. **4**, 427–433 (2023)
26. Shin, S.-H., et al.: Polymer-based 3-d printed Ku-band steerable phased-array antenna subsystem. IEEE Access **7**, 106662–106673 (2019)
27. Wu, L., Chu, H., Cao, D., Peng, S., Guo, Y.: 3-d printed antenna subsystem with dual-polarization and its test in system level for radiometer applications. IEEE Access **8**, 127856–127865 (2020)

28. Kempel, L.C., Volakis, J.L., Sliva, R.: Radiation by cavity-backed antennas on a circular cylinder. Proc. Inst. Elect. Eng., pt. H, 233–239 (1995)
29. Jun, S.Y., Elibiary, A., Sanz-Izquierdo, B., Winchester, L., Bird, D., McCleland, A.: 3-d printing of conformal antennas for diversity wrist worn applications. IEEE Trans. Components, Packag. Manuf. Technol. **8**(12), 2227–2235 (2018)
30. Nate, K., Tentzeris, M.M.: A novel 3-D printed loop antenna using flexible NinjaFlex material for wearable and IoT applications. In: Proceedings 24th IEEE Electrical Performance of Electronic Packaging and Systems, pp. 171–174 (2015)
31. Wang, R., Klionovski, K., Shamim, A.: Fully printed 3-d antennas with wideband radiation isotropy based on annular currents models. IEEE Open J. Antennas Propag. **4**, 484–491 (2023)
32. Rizwan, M., Khan, M. W. A., He, H., Virkki, J., Sydänheimo, L., Ukkonen, L.: Flexible and stretchable 3D printed passive UHF RFID tag. Electron. Lett. (2017)
33. Kiziltas, G., Psychoudakis, D., Volakis, J.L., Kikuchi, N.: Topology design optimization of dielectric substrates for bandwidth improvement of a patch antenna. IEEE Trans. Antennas Propag. **51**(10), 2732–2743 (2003)
34. Njoku, C.C., Whittow, W.G., Vardaxoglou, J.C.: Simulation methodology for synthesis of antenna substrates with microscale inclusions. IEEE Trans. Antennas Propag. **60**(5), 2194–2202 (2012)
35. Njoku, C.C., Whittow, W.G., Vardaxoglou, J.C.: Effective permittivity of heterogeneous substrates with cubes in a 3-D lattice. IEEE Antennas Wirel. Propag. Lett. **10**, 1480–1483 (2011)
36. Lewin, L.: The electrical constants of a material loaded with spherical particles. Electr. Eng., Part III: Radio Commun. Eng. **94**(27), 65–68 (1947)
37. Tribe, J., Whittow, W.G., Kay, R.W., Vardaxoglou, J.C.: Additively manufactured heterogeneous substrates for three-dimensional control of local permittivity. Electron. Lett. **50**(10), 745–746 (2014)
38. Monkevich, J.M., Le Sage, G.P.: Design and fabrication of a custom-dielectric fresnel multi-zone plate lens antenna using additive manufacturing techniques. IEEE Access **7**, 61452–61460 (2019)
39. Zhang, S., Cadman, D., Vardaxoglou, J.Y.C.: Additively manufactured profiled conical horn antenna with dielectric loading. IEEE Antennas Wirel. Propag. Lett. **17**(11), 2128–2132 (2018)
40. Paraskevopoulos, A., Maggiorelli, F., Gashi, I., Giovampaola, C.D., Albani, M., Maci, S.: 3-D printed all-dielectric GRIN lens antenna with an integrated feeder. IEEE Open J. Antennas Propag. **4**, 528–536 (2023)
41. Alkaraki, S., et al.: Compact and low-cost 3-d printed antennas metalized using spray-coating technology for 5G mm-wave communication systems. IEEE Antennas Wirel. Propag. Lett. **17**(11), 2051–2055 (2018)
42. Zhu, Y.-J., et al.: A 3-d printed spherical antenna with bandwidth enhancement under operation of dual resonance. IEEE Access **8**, 19345–19352 (2020)
43. Le Sage, G.P.: 3D printed waveguide slot array antennas. IEEE Access **4**, 1258–1265 (2016)
44. Cosker, M., Lizzi, L., Ferrero, F., Staraj, R., Ribero, J.-M.: Realization of 3-d flexible antennas using liquid metal and additive printing technologies. IEEE Antennas Wirel. Propag. Lett. **16**, 971–974 (2017)
45. Alkaraki, S., Gao, Y., Munoz Torrico, M.O., Stremsdoerfer, S., Gayets, E., Parini, C.: Performance comparison of simple and low-cost metallization techniques for 3d printed antennas at 10 GHz and 30 GHz. IEEE Access **6**, 64261–64269 (2018)
46. Bharambe, V., Parekh, D.P., Ladd, C., Moussa, K., Dickey, M.D., Adams, J.J.: Liquid-metal-filled 3-d antenna array structure with an integrated feeding network. IEEE Antennas Wirel. Propag. Lett. **17**(5), 739–742 (2018)
47. Yu, X., et al.: 3-d printed parts for a multilayer phased array antenna system. IEEE Antennas Wirel. Propag. Lett. **17**(11), 2150–2154 (2018)
48. Ketterl, T.P., et al.: A 2.45 GHz phased array antenna unit cell fabricated using 3-D multi-layer direct digital manufacturing. IEEE Trans. Microw. Theory Tech. **63**(12), 4382–4394 (2015)

49. Hwang, S., Reyes, E.I., Moon, K.-S., Rumpf, R.C., Kim, N.S.: Thermo-mechanical characterization of metal/polymer composite filaments and printing parameter study for fused deposition modeling in the 3D printing process. J. Electron. Mater. **44**, 771–777 (2015)
50. Goode, I., Saavedra, C.E.: 3-D printed dually symmetric orthomode transducer and horn antenna at X-band. IEEE Open J. Antennas Propag. **4**, 383–391 (2023)
51. Onuike, B., Heer, B., Bandyopadhyay, A.: Additive manufacturing of Inconel 718—copper alloy bimetallic structure using laser engineered net shaping (LENS™). Addit. Manuf. **21**, 133–140 (2018)
52. Liu, J., Yu, H., Chen, C., Weng, F., Dai, J.: Research and development status of laser cladding on magnesium alloys: a review. Opt. Lasers Eng. **93**, 195–210 (2017)
53. Wheat, E., Vlasea, M., Hinebaugh, J., Metcalfe, C.: Sinter structure analysis of titanium structures fabricated via binder jetting additive manufacturing. Mater. Des. **156**, 167–183 (2018)
54. Rojas-Nastrucci, E.A., Nussbaum, J.T., Crane, N.B., Weller, T.M.: Ka-band characterization of binder jetting for 3-d printing of metallic rectangular waveguide circuits and antennas. IEEE Trans. Microw. Theory Tech. **65**(9), 3099–3108 (2017)
55. Jun, S.Y., Sanz-Izquierdo, B., Parker, E.A., Bird, D., McClelland, A.: Manufacturing considerations in the 3-d printing of fractal antennas. IEEE Trans. Components, Packag. Manuf. Technol. **7**(11), 1891–1898 (2017)
56. Zhang, B., et al.: Metallic 3-d printed antennas for millimeter- and submillimeter wave applications. IEEE Trans. Terahertz Sci. Technol. **6**(4), 592–600 (2016)
57. Yuan, H., et al.: SLM printed wideband circularly polarized multilayer antenna array with reduced impact by the manufacturing imperfection. IEEE Access **11**, 105368–105378 (2023)
58. Zhang, Y.-X., Jiao, Y.-C., Zhang, L., Wen, J.-X.: Wideband 2-d monopulse antenna array with higher-order mode substrate integrated waveguide feeding and 3-d printed packaging. IEEE Trans. Antennas Propag. **68**(4), 3259–3264 (2020)
59. Zhang, Y.-X., Jiao, Y.-C., Liu, S.-B.: 3-D-printed comb mushroomlike dielectric lens for stable gain enhancement of printed logperiodic dipole array. IEEE Trans. Antennas Propag. **17**(11), 2099–2103 (2018)
60. Chen, Z., et al.: Circular polarized 3-d-printed dielectric loaded antenna using inset waveguide-to-dielectric transition for 5G millimeter-wave application. IEEE Antennas Wirel. Propag. Lett. **19**(11), 1929–1932 (2020)
61. Sharma, Y., Chen, X., Wu, J., Zhou, Q., Zhang, H.H., Xin, H.: Machine learning methods-based modeling and optimization of 3-d-printed dielectrics around monopole antenna. IEEE Trans. Antennas Propag. **70**(7), 4997–5006 (2022)
62. Cho, S.-S., Yoon, S.-H., Hong, I.-P.: Design of three-dimensional frequency selective structure with replaceable unit structures using a 3-d printing technique. IEEE Antennas Wirel. Propag. Lett. **17**(11), 2041–2045 (2018)
63. Muntoni, G., et al.: A curved 3-D printed microstrip patch antenna layout for bandwidth enhancement and size reduction. IEEE Antennas Wirel. Propag. Lett. **19**(7), 1118–1122 (2020)
64. Ahmed, F., Hayat, T., Afzal, M.U., Zhang, S., Esselle, K.P., Whittow, W.G.: 3-D printable synthetic metasurface to realize 2-D beam-steering antenna. IEEE Open J. Antennas Propag. **4**, 506–519 (2023)
65. Colella, R., Michel, A., Catarinucci, L.: Compact 3-D-printed circularly polarized antenna for handheld UHF RFID readers. IEEE Antennas Wirel. Propag. Lett. **17**(11), 2021–2025 (2018)
66. Farooqui, M.F., Kishk, A.: 3-D-printed tunable circularly polarized microstrip patch antenna. IEEE Antennas Wirel. Propag. Lett. **18**(7), 1429–1432 (2019)
67. Oh, Y., Kordsmeier, N., Askari, H., Adams, J.J.: Low profile GRIN lenses with integrated matching using 3-d printed ceramic. IEEE Open J. Antennas Propag. **4**, 12–22 (2023)
68. Wang, S., et al.: Radar cross-section reduction of helical antenna by replacing metal with 3-D printed zirconia ceramic. IEEE Antennas Wirel. Propag. Lett. **19**(2), 350–354 (2020)
69. Mu, C., Fang, S., Liu, H., Wang, Z., Fu, S.: 3-D-printed dielectric lens with cone-shaped cavity for axial ratio beamwidth enhancement of circularly polarized patch antenna. IEEE Access **7**, 105062–105071 (2019)

# Index


**A**

Access Point, 1, 4, 9, 13, 15, 17, 18, 23, 31, 33, 35–37, 40, 235, 265, 276
Acrylonitrile Butadiene Styrene, 286, 290, 293
Active Frequency Selective Surfaces, 116
Active Phase Shifter, 79
Adaptability, 1, 6, 316
Additive Manufacturing, 172, 283, 285–288, 295, 310, 314, 316
Advanced Mobile Phone System, 25
Advanced Mobile Telephone System, 25
Air Cavity, 107, 141, 146, 253, 262, 269, 295
American Telephone and Telegraph (AT&T), 8
Angle of Arrival, 242
Anisotropic Metasurface, 113
Anisotropic Zero-Index Metamaterials, 248, 249
Anisotropy, 87, 89, 108
Antenna, 19, 25, 28, 32–37, 39, 45, 47–67, 71–79, 81–86, 88–90, 96–103, 107–113, 115–117, 123–125, 127–148, 153, 154, 157–185, 189–216, 218, 220–238, 241–260, 262, 264–272, 274–278, 287–303, 305–312, 314–316
Antenna Array, 33, 39, 59, 60, 73, 77–79, 102, 124, 127, 129, 143–146, 154, 172–174, 180, 182, 242, 243, 252, 255, 257, 259–262, 278, 303, 307, 308
Antenna Decoupling Surface, 127
Antenna Pattern Diversity, 49, 50, 265
Antipodal Tapered Slot Antenna, 163
Antireflection, 105
Aperture, 33, 46, 57, 64, 74, 82, 86, 88, 89, 97–99, 102, 104, 105, 110, 113, 117, 124, 126, 127, 154, 157, 163, 170, 172, 174–176, 179, 182, 193, 196, 199, 201, 204, 206, 207, 222, 228, 231, 232, 236–238, 248, 259, 296, 298, 307, 310, 313
Array Feeding Region, 103, 104
Artificial Magnetic Conductor, 128, 165, 167
Augmented Reality, 29
Axial Ratio, 221, 222, 315
Azimuth, 33, 54, 82, 99, 101, 106, 107, 116, 181, 211, 249, 311

**B**

Bandwidth Enhancement, 299, 310
Barium Strontium Titanate, 85
Basic Service Set, 13, 19
Beam Scanning, 71–80, 82–84, 86, 88–90, 95, 96, 98–101, 103, 105, 108, 109, 111, 113–115, 117, 118, 123–126, 129–133, 144, 145, 147, 148, 243, 310
Beam Steering, 78, 82–85, 88, 89, 95, 98–100, 109–113, 116, 123, 132, 248, 310, 316
Beam-Switching, 74, 81, 82, 89, 275
Beamwidth Enhancement, 315
Bill of Materials, 31
Binder Jetting, 286, 305, 306
Bluetooth, 2–4, 9, 10, 16, 246
Broadband, 10, 11, 28, 29, 38, 39, 55, 77, 85, 109, 172, 173




S. K. Koul et al., *Antenna Systems for Modern Wireless Devices*, Signals and Communication Technology, https://doi.org/10.1007/978-981-97-3369-9

Broadside, 55, 57–59, 64, 66, 76–78, 98, 100–102, 125, 127, 128, 130, 131, 133, 134, 141, 145, 174, 177, 184, 189, 209, 212, 213, 215, 222, 226, 227, 236, 237, 246, 252, 255, 257, 259, 262, 265, 266, 268, 271, 276, 278, 311

**C**

Capacitive, 58, 79, 85, 86, 311
Cellular Communication, 4, 6, 38, 72
Cellular Generations, 24
Circular Patch Antenna, 57, 89, 214, 222
Circular Polarization, 61, 62, 64, 66, 259
Client, 7, 9, 13, 16, 20–22, 32
Code Division Multiple Access, 26
Code Division Multiplexing Access, 26
Coefficient of Thermal Expansion, 301
Commercial Off-The Shelf, 310
Compact Structure, 115, 195, 311
Complementary Code Keying, 16
Complementary Split-Ring Resonator, 112, 165, 243
Computer Numerical Control, 288
Conductive Spray, 299
Conformable Substrate, 292
Conventional, 22, 29, 55, 58, 74, 75, 81, 85, 89, 95, 96, 99, 100, 103, 112, 124, 130, 134, 156, 160, 161, 163, 167, 172, 173, 182, 183, 212, 242, 248, 252, 283, 287, 288, 290, 291, 300, 309, 312–315
Coplanar Waveguide, 62
Coplanar Waveguide Ground, 127
Corrugated SIW, 83
CRLH Transmission Line, 84
Cross Circular Loop Resonator, 162
Cylindrical Metasurface, 113

**D**

3-D Printing, 134, 262, 272, 283–285, 316
Delay Spread, 37, 46
Denial Of Service, 20
Dielectric, 77, 85, 87–89, 99, 101–103, 105, 107–109, 114, 115, 129, 134, 140–142, 147, 156, 163, 168–170, 172–175, 178, 182, 183, 185, 194, 196, 204, 212, 228, 229, 253, 262–265, 267–274, 276–278, 286–290, 295–298, 303, 306–311, 314, 315
Dielectric Lens, 103, 105, 110, 132, 162, 172, 173, 268, 315
Dielectric Loss, 88, 110, 172, 183, 194, 296, 303, 314
Dielectric Slab Superstrate, 174
Dipole, 33, 34, 58, 60, 64, 82, 89, 115, 128, 131, 160, 162, 166, 180–182, 189, 193–195, 197, 199–202, 204–216, 218–222, 246, 247, 276, 292–294, 309
Directed Energy Deposition, 304
Direction, 27, 33, 38, 41, 46, 51, 53, 54, 56, 57, 59–61, 63, 64, 66, 71–73, 75, 76, 78–80, 82, 84, 86, 87, 89, 90, 97–102, 105, 107, 109–116, 127–132, 141, 147, 153, 154, 159, 161–167, 176, 180, 183, 184, 189, 194, 196, 197, 199, 200, 204, 206, 209, 212, 216, 219, 222, 227, 229, 230, 232, 236, 237, 242, 248–250, 252, 255, 259, 262, 264–266, 268, 271, 276, 278, 310
Direct Metal Laser Sintering, 286, 304
Direct Sequence Spread Spectrum, 15
Discrete, 81, 83, 84, 89, 90, 95, 105, 310
Dispersive Medium, 46
Distributed Phase Shifter, 79
Distributed System, 13, 98
Distribution System Service, 14
Diversity, 34, 35, 37, 45, 48–56, 58, 59, 63, 66, 72, 202, 209, 222, 235, 241–243, 245, 246, 248, 265, 272
Diversity Techniques, 48, 55, 66
Double Negative Materials, 157
Dynamic Frequency Selection, 17

**E**

Effective Isotropic Radiation Power, 242
802.11, 8, 9, 12–24, 30, 32, 40
Electromagnetic, 7, 41, 45, 73, 74, 76, 77, 101, 111, 114, 125, 127, 129, 155, 159, 165, 168–170, 172, 175, 201, 242, 288, 299, 304, 306
Electromagnetic Band Gap Structures, 168
Electromagnetic Spectrum, 2
Electromagnetic Waves, 41, 46, 47, 72, 75, 77, 89, 105, 114, 154–157, 159, 160, 162, 164, 167, 168, 177, 183, 184, 230, 234, 307
Electron Beam Melting, 286
Electronic Scanning, 72–74, 81, 118
Electroplating, 292, 293, 299, 300, 306

Energy, 25, 26, 29, 32, 40, 41, 46, 50, 54, 72, 75, 76, 114, 123, 129, 136, 153, 163, 183, 199, 222, 292, 304, 307
Enhanced Data Rates for GSM Evolution, 26
Enhanced Mobile Broadband, 29
Envelope Correlation Coefficient, 51, 202, 272
Epsilon-Near-Zero, 160
Epsilon Negative, 157
European Radiocommunications Office, 2
European Telecommunications Standards, 8
Even Mode, 58
Extended Service Set (ESS), 14, 23

**F**

Fabry–Perot cavity, 227, 228, 231, 264, 268, 270, 276
Fading, 45–48, 51, 52, 54, 55, 61, 62, 241, 243
Federal Communications Commission, 2, 8, 12
Feed Network, 56, 58, 65, 75, 77, 82, 98, 115, 123, 133, 154, 255, 302, 305, 307, 311
Ferrite, 73, 86, 89, 90, 183
Ferroscan, 89
Fifth Generation (5G), 2, 6, 24, 29, 30, 38, 72, 123, 168, 172, 245, 246, 298, 308
Filament, 253, 285, 286, 289, 290, 293, 294, 297, 304, 310, 311
Flexibility, 1, 7, 19, 23, 59, 65, 74, 114, 116, 129, 183, 255, 283, 287, 293, 297, 309, 310
Flexible Substrates, 288, 293, 295
Focal To diameter ratio, 105
Fourth Generation (4G), 2, 6, 24, 28, 29, 55, 155
Free-Space Attenuation, 46
Frequency Diversity, 31, 49, 50
Frequency Hopping Spread Spectrum, 15
Frequency Range 2, 2
Frequency Scanning, 73, 74, 77, 95, 129
Frequency Scanning Antennas, 73, 75, 80
Frequency Selective Surface, 114, 116, 165, 169
Frequency Tuning, 84, 311
Front-To-Back Ratio, 137, 140, 145, 180–182, 192, 195, 196, 201, 204, 209, 212, 215, 220, 228, 233, 257, 260, 266, 271
Fused Deposition Modeling, 134, 284, 285, 290, 293, 303
Fused Filament Fabrication, 285, 309

**G**

2.5 Generation, 26
2.75 Generation, 26
3.5 Generation, 27
3.75 Generation, 28
Gain Enhancement, 153–155, 157–161, 163, 165–169, 171–175, 177, 178, 180–182, 185, 189–193, 199–201, 206, 207, 211, 218, 219, 222, 227, 228, 230, 234, 236–238, 257–260, 264, 268–271, 274, 276–278, 313
General Packet Radio Service, 26
Global Positioning System, 27
Global System for Mobile, 133
Gradient Index, 107, 108, 164, 287, 313
Group Special Mobile, 26
Gyromagnetic Resonance, 90

**H**

Half-Power Beamwidth, 33, 132, 137, 140, 145, 163, 228, 232, 260, 266, 272
High-Performance Radio Lan, 8, 31
High Refractive Index, 163
High-Speed Downlink Packet Access, 27
High-Speed Uplink Packet Access, 28
Highway Addressable Remote Transducer, 10
Horn Antenna, 99, 100, 142, 202, 223, 235, 262, 272, 296, 305, 306, 313
H-Plane, 84, 125, 128, 173, 182, 201, 209, 313
H-Shaped Resonator, 166, 193
Hybrid Partially Reflecting Surface, 178
Hybrid Surface, 178
Hybrid Wireless Mesh Protocol, 19

**I**

IEEE 802.11, 5, 8–10, 12–15, 17–20, 22–24, 133
Impedance, 37, 57, 77, 80, 84, 86, 105, 125, 127–130, 136, 137, 140, 143, 144, 159, 162, 164, 167, 174, 181–183, 185, 195, 201, 204, 207, 212, 214, 215, 220, 228, 231, 244, 246, 257, 260, 266, 271, 287, 292, 294, 297, 316
Impedance Modulation Surface, 84

Improved Mobile Telephone Service, 25
Independent, 11, 26, 49, 50, 55, 64, 103, 109, 113, 241, 243, 244, 248
Independent BSS, 13
Indoor Antenna, 33, 37
Industrial Scientific and Medical, 2
Infrastructure BSS, 13
Inhomogeneous Substrate, 288, 295
Integrated Lens Antennas, 110, 132, 172
Integrated Services LAN, 10
Intelligent Reflecting Surface, 39
Intelligent Transportation Systems, 19
Interconnections, 5, 86
Interim Standard 95, 26
Intermediate Frequency, 31
International Mobile Telephone 2000, 27
International Telecommunication Union, 2, 27
Internet of Nanothings, 41
Internet Protocol, 16
Inverted-F Antenna, 243, 301
ISM Bands, 2–4, 8, 12, 15, 59, 246
Isotropic, 34, 87, 89, 153, 154, 159, 179

**K**

Ku-Band, 298

**L**

LAN/MAN Standards Committee, 9
Leaky Wave Antenna, 74–76, 83, 90, 289
Left Hand Circular Polarization, 35
Left-Handed Materials, 156
Lens Antenna, 74, 88, 99, 104, 105, 107–109, 132, 298
Lightweight, 88, 293, 297, 300, 306
Linear Polarization, 61, 63–66, 130
Line-Of-Sight, 5, 15, 32, 33, 39, 46, 48, 62
Link Service Access Point, 11
Liquid Crystal Display, 87
Liquid-Crystal Reflectarray, 95
Liquid Crystals, 74, 86–88, 90, 95, 115
Liquid Metal, 115, 301, 302
Local Area Network, 8
Local Oscillator, 31
Logical Link Control, 11
Long Term Evolution, 28
Loss Tangent, 88, 134, 170, 253, 297, 303, 307, 315
Low-Cost, 6, 31, 32, 37, 59, 80, 85, 88, 103, 123, 124, 129, 158, 172, 173, 262, 283, 286, 288, 289, 298, 302, 304, 309
Low Earth Orbit, 123
Low-Index Materials, 159
Low-Noise Amplifier, 30, 77
Low Power Wide Area Network, 6
Low Temperature Cofired Ceramic, 182

**M**

Machine Learning, 40, 309
MAC Service Access Point, 11
Magnetic Field, 54, 87, 90
Magnetic Loss, 183
Magnetization, 90
Massive, 29, 38–40, 74, 248, 259
Massive Machine Type Communication, 29
Material Jetting, 286, 305
Maximum Ratio Combining, 241
Mechanical Beam Steering Array, 100
Medium Access Control, 11, 15
Medium Earth Orbit, 123
Metal-Dielectric Composite Metasurfaces, 310
Metal Extrusion, 304
Metallization, 158, 290, 298, 300, 302, 313
Metal Plating, 301, 304
Metal Printing, 298, 304, 305
Metamaterials, 41, 58, 59, 66, 108, 109, 111, 154, 155, 157, 158, 161, 162, 164–166, 185, 189, 190, 196–198, 200, 201, 205–207, 209, 211, 212, 216, 218, 227, 229, 230, 237, 238, 248, 264, 271, 272, 274, 276–278, 316
Microstrip, 36, 55, 59, 60, 62, 64, 66, 76, 77, 83, 86, 88, 89, 107, 108, 110, 113, 127, 128, 166, 167, 175, 180, 183, 184, 194, 222, 243, 248, 250, 254, 302, 310
Microstrip Patch Antenna, 55, 172, 175, 183, 249, 311
Microwave, 8, 9, 15, 16, 35, 41, 80, 85–88, 98, 99, 123, 132, 154, 155, 164, 172, 173, 197, 216, 229, 287, 290, 292, 306, 307
Mm-Wave, 307, 308
Mobile Telephone System, 25
Mobility, 6, 28, 29, 32, 46
Monolithic Microwave Integrated Circuit, 110
Monopole, 33, 35, 36, 55–57, 62–64, 164, 165, 180, 181, 196, 245, 246, 248, 309, 311
Multibeam Scanning, 72

Multilayer Antenna, 307
Multimedia Messaging Service, 26
Multipath, 33, 37, 38, 47, 48, 51–53, 55, 61, 62, 65, 71, 77, 241–243
Multipath Fading, 47, 48, 51, 54, 66, 247, 278
Multiple Beams, 82, 243, 248, 250, 252
Multiple-Input, Multiple-Output, 39
Multiple Modes, 57
Multiple Radiators, 53, 55, 56, 248
Multiple-Unit Steerable Antenna, 54
Mu-Near-Zero, 160, 161
Mu-Negative, 157
Mutual Coupling, 51, 54, 62, 81, 125–129, 196, 201, 202, 204, 207, 208, 214, 225, 228, 234, 235, 242–245, 262, 272, 274

**N**
Nanoparticle Jetting, 314
Near Field Communication, 5
Near-Field Meta Steering, 99
Near Zero-Index Materials, 159
Negative Refractive Index, 111, 164, 165
Next-Gen Networks, 29
Nippon Telephone and Telegraph, 25
Nordic Mobile Telephone, 25
Null, 16, 37, 38, 55, 59, 65, 75
Null Scanning, 72, 75

**O**
Odd Mode, 58
Omnidirectional, 32–34, 53, 55, 58, 59, 64, 82, 96, 116, 133, 180, 181, 226, 248
Omnidirectional Antenna, 33, 35, 248
Open-Stop Band, 76, 130
Open System Interconnection, 11
Operating Frequency, 17, 25, 41, 66, 75, 89, 105, 127, 129, 159, 161, 177, 181, 182, 194–197, 199, 200, 203, 206, 209, 212, 216, 220–222, 229, 233, 238, 249, 250, 273
Optical Beam Scanning, 74
Optimization, 96, 110, 206, 309
Orthogonal, 33, 50, 55, 57, 60–62, 64, 66, 113, 179, 183, 201, 211, 246, 294
Orthogonal Frequency Division Multiplexing, 15, 22
Orthogonal-Mode Transducer, 290
Orthogonal Radiating Choke, 184
OSI Reference Model, 11
Outdoor Antenna, 33

**P**
Panel Antenna, 34
Parabolic Grid, 35
Parallel-Plate Waveguide, 88, 90
Partially Reflecting Surface, 154, 238
Passive Phase Shifter, 79
Patch Antenna, 35, 36, 58, 65, 82–84, 112, 128, 134, 173–175, 177–179, 183, 211, 212, 214–216, 218, 220–222, 228, 230, 248, 249, 253–255, 257, 258, 265, 266, 278, 310, 315
Pattern Diversity, 45, 50–67, 189, 190, 193, 194, 204, 226, 237, 238, 243, 252, 264, 278
Perfect Electric Conductor, 167, 197
Perfect Magnetic Conductor, 167, 197
Permeability, 89, 108, 109, 155, 156, 158, 183, 197, 198, 219, 250, 310
Permittivity, 62, 87–89, 102, 104, 106–109, 115, 155–157, 159, 173–175, 178, 183, 197, 198, 215, 219, 250, 254, 287, 289, 294–297, 307, 310, 311, 313, 314
Phase, 39, 47, 51, 53, 56–59, 61, 63, 66, 73, 75–83, 87–90, 95, 96, 99–102, 105, 106, 109–114, 127, 129, 130, 132, 133, 144, 154, 156, 159, 160, 167, 170, 172, 173, 175–180, 197, 218–220, 231, 242, 248, 260, 262, 265–267, 270, 274, 276, 293, 294, 310, 311, 313
Phased Array, 73, 74, 77, 78, 80, 81, 89, 90, 123, 126–128, 242, 248, 303
Phased Array Antennas, 53, 73, 75, 77, 80, 81, 90, 98, 111, 123–125, 128, 129, 132, 248
Phase Matching Method, 105
Phase Scanning, 73, 74
Phase Shifter, 53, 73, 74, 77–80, 83, 85, 90, 102, 113, 123, 129, 133, 134, 146, 242, 264, 290, 293, 294
Phase-Shifting Surface, 95, 96, 99, 100
Physical Service Access Point, 11
Physical Specification, 9
Pin Diode, 65, 66, 74, 79, 81–85, 88, 90, 95, 111, 115, 116
Planar Inverted F Antenna, 36, 243
Plasma, 159
Point-To-Multipoint, 32, 33
Point-To-Point, 32, 33, 35, 76, 154
Polarization Diversity, 33, 45, 49, 50, 52, 55, 61–64, 66, 211, 243, 245
Polarization Rotator, 113

Polycarbonate, 303
Polylactic Acid, 134, 190, 235, 268, 286, 290, 296
Polymer, 235, 268, 283–286, 289, 290, 292, 304
Portable Stations, 23
Power Amplifier, 30, 31, 77, 108, 123
Power Divider, 57, 66, 74, 307
Printed Log-Yagi Dipole Array Antenna, 181
Push To Talk, 25

**Q**
Q-Factor, 85, 86, 134
Quality Of Service, 16, 22, 29, 40

**R**
Radiation Efficiency, 90, 125, 134, 146, 175, 201, 202, 208, 209, 222, 223, 225, 226, 252–255, 272, 275, 277, 297, 303, 310, 314
Radiation Pattern, 33–36, 45, 50, 51, 53, 55–60, 63–66, 71, 72, 77, 82, 86, 90, 116, 125, 127, 128, 133, 135, 137, 138, 140–142, 145, 154, 158, 161, 166, 169, 173, 181, 189, 192, 193, 195, 196, 201, 203, 204, 206, 209, 212, 220–222, 232, 233, 242, 246, 248, 250, 251, 255, 256, 260, 263, 266, 271–273, 278, 292, 300
Radio, 1, 2, 4, 7, 8, 11, 16, 18, 21, 22, 25, 30–33, 35, 37, 38, 40, 45–48, 52–54, 63, 81, 154, 247, 292, 304
Radio Frequency, 1, 2, 39, 77, 123, 287, 293, 312
Radio Frequency Identification, 5
Radio Frequency Interference, 299
Radio Propagation, 45–48
Radio Spectrum, 1, 2, 12, 22
Rapid Prototyping, 283, 303
Rayleigh Fading, 47, 61
Reconfigurability, 45, 65, 66, 84, 86, 98
Reconfigurable Antennas, 65, 82, 86, 88
Rectangular Waveguide, 76, 89, 305
Reflectarray, 74, 95, 124, 287
Reflections, 38, 46, 47, 50, 61, 66, 77, 79, 88, 99, 105, 116, 117, 126, 127, 130, 142, 155, 164, 167, 170, 174–180, 183, 185, 195, 197, 202, 214, 216, 220, 222, 227–232, 234, 238, 253, 266, 267, 271, 272
Reflective Surface, 175, 178, 227, 229
Refractive Index, 112, 113, 157, 159, 160, 162–166, 193, 197, 216
Resonant, 58, 59, 66, 75, 80, 114, 127, 128, 133, 135, 164, 175, 176, 178, 179, 183, 184, 192, 194, 195, 197, 213–215, 219, 220, 222, 231, 253, 255, 257, 266, 272, 292, 299
Resonant Cavity Antenna, 102, 174
RF Mems Switch, 74, 85, 86, 90
Right Hand Circular Polarization, 35
Risley Prism Antenna, 99
Robust Security Network, 18

**S**
Satellite Communication, 2, 75, 77, 123, 126, 154
Saturation Voltage, 88
Scalability, 6, 23
Scan Loss, 100, 107, 110, 124, 131–133, 148, 264
Scattering, 38, 46, 50, 55, 61, 66, 234, 247
Second Generation (2G), 6, 25
Selective Laser Melting, 285, 286, 306
Selective Laser Sintering, 284, 285
Sensor, 3–5, 22, 72, 74
Short Message Service, 25, 30
Sidelobe Level, 80, 96, 132, 255
Signal Noise Ratio, 132
Signal-To-Noise Plus Interference Ratio, 242
SIW Fed-Slot Antenna, 162
Sixth Generation (6G), 24, 29, 38, 40, 123, 308
Slot Array Antenna, 96
Solid, 87, 110, 254, 268, 283, 296, 299–301, 304, 305, 308
Solid-Freeform, 283
Space Diversity, 49, 50, 243
Split-Ring Resonators, 155, 157, 161, 165
Spoof Surface Plasmon Polaritons, 131, 248
Stations, 6, 11–14, 21, 23, 32, 33, 35, 39, 45–50, 61, 72, 81, 133, 154
Station Service, 14
Stereolithography, 284
Substrate Integrated Waveguide, 58, 77, 82, 162, 172
Surface-Wave Launchers, 77, 297
Surface Waves, 77, 127, 183, 194, 212, 297, 310
Symmetric, 76, 96, 105, 108, 131, 266
Symmetry, 55, 64, 66, 83, 96, 97

**T**
Telecom Regulatory Authority of India, 2
Terahertz, 95
Third Generation (3G), 6, 24, 27, 28, 55
Three-Dimensional (3-D), 172, 283, 288, 302
Three-Port Antenna, 63, 189, 204, 211, 212, 227, 237, 253, 259, 276
THz Bands, 39, 41, 88
Time Diversity, 49, 51
Time Division Multiple Access, 26, 30
Total Access Communications Systems, 25
Transceiver, 25, 30–32, 41
Transmission Line, 59, 73, 79, 80, 111, 165, 228, 302
Transmitarray, 74, 95, 99
Transmit Power Control, 17
Tunability, 85, 88, 114, 115
Tunable, 74, 83–85, 87, 88, 100, 115, 290, 311, 312, 316
Two-Dimensional (2-D), 99, 100, 109, 110, 113, 126, 128, 175, 182, 307, 310
Two-Port Antenna, 194, 195, 198

**U**
Ultra-High Frequency, 38, 294
Ultra Reliable Low Latency, 29
Ultraviolet, 283, 289
Ultrawideband, 39, 64, 196
Unit Cell, 84, 96, 99, 105, 114, 116, 130, 155, 158, 161–164, 166, 167, 171, 177, 179, 180, 197–201, 207, 216, 218, 249, 314
Universal Mobile Telecommunications, 27
Unmanned Aerial Vehicles, 72, 124

**V**
Vacuum Filling, 302
Varactor Diodes, 74, 83, 84, 88, 90, 95, 115
Variable Gain Amplifier, 31
Vertically Polarized, 61–64
Vertical Split-Ring Resonator, 182
Very Large Epsilon, 162
Virtual Reality, 29
Voice Over Internet Protocol, 15
Volume, 74, 89, 98, 115, 132, 173, 278, 295, 303, 308

**W**
Wave Guide, 47, 75, 76, 88–90, 96, 107, 108, 124, 132, 142, 147, 174, 175, 182, 202, 223, 235, 272, 287, 288, 290, 300, 301, 304, 305, 307, 308
Wide-Angle Beam Scanning, 103, 124–126, 148, 266
Wide-Angle Impedance Matching, 127, 130
Wideband Code Division Multiple Access, 27, 28
Wide Scan, 74, 107, 108, 127
Wi-Fi, 2, 4, 5, 8–10, 22–24, 28, 32–34, 37, 38, 40, 246
Wi-Fi 8, 40
Wi-Fi Protected Access, 18
Wired Equivalent Privacy, 17
Wireless, 1, 3–13, 15, 16, 19–25, 27, 29, 30, 32, 34, 35, 38–41, 45–50, 53, 59, 65, 66, 71, 72, 80, 81, 83, 85, 90, 95, 112, 123, 133, 155, 172, 178, 211, 241, 242, 246–248, 250, 278, 293, 315, 316
Wireless Access Point, 5, 13, 17, 23, 32
Wireless Application Protocol, 26
Wireless Fidelity, 31
Wireless Local Area Network, 1, 31
Wireless Metropolitan Area Network, 4
Wireless Networks, 1, 3, 6, 7, 9, 13, 14, 17, 20, 21, 23, 31, 37, 38, 123, 154, 308
Wireless Personal-Area Network, 3
Wireless Power Transfer, 40, 41, 72
Wireless Technology, 1, 4–7, 25, 29, 38
Wireless Wide Area Network, 4
Wire Mesh Embedding, 253, 303
WLAN Standards, 9, 15, 32

**Z**
Zero-Index Material, 156, 159
Zigbee, 3–5, 10
Z-Wave, 4, 5

MIX
Papier aus verantwortungsvollen Quellen
Paper from responsible sources
FSC® C105338

If you have any concerns about our products,
you can contact us on
**ProductSafety@springernature.com**

In case Publisher is established outside the EU,
the EU authorized representative is:
**Springer Nature Customer Service Center GmbH**
**Europaplatz 3, 69115 Heidelberg, Germany**

Printed by Libri Plureos GmbH
in Hamburg, Germany